Tumor Antigens Recognized by T Cells and Antibodies

The tumor immunology and immunotherapy series
A series of books exploring the multidisciplinary nature of the field of tumor immunology
Edited by Giorgio Parmiani
National Cancer Research Institute, Milan, Italy and
Michael T. Lotze
University of Pittsburgh Cancer Institute, Pittsburgh, USA

Volume One
Tumor Immunology: Molecularly Defined Antigens and Clinical Applications
Edited by Giorgio Parmiani and Michael T. Lotze

Volume Two
Mechanisms of Tumor Escape from the Immune Response
Edited by Augusto C. Ochoa

Volume Three
Tumor Antigens Recognized by T Cells and Antibodies
Edited by Hans J. Stauss, Yutaka Kawakami and Giorgio Parmiani

Tumor Antigens Recognized by T Cells and Antibodies

Edited by
Hans J. Stauss, Yutaka Kawakami and Giorgio Parmiani

LONDON AND NEW YORK

First published 2003
by Taylor & Francis
11 New Fetter Lane, London EC4P 4EE

Simultaneously published in the USA and Canada
by Taylor & Francis Inc,
29 West 35th Street, New York, NY 10001

Taylor & Francis is an imprint of the Taylor & Francis Group

Typeset in Baskerville by
Newgen Imaging Systems (P) Ltd, Chennai, India
Printed and bound in Great Britain by
TJ International Ltd, Padstow, Cornwall

British Library Cataloguing in Publication Data
A catalogue record for this book is available
from the British Library

Library of Congress Cataloging in Publication Data
Tumor antigens recognised by T cells and antibodies/edited by Hans Stauss, Yutaka Kawakami, and Giorgio Parmiani.
p. cm. – (Tumour immunology and immunotherapy series; v. 3)
Includes bibliographical references and index.
1. Tumor antigens. [DNLM: 1. Antigens, Neoplasm – immunology. 2. Antibodies, Neoplasm – immunology. 3. T-Lymphocytes – immunology. QW 573 T925 2003] I. Stauss, Hans J. (Hans Josef), 1956– II. Kawakami, Yutaka III. Parmiani, Giorgio. IV. Series.

QR188.6 .T847 2003
616.99′20792–dc21 2002152712

ISBN 0–415–29698–6

Contents

Figures

Tables

Contributors

Susanne Beckebaum
University of Pittsburgh Cancer Institute
Division of Basic Research and the Department of Pathology
School of Medicine
University of Pittsburgh
Pittsburgh, PA 15213

Gavin Bendle
Department of Immunology
Imperial College of Science Technology and Medicine
Faculty of Medicine
Hammersmith Campus, Du Cane Road
London W12-0NN
United Kingdom

Claudia Bormann
Med. Klinik und Poliklinik
Innere Medizin
Saarland University Medical School
D-6642 Homburg
F. R. Germany

Matteo Carrabba
Tumor Immunotherapy Unit
Instituto Nazionale Tumori
Via Venezian 1, 20133 Milano
Italy

Vito R. Cicinnati
University of Pittsburgh Cancer Institute
Division of Basic Research and the Department of Pathology
School of Medicine
University of Pittsburgh
Pittsburgh, PA 15213

Albert B. DeLeo
University of Pittsburgh Cancer Institute
Division of Basic Research and the Department of Pathology
School of Medicine
University of Pittsburgh
Pittsburgh, PA 15213

Mary L. Disis
Box 356527, Oncology
University of Washington
Seattle, WA 98195-6527

Liquan Gao
Department of Immunology
Imperial College of Science Technology and Medicine
Faculty of Medicine
Hammersmith Campus, Du Cane Road
London W12-0NN
United Kingdom

Roopinder Gillmore
Department of Immunology
Imperial College of Science Technology and Medicine
Faculty of Medicine
Hammersmith Campus, Du Cane Road
London W12-0NN
United Kingdom

John B. A. G. Haanen
Department of Immunology and Medical Oncology,
The Netherlands Cancer Institute
Plesmanlaan 121 1066 CX Amsterdam
The Netherlands

Mamoru Harada
Department of Immunology
Kurume University School of Medicine
67 Asahi Machi, Kurume 830-0011
Japan

Angelika Holler
Department of Immunology
Imperial College of Science Technology and Medicine
Faculty of Medicine
Hammersmith Campus, Du Cane Road
London W12-0NN
United Kingdom

Masaaki Ito
Department of Immunology
Kurume University School of Medicine
67 Asahi Machi
Kurume 830-0011
Japan

Kyogo Itoh
Department of Immunology
Kurume University School of Medicine
67 Asahi Machi, Kurume 830-0011
Japan

Dirk Jäger
II. Medizinische Klinik
Hämatologie – Onkologie
Krankenhaus Nordwest, Frankfurt
Germany

Elke Jäger
II. Medizinische Klinik
Hämatologie – Onkologie
Krankenhaus Nordwest, Frankfurt
Germany

Kazuko Katagiri
Department of Immunology
Kurume University School of Medicine
67 Asahi Machi, Kurume 830-0011
Japan

Yutaka Kawakami
Professor, Division of Cellular Signaling
Institute for Advanced Medical Research
Keio University School of Medicine
35 Shinanomachi, Shinjukuku, Tokyo
Japan 160-8582

Alexander Knuth
II. Medizinische Klinik
Hämatologie – Onkologie
Krankenhaus Nordwest, Frankfurt
Germany

Zihai Li
Center for Immunotherapy of Cancer and Infectious Diseases, MC 1601
University of Connecticut
School of Medicine
Farmington CT 06030-1601
USA

Takashi Mine
Department of Immunology
Kurume University School of Medicine
67 Asahi Machi, Kurume 830-0011
Japan

Frank Neumann
Med. Klinik und Poliklinik
Innere Medizin
Saarland University Medical School
D-6642 Homburg
F. R. Germany

Giorgio Parmiani
Unit of Immunotherapy of Human Tumors
Instituto Nazionale per lo Studio e la Cura dei Tumori
Via G. Venezian, 1 – 20133 Milan
Italy

Michael Pfreundschuh
Med. Klinik und Poliklinik
Innere Medizin
Saarland University Medical School
D-6642, Homburg
F. R. Germany

Lorenzo Pilla
Tumor Immunotherapy Unit
Instituto Nazionale Tumori
Via Venezian 1, 20133 Milano
Italy

Klaus-Dieter Preuss
Med. Klinik und Poliklinik
Innere Medizin
Saarland University Medical School
D-6642 Homburg
F. R. Germany

Francisco Ramirez
Department of Immunology
Imperial College of Science Technology and Medicine
Faculty of Medicine
Hammersmith Campus, Du Cane Road
London W12-0NN
United Kingdom

Licia Rivoltini
Tumor Immunotherapy Unit
Instituto Nazionale Tumori
Via Venezian 1, 20133 Milano
Italy

Lupe Salazar
1959 NE Pacific St, BB 1321
Box 356527, Seattle, WA
USA 98195-6527

Ton N. M. Schumacher
Department of Immunology and Medical Oncology
The Netherlands Cancer Institute
Plesmanlaan 121 1066 CX Amsterdam
The Netherlands.

Shigeki Shichijo
Department of Immunology
Kurume University School of Medicine
67 Asahi Machi, Kurume 830-0011
Japan

Hans J. Stauss
Department of Immunology
Imperial College of Science Technology and Medicine
Faculty of Medicine
Hammersmith Campus, Du Cane Road
London W12-0NN
United Kingdom

Catia Traversari
MolMed SpA and Cancer Immunotherapy and Gene Program
Scientific Institute HS Raffeale, Milan
Italy

Rong-Fu Wang
The Center for Cell and Gene Therapy
Baylor College of Medicine, 1 Balor Plaza
Houston, TX 77030

Shao-an Xue
Department of Immunology
Imperial College of Science Technology and Medicine
Faculty of Medicine
Hammersmith Campus, Du Cane Road
London W12-0NN
United Kingdom

Akira Yamada
Department of Immunology
Kurume University School of Medicine
67 Asahi Machi
Kurume 830-0011
Japan

Carsten Zwick
Med. Klinik und Poliklinik
Innere Medizin
Saarland University Medical School
D-6642 Homburg
F. R. Germany

Series preface

Tumor immunology has been a conflicting area of investigation for several decades, and has been characterized by a succession of excitements and disappointments. However, three major discoveries have been instrumental in causing a resurgence of interest in the field. First, the understanding of molecular steps of antigen recognition, processing and presentation for both HLA classes I and II restricted antigens; second, the milestone event of cloning genes encoding the T-cell recognized human melanoma antigens; and third, the identification of stimulatory and now inhibitory receptors of NK and T lymphocytes. Furthermore, the availability of vectors that allow the genetic engineering of most immune cells and of tumor cells significantly widened the possibility of understanding mechanisms of immune recognition and of manipulating, for therapeutic purposes, the immune system of tumor-bearing individuals. But also previous reagents, like monoclonal antibodies, apparently inefficient as a magic bullet in early therapeutic approaches, have now found new applications and remain the focus of intensive research in tumor immunology.

Tumor immunology is therefore, once again, enjoying a remarkable popularity and could lead to future successes in the immunotherapy of cancer, though several crucial questions need to be answered that require a concomitant effort of both pre-clinical and clinical investigators. We are not only continuing our quest for molecules that make tumor cells diverse from normal counterparts and foreign to the body but we have now to face the unexpected finding and understand how normal proteins and peptides can be recognized by the immune system and whether they can serve as targets of the immune response against growing neoplastic cells.

This new series of books in tumor immunology reflects the increased interest in this area which requires a multidisciplinary approach. It will attract the attention of molecular biologists, immunologists, gene therapists, and experimental and clinical oncologists. It intends to offer a forum of discussion in tumor immunology covering the latest results in the field.

Giorgio Parmiani and Michael T. Lotze

Preface

Historic immunization experiments with irradiated tumor cells have clearly demonstrated the existence of rejection antigens expressed by carcinogen induced murine tumors (Foley, 1953). This led Burnett to postulate that a major function of the immune system was early recognition and elimination of transformed cells, thus preventing the development of overt malignancies (Burnet, 1970). This concept of immunological surveillance is clearly relevant for virus-associated malignancies, such as EBV lymphomas, which frequently develop in immune suppressed individuals. Furthermore, recent experiments demonstrated a hugely increased incidence of adenomas and adenocarcinomas in mice lacking functional RAG recombinase required for rearrangement of antigen receptors of B and T lymphocytes (Shankaran *et al.*, 2001). These mice are devoid of mature B and T lymphocytes, resulting in an inability to recognize and destroy tumors at an early stage of development. These observations indicate that antigen-specific immune recognition by B and T cells are critically important for protective tumor immunity.

A major focus of this book is the molecular nature of tumor antigens that can be recognized by antibodies, helper T lymphocytes and cytotoxic T cells in humans (Chapters 3–12). In addition Chapters 1 and 2 describe murine tumor models suitable to explore mechanisms underlying immune responses against tumor antigens.

Although helper T cells are critically important in the generation and maintenance of effective tumor immunity, only a relatively small number of "helper antigens" have been molecularly cloned to date. Unlike the well-established technologies for identification of antigens recognized by antibodies and CTL, the strategies for cloning helper T cell-recognized antigens are relatively new and have been used successfully in a limited number of research laboratories. It is anticipated that the technologies will mature and lead to the identification additional "helper antigens" in the next few years. Chapters 8 and 9 of this book provide an overview of currently identified tumor antigens recognized by helper CD4 T cells in humans.

Chapters 10–12 describe a powerful technology, SEREX, that has been developed recently and led to the identification of tumor antigens recognized by antibodies present in the serum of cancer patients. Interestingly, some of the antibody-recognized antigens are also targets for tumor-reactive cytotoxic T cells (Chapter 12).

The cloning of tumor antigens recognized by autologous CTL in melanoma patients was successfully achieved for the first time in 1989 (van der Bruggen *et al.*, 1991). This seminal work was followed by the identification of a large number of CTL-recognized human tumor antigens, some of which are now in clinical vaccination trials. Tetrameric MHC class I molecules folded around CTL-recognized peptide epitopes have revolutionized the analysis of tumor-specific immune responses in cancer patients at the single cell level (see Chapter 7).

The immunological properties of CTL-recognized antigens in melanoma and squamous cell carcinoma are described in Chapters 4 and 5, and the exciting observation that heat shock proteins isolated from tumor cells can be used as vaccine to stimulate tumor-protective CTL responses is discussed in Chapter 2.

In most cases the CTL-recognized tumor antigens are encoded by genes expressed at high levels in tumor cells, but also at lower levels in normal tissues. The expression in normal tissues is likely to establish immunological tolerance, affecting primarily high avidity CTL specific for these antigens. This may explain why tumor-reactive CTL isolated from melanoma patients are frequently of low avidity. One strategy to improve the quality/avidity of the CTL response is based on altered peptide ligands that differ by single amino acid substitutions from the native peptides derived from the tumor antigens (Chapter 6). Another strategy is the circumvention of immunological tolerance by directing allogeneic CTL of HLA-mismatched donors against peptide epitopes that are expressed at elevated levels in tumor cells (Chapter 3).

The knowledge of tumor antigens recognized by all component of the adaptive immune system, together with the ability to generate high avidity CTL responses against specific antigens provides an exciting platform for innovative clinical studies in the next decade.

Hans J. Stauss

References

Burnet, F. M. (1970) The concept of immunological surveillance. *Prog. Exp. Tumor Res.*, **13**, 1–27.

van der Bruggen, P., Traversari, C., Chomez, P., Lurquin, C., De Plaen, E., Van den Eynde, B., Knuth, A. and Boon, T. (1991) A gene encoding an antigen recognized by cytolytic T lymphocytes on a human melanoma. *Science*, **254**, 1643–7.

Foley, E. J. (1953) Antigenic properties of methylcholanthrene-induced tumours in mice of the strain of origin. *Cancer Res.*, **13**, 835–7.

Shankaran, V., Ikeda, H., Bruce, A. T., White, J. M., Swanson, P. E., Old, L. J. and Schreiber, R. D. (2001) IFNgamma and lymphocytes prevent primary tumour development and shape tumour immunogenicity. *Nature*, **410**, 1107–11.

Part 1

Animal models

Chapter 1

Mouse models in the recognition of tumor antigens

Vito R. Cicinnati, Susanne Beckebaum and Albert B. DeLeo

Summary

Accompanying the clinical introduction of tumor vaccines has been an increasing awareness for the need of suitable preclinical mouse tumor model systems to guide some of the potentially critical analogous issues and concerns that arise in the design and implementation of clinical protocols. Due to the increased ability to manipulate the murine germline, the mouse has become a primary organism in which to investigate some of the concepts and mechanisms involved in tumor immunology. Several murine homologs of human tumor antigens have been identified, but there is a noticeable lack of comparable naturally processed murine T-cell defined tumor antigens for a major class of currently defined human tumor antigens, namely the cancer/testis (CT) antigens. For the past 50 years, carcinogen-induced or spontaneous arising tumors in inbred mice were the major source for the identification of murine tumor antigens. More recently, advances in genetic engineering technologies have opened up new possibilities to study and discover new tumor antigens using transgenic and/or knockout mouse models with translational potential for the human system. This chapter focuses on preclinical murine tumor models involving murine antigen homologs of human tumor antigens, and the relevancy of their use in the development of cancer vaccines and strategies for immunotherapy of cancer.

Introduction

The studies of the immunogenicity of experimentally induced tumors of inbred mice provided the basis for many of the fundamental concepts of tumor immunology, in particular, tumor antigens and the their use in vaccines for immunotherapy of cancer (Old, 1981). Following a nearly twenty-year quest to confirm the existence of these determinants, the demonstrated immunogenicity of a chemically induced tumor in the primary host provided the foundation for pursuing the molecular identification of these determinants (Klein *et al.*, 1960). The intention was that this information would facilitate the molecular characterization of human homologs.

Advances in molecular biology and immunology over the past two decades revolutionized tumor immunology. The demonstrated role of T cells in tumor rejection coupled with a better understanding of the mechanisms involved in T lymphocyte recognition of antigenic determinants have enhanced the potential of a vaccine approach to the treatment of cancer (Sogn, 1998). The methodologies developed for identification of the first tumor rejection antigen, the P1A determinant of the murine P815 mastocytoma, provided the basis for the subsequent identification of T-cell defined human tumor antigens (Van den Eynde *et al.*, 1991).

The subsequent rapid pace in successfully identifying human cancer antigens and their clinical introduction have reduced the need, *per se*, to identify murine tumor antigens in order to facilitate the identification of their human homologs. Instead, the emphasis has been on identifying relevant preclinical murine tumor model systems for use in guiding the development of cancer vaccines and clinical strategies for their use. The "ideal" murine tumor models being sought need to reflect the various classes of cancer antigens recognized by the host's immune system, as well as the genetic events associated with malignant transformation at specific tissue/organ sites. In those instances where homologs of genes encoding human tumor antigens are not expressed in inbred strains of mice, basic molecular biological technologies have been applied to develop genetically modified strains of mice that can express the targeted gene product.

A recent listing of the human tumor antigens defined by T cells based on their pattern of distribution rather than origin or function classified these determinants into four groups, three of detailing class I HLA-restricted epitopes derived from non-mutated, "self" epitopes, while the fourth comprizes mutated epitopes (Renkvist *et al.*, 2001). The largest group of "self": tumor antigens is the "CT" antigen group. These determinants are derived from proteins expressed in tumors but not normal tissues, with the exception of testis. The other groups of "self" tumor antigens are derived from lineage-specific or differentiation antigens or proteins overexpressed in tumors relative to normal cells. Regardless of their category, these "self" determinants are shared, tumor-associated antigens and suitable for use in the development of broadly applicable vaccines. In contrast, the "mutant tumor-specific" epitopes (Mumberg *et al.*, 1996) are expressed on tumors derived from individual patients. In a sense, these antigens correspond to the highly restricted or "unique" tumor-rejection antigens expressed by most murine tumors (Old, 1981). Most of these antigens have been identified as mutant epitopes derived from unrelated gene products. Due to their restricted nature, "custom made" vaccines would be required for targeting this class of tumor antigen. Although, these vaccines would probably be the most effective in inducing tumor rejection, identifying members of this class of tumor antigens is most difficult and considered impractical due to their limited applicability.

The "self" nature of most T-cell defined human tumor antigens has linked tumor immunology to autoimmunity to the extent that a critical concern of immunotherapy is whether the induction of effective anti-tumor immune responses can be achieved in the absence of deleterious autoimmune side effects (Gilboa, 2001). This review focuses on studies utilizing transplantable as well as primary murine tumor antigen model systems, involving wild type (wt) and genetically modified mice, in which the targeted epitopes are homologous to defined human tumor antigens, and the conditions under which they are expressed are relevant to the clinical situation and, therefore, have optimal translational relevancy.

Animal tumor models

Transplantable tumor models

Murine homologs of human tumor antigens are listed in Table 1.1. All of the listed determinants are being used in preclinical murine studies that focus on development of vaccine vehicles, as well as immunization strategies. Most involve the use of stable, transplantable tumors to evaluate the efficacies of immunization protocols in the protection as well as therapy settings. Although murine tumor models have and can continue to provide investigators with valuable insights into tumor immunotherapy, they do have critical shortcomings relative to

Table 1.1 Murine homologs of human tumor associated antigens

Class I MHC-restricted tumor antigens	Murine tumor antigens with existing human counterparts
Cancer/testis	MAGE, NY-ESO
Widely expressed	*p53, mdm2, cyclin DI, Her-2/neu, ras, WT1 CEA, AFP, muc-1, EGP-2*
Melanoma differentiation	Gp100, TRP1, TRP2, MART-1

the clinical situation (Sogn, 1998). Primarily, these are the distinct differences in tumor heterogeneity and immune status of the hosts that exist between murine tumor model systems and the clinical presentations of human cancers. As a result, interpretations and conclusions drawn from the results of vaccination studies employing murine tumor models, in particular, transplantable tumors, need to be guarded.

Cancer/testis tumor antigens. The human CT antigens comprise a large group of determinants expressed on tumors, as well as testis. Included in this group are the MAGE and NY-ESO antigens. The murine homologs of MAGE genes (Jurk *et al.*, 1998; De Plaen *et al.*, 1999; Lee *et al.*, 2000; Osterlund *et al.*, 2000), have been identified, along with an NY-ESO homolog (Alpen *et al.*, 2002). Relative to murine CT models, Bueler and Mulligan (1996) and Van Pel *et al.* (2001) used MAGE-transfected tumors to evaluate vaccines in mice. Recently, Sypniewska *et al.* (2002) identified Mage b gene expression in mammary carcinomas in transgenic mice, a finding that should greatly facilitate murine studies of CT-based vaccines and strategies for their use.

Widely expressed murine tumor antigens. The second class of shared tumor antigens listed in Table 1.1 are non-mutated "self" epitopes expressed by a wide range of tumors. Most of these determinants are derived from either oncogenes, such as *mdm2, cyclin D1, Her-2/neu, ras* and *WT1*, the *p53* tumor suppressor gene, or oncofetal genes, such as *AFP* and *CEA*.

The immunotherapeutic targeting of p53 in mice has been and continues to be a particularly active area of investigation. p53 was initially identified as a transformation-related antigen in chemically induced sarcomas and other transformed cells of the mouse (DeLeo *et al.*, 1979). Its subsequent classification as a tumor suppressor gene product, which is frequently mutated in human tumors made it an attractive candidate for development of cancer vaccines (Hollstein *et al.*, 1991). While the mutations represented ideal targets for tumor-specific vaccines (Yanuck *et al.*, 1993), the constraints imposed by antigen presentation highlighted the limited applicability of such vaccines. The frequent accumulation or "overexpression" of mutant p53 molecules in tumors, however, suggested that non-mutated epitopes derived from the mutated p53 molecules accumulating in tumors might result in their enhanced presentation by the tumor for immune recognition (Houbiers *et al.*, 1993). To date, a number of CTL and Th-defined wt sequence epitopes of murine p53 have been identified.

The identification of H-2^d and H-2^b restricted wt and mutant mouse p53 CTL-defined epitopes permitted studies aimed at developing peptide-based as well as DNA vaccines and immunization strategies targeting wt p53 epitopes. Noguchi *et al.* (1994) first characterized p53 epitopes recognized by $CD4^+$ and $CD8^+$ T cells. They immunized BALB/c or (BALB/c × C57BL/6)F1 mice with p53 peptides using complete Freund's adjuvant (CFA)-based vaccines and tested the peptide-specificity of splenocytes obtained from these mice. They could show that cytotoxic $CD8^+$ as well as $CD4^+$ T cells recognizing the H2-K^d-restricted $p53_{232-240}$ epitope harboring a mutation at codon 234, but not the wt sequence epitope could be induced. The CTL were able to lyse mutant p53 transfected target cells, whereas $CD4^+$ T cells showed a proliferative response with broader specificity to this epitope region. The use of these vaccines in the protection setting, however, induced only a marked, but not *a* significant inhibition of tumor growth. A Meth A mutant p53 peptide-based vaccine using the QS-21 adjuvant co-administered with IL-12 was subsequently shown to induce rejection of Meth A in protection and therapy settings, as well as induced higher mutant peptide-specific CTL activity (Noguchi *et al.*, 1995). Mayordomo *et al.* (1996) were able to show that vaccines consisting of either the wt or Meth A mutant $p53_{232-240}$ peptide pulsed onto bone marrow-derived dendritic cell (DC) induced CTL specific these epitopes, as well as rejection of tumors expressing either the mutant or wt p53 epitope in the prevention and therapy settings without inducing autoimmunity. In addition to DC-based vaccines, a variety of p53-based DNA vaccines employing viral as well as non-viral plasmid vectors have also been evaluated. Tüting *et al.* (1997) transfected DC with a non-viral plasmid encoding a wt p53 sequence and were able to inhibit tumor growth of the chemically induced CMS4 sarcoma in BALB/c mice. A bioballistic "gene gun" immunization using cDNA encoding the Meth A mutant $p53_{232-240}$ epitope was also shown to be effective in protecting mice from subsequent tumor challenge (Tüting *et al.*, 1999a). Ishida *et al.* (1999) showed that immunization of mice with DC transduced with an adenoviral construct encoding the highly homologous human wt p53 increased the resistance of the mice to a subsequent lethal Meth A sarcoma challenge.

A major concern related to targeting the widely expressed "self" tumor antigens, such as the wt p53 epitopes, is the potential of inducing deleterious autoimmune responses in addition to anti-tumor responses. In the case of p53, which is expressed by all nucleated cells, modeling studies comparing the affinities of the anti-p53 CTL responses induced in p53 knockout (null) mice with those induced in normal mice have demonstrated that high affinity anti-p53 CTL are induced in the null mice but tolerization or deletion of these CTLs occurs in normal mice. As a result, the anti-p53 CTL response in normal mice consists of intermediate/low affinity anti-p53 CTL (Theobald *et al.*, 1997). This observation raises the question of whether the "intermediate/low" affinity anti-p53 CTLs that survive in normal individuals are capable of inducing/participating in effective anti-tumor immune responses. The administration of anti-CTLA-4 monoclonal antibodies (mAb) was recently employed by Hernandez *et al.* (2001) as a means of up-regulating the anti-p53 immune response as well as reduce tolerance to this antigen in HLA-A2.1 transgenic mice. This antibody has been shown to significantly reduce tolerance in mice, presumably by blocking the down-regulation of T cells following their activation. The anti-CTLA-4 mAb treatment resulted in an enhanced expansion of anti-p53 CTL in the mice, but the avidity of these T cells for the HLA-A2-restricted mouse $p53_{261-269}$ epitope was not increased. Vierboom *et al.* (1997) showed tumor eradication in normal C57BL/6 mice following adoptive transfer of high affinity CTL specific for the H2 K^b-restricted wt $p53_{158-166}$ epitope that were generated in $H2^b$ p53 knockout mice. The potential of using high affinity anti-human p53 CTL generated in mice for immunotherapy of humans was also studied by Liu *et al.* (2000). Following the generation of anti-human $p53_{149-157}$ CTL in HLA-A2.1 transgenic mice, human Jurkat (TCR^-) T cells were transfected with cDNAs of the TCR α/β chains expressed by the murine CTL. The transfected Jurkat cells recognized a variety of $HLA\text{-}A2.1^+$ human tumor cells, but not normal cells.

Preclinical murine tumor models are also being recruited to study the processing and presentation of wt p53 epitopes in murine and human tumors, which is more complicated than originally thought. The mere accumulation of genetically altered p53 molecules in a tumor does not appear to be the sole prerequisite for processing and presentation. Apparently, the site of a missense mutation or extent of deletion/alterations in *p53* impact on processing and presentation of epitopes derived from the altered gene products in the tumor as well as the product of the wt gene, when it too is present in that tumor. In this regard, the high affinity anti-p53 induced in p53 null mice have been used as a probe for investigating the molecular characteristics of altered p53 which may influence the processing of epitopes derived from these molecules (Vierboom *et al.*, 2000).

The cyclin D1 and mdm2 oncogenes are "overexpressed," primarily as a result of gene amplification, in a high percentage of human cancers, such as melanoma, leukemia, sarcoma and epithelial tumors (Watanabe *et al.*, 1996; Donnellan and Chetty, 1998). Dahl *et al.* (1996) investigated whether class I MHC-restricted mdm2 and cyclin Dl epitopes could represent potential targets for tumor-reactive CTL. Murine dendritic cells pulsed with several H2-D^b or H2-K^b-binding wt cyclin Dl peptides were shown capable of *in vitro* induction of peptide-specific CTL, only the H2-K^b-restricted $mdm2_{441–449}$ peptide was shown to be a naturally presented epitope. In addition, mdm2 epitopes have been used to demonstrate the potential of allogenic-derived anti-tumor peptide-specific CTL in tumor eradication (Sadovnikova and Stauss, 1996). The H2-K^b-restricted $mdm2_{100–107}$ peptide was shown capable of inducing allogeneic BALB/c CTL that had specificity for H2-K^b-melanoma and lymphoma cells. Upon engraftment in bone marrow-transplant recipients, these CTL were viable up to 14 weeks following administration and did not induce deleterious side effects.

The *ras* p21 proto-oncogenes encode a family of cellular guanosine triphosphate (GTP) binding proteins important for cellular differentiation (Grand and Owen, 1991; Satoh and Kaziro, 1992). In mammalian cells, the *ras* proto-oncogenes consist of three highly homologous members, K-*ras*, H-*ras* and N-*ras* (Bos, 1989). A key genetic event in the ontogeny of many tumors is mutation of *ras* at codons 12, 13 or 61. In mice, the majority of the ras mutations are found at codon 61, while in humans, they occur primarily at codon 12. Because of the limited spectrum of ras "hot spot" mutations coupled with the lack of "accumulation" of altered ras products in tumors, the development of ras-based vaccines has focused primarily on the targeting of mutant tumor specific epitopes rather than wt epitopes. However, since the mutant ras epitopes are "widely expressed," vaccines targeting them have a potentially broader applicability than a vaccine targeting an infrequently occurring mutant tumor-specific epitope (Disis and Cheever, 1996). Nonetheless, the applicability of mutant ras-based vaccines is limited by the inherent constraints of antigen presentation, namely that the mutations at codons 12, 13 or 61 occur within or create epitopes capable of being presented by the host. As a result, preclinical murine studies targeting ras have required the selective use of inbred strains of mice as well as modification of ras-derived peptides to generate T-cell defined epitopes.

Peace *et al.* (1991) were the first to show that immunization of mice with mutant ras peptide elicited ras-specific T-cell responses. Immunization of C57BL/6 ($H2^b$) mice the $ras_{5–16}$ peptide containing the G12R mutation induced ras-specific $CD4^+$ T cells. These T cells recognized the mutated but not wt sequence ras peptide *in vitro*. In a follow up study, Peace *et al.* (1993) demonstrated that immunization of mice of a different haplotype, C3H/HeN($H2^k$), with a mutant ras protein expressing the Q61L mutation induced a ras mutation-specific T-cell response. These investigators were also able to induce in C57BL/6 mice an H2-K^b-restricted-CTL response to the ras Q61L mutation using a combination of 6–17 mer peptides incorporating this missense mutation (Peace *et al.*, 1994). The most effective H2-K^b binding target peptide was defined as $ras_{60/61-67}$, and these effectors were reactive against fibroblast cell line transformed with ras expressing the Q61L mutation indicating that the epitope was naturally presented. Fenton *et al.* (1993) showed the immunogenicity of the ras G12R mutation in BALB/c (H-2^d) mice. Mice were immunized with either the wt or mutant ras proteins encoding either the G12R or G12V missense mutations, but only the G12R mutant ras-immunized mice showed an increase resistance to a challenge with a tumor expressing that ras mutation. This response correlated with the induction of mutant ras specific-CD8$^+$ T cells in these mice that were reactive against tumor cells expressing the mutation. Abrams *et al.* (1995, 1996) demonstrated the immunogenicity of the ras G12V mutation in BALB/c mice by immunizing them with a 13-mer ras_{4-16} peptide containing the G12V mutation. Since Ia^d-restricted CD4$^+$ as well as H2-K^d-restricted CD8$^+$ mutant ras-specific T-cell lines were obtained from splenocytes of these immunized mice, this peptide sequence apparently expresses overlapping class I as well as class II MHC epitopes. In a subsequent study from this group, Bristol *et al.* (2000) showed that immunization of BALB/c mice with mutant ras_{4-16} peptide containing the G12V mutation or a non-viral plasmid based DNA vaccine encoding this sequence was effective in inducing class I and class II restricted mutant ras specific T cells. Mice immunized with the wt ras_{4-16} peptide did not generate a T-cell response. Skipper and Stauss (1993) is the only report identifying mutant as well as wt ras epitopes. In this study, two CTL-defined ras epitopes were identified following immunization of C57B1/10 (H-2^b) mice with recombinant vaccinia viruses expressing either wt ras or ras expressing the Q61K missense mutation. Interestingly, CTL from mice immunized with the mutant ras lysed only cells expressing the mutated ras_{4-16} epitope, whereas T cells induced with virus encoding wt ras were able to recognize transfected tumor cells overexpressing wt ras. A second CTL-defined epitope, wt $ras_{152-159}$, was also identified in this study.

The Wilm's tumor gene encoded transcription factor (WT1) is primarily expressed during embryogenesis and at low levels in some adult cells. It is, however, critically involved in leukemogenesis. It is overexpressed in nearly all types of leukemia in humans, and many types of solid tumors as well. In humans, WT1 can serve as a target for CTL and these cells have a high specificity for leukemic progenitor cells (Gao *et al.*, 2000). Recently, the results of two preclinical WT1 immunization studies in mice were reported. The results reflect the variances and subtleties that differences in the immunogenicities of peptides and polypeptides together with the choice of adjuvant can have on inducing T-cell mediated anti-tumor responses, and could have clinical relevancy. Gaiger *et al.* (2000) demonstrated that the wt $WT1_{117-139}$ peptide admixed with CFA as the adjuvant induced in C57BL/6 mice CTL with anti-tumor reactivity against SV40-transformed mouse cells, as well as an anti-WT1 antibody response. In this study, $WT1_{130-138}$ sequence was concluded to encode the immunodominate H2-D^b-restricted CTL epitope. This vaccine, however, did not increase tumor resistance in mice. In contrast, Oka *et al.* (2000) demonstrated that a vaccine consisting of the $WT1_{126-134}$ peptide pulsed onto LPS-stimulated splenocytes induced H2-D^b-restricted, anti-tumor CTL in C57B1/6 mice as well as tumor rejection in the protection setting. Furthermore, neither immunization protocol induced any observable evidence of autoimmunity. In these studies, reflecting the choices that need to be made in devising clinical protocols, the choice of peptide, vaccine vehicle as well as tumor target cells selected for analysis impacted on whether tumor resistance could be achieved. The use of DNA vaccines in the WT1 model has also been evaluated. Tsuboi *et al.* (2000) immunized C57BL/6 mice with plasmid DNA encoding murine full-length WT1. The immunized mice generated CTL that specifically killed WT1-expressing tumor cells in an MHC class I-restricted manner, rejected WT1-expressing tumor cells and survived with no signs of autoimmunity.

Most of the preclinical murine studies targeting the *Her-2/neu* oncogene involve the use of transgenic mice (see later section), but two interesting reports by Shiku and colleagues have investigated the immunogenicity of the endogenously expressed murine Her-2/neu. Nagata *et al.* (1997) identified H2-K^d-restricted CTL-defined peptides derived from human as well as murine Her-2/neu. The peptides, human Her-2/neu_{63-71}, mouse Her-2/neu_{63-71} and the common Her-2/$neu_{780-788}$, were shown to be naturally presented by murine sarcomas transfected with the appropriate *Her-2/neu* gene. Furthermore, growth of a murine *Her-2/neu* transfected sarcoma was inhibited in mice immunized with either of the murine Her-2/neu peptides encoded by codons 63–71 or 780–788. Recently, Ikuta *et al.* (2000) and Okugawa *et al.* (2000) demonstrated that the DC pulsed with mouse Her-2/neu_{63-71} or Her-2/$neu_{780-788}$ peptide were capable of inducing HLA-A24-restricted CTL from PBMC obtained from healthy donors and cancer patients as well. These effectors were cytolytic against human ovarian carcinoma cell line transfected to express the HLA-A24 restriction element. This finding represents a unique preclinical/clinical system for developing Her-2/neu-based vaccines and strategies for their use.

The other group of widely expressed, non-lineage-specific epitopes to be discussed consists of MUC-1 and three "oncofetal" proteins, namely alpha-fetoprotein (AFP), carcinoembryonic antigen (CEA) and the epithelial cell adhesion molecule, EGP-2. Murine studies involving these tumor-associated antigens included immunization of mice with cDNA encoding the human proteins or recombinant human proteins combined with the use of mouse tumor target cells genetically modified to express the human proteins. Therefore, the immune responses of the mice to these immunogens were "non-self" rather than "self." These reports are discussed here because they did provide the basis for subsequent preclinical murine studies that involve the mouse gene products or mice transgenic for the human genes.

The AFP is produced in fetal liver cells and is the major serum protein produced during fetal and embryonal development. In adults, it is produced only by hepatocellular carcinomas and cancers derived from embryonal cells, such as testicular carcinoma. Grimm *et al.* (2000) demonstrated induction of AFP-specific CTL in C57L/J (H-2^b) mice following immunization of these mice with vaccinia virus encoding murine AFP co-administered with plasmids encoding cytokines, such as IL-12, GM-CSF and interleukin 18 (IL-18). Immunization of mice bearing the syngeneic Hepa 1–6 hepatoma induced significant tumor regression as compared to relative controls. To date no MHC-restricted AFP epitopes have been identified.

The epithelial cell adhesion molecule, EGP-2 (Ep-CAM), is also identified as the 17-1A antigen, based on its identification and characterization by the 17-1A mAb. EGP-2 is a 40-kDa glycoprotein that mediates Ca^{2+}-independent homotypic cell–cell adhesions. This oncofetal antigen is a human pan carcinoma-associated antigen that is abundantly expressed in colorectal carcinomas. Its expression is related to increased epithelial proliferation and negatively correlates with cell differentiation. A regulatory function of EGP-2 in the morphogenesis of epithelial tissue has been demonstrated for a number of tissues, in particular pancreas and mammary gland. Nelson *et al.* (1996) demonstrated that gp40, a molecule previously shown to be expressed by thymic epithelial cell lines *in vitro* and by thymic epithelial cells *in vivo*, is the murine homolog of human EGP-2. In a murine system, Xiang *et al.* (1997) showed recruitment of $CD8^+$ T cells as well as natural killer (NK) cells after immunization with an anti-human EGP-2 antibody-IL-2 fusion protein (huKSl/4-IL2) in BALB/c mice bearing the human EGP-2-transfected CT26 colon cancer, CT26-EpCAM. Recently, Imboden *et al.* (2001) used the cell line, CT26-Ep21.6, a subclone of CT26-EpCAM expressing low levels of MHC class I, and reported that the anti-tumor activity involved recruitment of NK cells rather than T cells, an observation consistent with the "no-self" nature of the functional activities of NK cells (Karre, 1995).

Elevated levels of the third oncofetal antigen listed, CEA, as well as the underglycosylated form of MUC-1 are expressed on gastrointestinal cancers as well as breast and non-small-cell lung cancers. Although the murine homologs from CEA and the *MUC-1* genes have been cloned (Beauchemin *et al.*, 1989; Vos *et al.*, 1991), most immunization studies in mice have employed the human genes and proteins. Kantor *et al.* (1992) were first to show anti-CEA cellular immune responses in mice immunized with a vaccinia virus vector encoding human CEA. They could show tumor growth inhibition of a murine colon carcinoma cell line transfected to express human CEA as well as concomitant humoral and cellular human CEA-specific responses.

Graham *et al.* (1996) immunized C57BL mice intramuscularly with naked *MUC1* cDNA and were able to achieve protection against a syngeneic MUC-1 expression tumor in a dose-dependent manner in 80% of mice. They reported the induction of a humoral as well as a cellular anti-MUC-1 immune response, but the antibody response did not correlate with tumor rejection. Recently, Xing *et al.* (2001) studied the effect of murine MUC1 in a particular mouse strain (C3H/HeOuj) prone to develop breast tumors expressing this antigen. The MUC1 was administered in the form of a mannan–muc1 fusion protein containing 10 tandem repeats. The murine muc1 immunization together with administration of cyclophosphamide, was required to enhance the anti-tumor immune response in these mice. Koido *et al.* (2000) immunized mice with MUC1 RNA-transfected DC and showed effective prevention as well as therapy against tumors expressing MUC 1. In transgenic mice tolerant to this antigen, however, the same vaccine was only successful when co-administered with IL-12.

Melanoma differentiation antigens. A large number of T-cell defined tumor antigens identified in humans over the past decade are shared determinants derived from lineage-specific or tissue-differentiation antigens expressed on melanomas. These are gp100, MAR-1, tyrosinase, TRP-1 and TRP-2 and the expression of their murine homologs has been confirmed. Nearly all preclinical murine studies involving these antigens have employed the spontaneously arising C57BL/6 BL6 melanoma and sublines derived from it, such as B16 and F16F10. In general, these tumor cells expressed little to no class I MHC (Gorelik *et al.*, 1991) and most of the immunization studies involving CTL-defined epitopes focus on inhibiting the growth of experimentally induced lung metastases rather than a locally growing subcutaneous tumor. A process highly dependent on NK-cell activity, furthermore, the level of pigmentation and other parameters of melanoma differentiation, including *in vivo* metastatic potential, of B16 cell lines have been shown to be influenced by the tissue culture medium used for its *in vitro* culture; tumor cells grown in DMEM medium showed an increased level of pigmentation and potential to colonize compared to cells grown in RPMI-1640 medium (Prezioso *et al.*, 1993). The use of RPMI-1640 medium to culture B16, therefore, may also influence its initial immunogenicity and antigenicity when used to challenge mice. Bloom *et al.* (1997) were first to clone the tissue-differentiation antigen tyrosinase-related protein 2 (TRP-2) in the mouse and identified the TRP-$2_{181-188}$ peptide as an H2-K^b-restricted epitope. They could show that CTL recognizing this epitope were able to eradicate established B16 pulmonary metastasis in the mouse. Similarly, Zeh *et al.* (1999)

generated anti-TRP-2 CTL from splenocytes of mice immunized with GM-CSF transduced B16 melanoma cell line and resistant to a parental B16 challenge. *In vitro* re-stimulation of these splenocytes in the presence of a low concentration of TRP-$1_{180-188}$ peptide (1×10^{-9}M) yielded higher avidity CTL than those generated in the presence of 1×10^{-5}M peptide. Overwijk *et al.* (1999) immunized normal mice with recombinant vaccinia virus (rVV) encoding murine TRP-2, gp100, MART-1 or tyrosinase and detected no autoimmune responses, in particular no loss of pigmentation. In contrast, immunization of normal mice with rVV encoding TRP-1 did induce autoimmune vitiligo. This vaccine was also effective in inducing immunity against the B16 melanoma. T-cell depletion experiments and immunization of MHC class II knockout mice revealed that the anti-tumor effect was $CD4^{+}$T cell-dependent. In contrast, Bronte *et al.* (2000) immunized mice (i.m.) with plasmid DNA encoding TRP-2 and showed that an anti-B16 melanoma response was induced in the absence of vitiligo. This response was shown by immunodepletion to be mainly dependent on CTL and NK-cell activity. Xiang *et al.* (2000) showed that oral application of a plasmid DNA vaccine consisting of the murine ubiquitin gene fused to mini genes encoding H2-D^{b}-restricted murine $gp100_{25-33}$ and H2-K^{b}-restricted TRP-$2_{181-188}$ epitopes was able to confer protection in mice and rejection of a B16-tumor growing subcutaneously. The ubiquitination of the antigens seems to be crucial for the efficacy of the construct, since a construct expressing only the epitopes was ineffective. Tüting *et al.* (1999b) have used a variety of viral and non-viral plasmid DNA vaccines encoding TRP-2 and delivery systems for preclinical murine tumor model studies. Tüting *et al.* (1999b) used "gene gun" immunization with plasmid DNA encoding TRP-2 to delay the outgrowth of B16 melanoma. Co-administration of a plasmid encoding IL-12 further enhanced the anti-tumor effect of the gene gun TRP-2 vaccine. In a related later study, gene gun immunization of mice with a plasmid encoding the highly homologous human *TRP-2* gene was shown to be more effective than the murine *TRP-2* gene immunization (Steitz *et al.*, 2000). The human gene-based vaccine induced depigmentation as well as significant protection against metastatic growth of B16. The enhanced immunogenicity of human TRP-2 relative to murine TRP-2 in mice was also evident when the efficacy of recombinant adenovirus encoding either murine or human *TRP-2* was studied (Steitz *et al.*, 2001). In other studies, a vaccine consisting of *ex vivo* genetically modified DC expressing TRP-2 was employed in immunization studies. Mice were immunized with DC transduced with an adenoviral vector expressing murine TRP-2. They were able to achieve prevention of metastatic disease in the B16-melanoma model and demonstrated that the anti-tumor effect was due to both CD8 and CD4 T-cell activity (Tüting *et al.*, 1999c).

Genetically engineered mouse models

The development of genetically altered mice for studying and manipulating the primary host's immune response to tumor antigens and tumorgenesis has become an increasingly important research objective in recent years. The ability to manipulate the murine germline is based principally on the advances in genetic engineering technologies that permit deletion or insertion of genes. (Smithies *et al.*, 1985; Thomas *et al.*, 1986). Several distinct types of genetically modified mice are available for use in analyzing various aspects of vaccine development and strategies for their use. Mice expressing HLA allelic transgenes are being used to facilitate the identification of naturally presented T-cell defined tumor antigens. The most applicable of these strains express the HLA-A2.1 allele, the most common allele among Caucasians with nearly 50% of the population expressing this haplotype (Lee, 1990). Many new T-cell epitopes have been identified in this way, Bernhard *et al.* (1988) and Theobald *et al.* (1995) being two examples in a period spanning nearly a decade of how transgenic mice expressing HLA-A2.1 allele were used to identify T-cell defined tumor epitopes. The fact that, in general, these mice can generate high affinity T cells to human tumor associated "self" antigens provides a novel approach to overcome tolerance to "self" tumor antigens. As previously mentioned, Liu *et al.* (2000) generated a high affinity, HLA-A2-resticted, anti-p53 CTL by transducing a human T-cell line with cDNA encoding the TCR expressed by a high-affinity, anti-human p53 CTL that was induced in an HLA-A2.1transgenic mouse. Recently, Stanislawski *et al.* (2001) extended this approach for overcoming tolerance to another human "self" tumor associated antigen, the HLA-A2.1-restricted, human $mdm2_{81-88}$ epitope.

The more prevalent types of transgenic mice used for developing immunotherapy of cancer are those expressing human tumor associated antigens, such as CEA and Her-2/neu. Although these transgenes are foreign to the mouse, the fact that they are not expressed in a "time and tissue specific" manner in these mice permits them to be "self" in nature. Adult mice that are transgenic for the human CEA gene express the transgene in the tongue, esophagus, stomach, small intestine, cecum, colon and trachea, and at low levels in the lung, testis and uterus. This model makes it possible to analyze negative side effects due to immunization against human CEA (Hasegawa *et al.*, 1991; Eades-Perner *et al.*, 1994; Clarke *et al.*, 1998). Kass *et al.* (1999) administered mice CEA as a whole protein in adjuvant or immunized mice with a recombinant vaccinia virus encoding CEA. Only mice immunized with rVV expressing CEA generated relatively strong anti-CEA IgG antibody titers and demonstrated evidence of immunoglobulin class switching, whereas the whole protein immunization failed to elicit an immune response. The development of Th 1-type CEA-specific $CD4^+$ T-cell responses and a CEA peptide-specific cytotoxicity correlated with protection of the transgenic CEA mice against a challenge with CEA-expressing tumor cells. They could not observe any autoimmunity related to CEA-based immunization. Xiang *et al.* (2001) showed that peripheral T-cell tolerance toward CEA could be broken in CEA-transgenic C57BL/6J mice by an oral CEA-based DNA vaccine delivered by the live, attenuated AroA-strain of *Salmonella typhimurium*. The vaccine-induced anti-tumor protection mediated by MHC class I-restricted $CD8^+$ T cells after a lethal challenge with murine MC38 colon carcinoma cells double transfected with CEA and the human epithelial cell adhesion molecule (Ep-CAM)/KSA. Boosts with the antibody–IL2 fusion protein KS1/4–IL2 markedly increased the efficacy of the tumor-protective immune response resulting in more effective tumor rejection. Activation of T cells was indicated by increased secretion of proinflammatory cytokines IFN-gamma, IL-12 and granulocyte/macrophage-colony stimulating factor, as well as specific tumor rejection and growth suppression in vaccinated CEA-transgenic mice.

Another type of transgenic mice useful in vaccine development are mice in which a transformation agent, such as simian virus 40 (SV40) large T, has been targeted for expression in a "time and tissue specific" manner, usually in breast and prostate tissues. Originally conceived as a useful tool for studying tumorigenesis, these mice also represent ideal models for evaluating therapies that target widely expressed transformation related antigens, such as p53, as well as lineage/tissue specific tumor associated antigens. In the case of the large T antigen, one must consider that the T antigen does represent a "non-self" antigen that may be contributing to the overall host's mediated tumor immune response.

Her-2/*neu* transgenic mice are being employed to evaluate a wide range of vaccines and immunization strategies for immunotherapy of breast cancer. In these mice, the oncogene is under the control of the MMTV promoter to induce its expression in a "time and tissue specific" manner in breast tissue. Amici *et al.* (1998) immunized transgenic FVB/*neu* mice with a plasmid DNA encoding neuNT leading to reduced outgrowth of mammary neoplasms as well as their metastatic penetrance. Booster immunizations skewed to a Th1 phenotype immune response and led to necrosis of established tumors. Subsequently, they immunized FVB/N *neu* transgenic mice by i.m injection of DNA encoding either full length rat *neu* oncogene or neu extracellular domain or neu extracellular and transmembrane domains (Amici *et al.*, 2000). They found that immunization against the transmembrane domain conferred the best protection against spontaneously arising mammary tumors. The co-administration of IL-12 encoding DNA enhanced this anti-tumor effect. In recent reports, Rovero *et al.* (2000, 2001) showed that the plasmid Her-2/neu DNA vaccination of transgenic BALB/c mice was more effective in blocking the spontaneous induction of tumors in these mice than the growth of a transplanted breast carcinoma, and that this effect was augmented by co-administration with a nonapeptide derived from IL-1β. Chen *et al.* (1998) immunized FVB/N *neu* transgenic mice with DNA expression vectors encoding either the full-length *neu* cDNA, the neu extracellular domain or the neu extracellular and transmembrane domains. Although all of these plasmids could induce protective immunity in FVB/N mice against Tgl-1 cells, a neu-expressing tumor cell line generated from a mouse mammary tumor that spontaneously arose in an FVB/N *neu* transgenic mouse, the full length Her-2/neu p185 cDNA was the least effective. The transmembrane expression plasmid was the most effective in inducing a humoral response, but induction of a humoral response did not correlate with tumor protection in this model system.

Cefai *et al.* (1999) induced an anti-Her-2/neu immune response in transgenic FVB mice using a cell-based vaccine consisting of allogeneic fibroblasts expressing Her-2/neu. The vaccination induced an anti-tumor immune response in these mice that prevented tumor formation of transplanted breast-tumor cells, and also protected the mice from spontaneous tumor formation. T-cell-mediated and humoral immune responses were detectable in the vaccinated mice. In contrast, the vaccine failed to protect against established spontaneous tumors.

Esserman *et al.* (1999) immunized *neu*-transgenic mice with a vaccine consisting of the recombinant extracellular domain of p185neu admixed with CFA. Immunized mice developed Her-2/neu-specific humoral immune responses, as measured by circulating anti-Her-2/neu antibodies in their sera, and cellular immune responses, as measured by lymphocyte proliferation to the antigen *in vitro*. In addition, the subsequent development of mammary tumors was significantly lower in immunized mice than in controls and vaccine treatment was associated with a significant increase in median survival.

The final type of genetically engineered mice to be discussed consists of mice that have been modified to yield tumors at specific sites, namely gastrointestinal carcinomas. All p53 homozygous mutant mice (p53−/−) develop tumors about nine months of age (Donehower *et al.*, 1992), with lymphomas accounting for nearly 75% of observed tumors, with a variety of solid tumors comprising the rest. Heterozygous mutant mice (p53+/−) usually develop tumors at a later timepoint, similar to the Li–Fraumeni syndrome in humans, in which patients heterozygous for constitutional mutations of p53 have sarcomas and tumors of the brain and breast (Malkin *et al.*, 1990; Jacks *et al.*, 1994). Recently, a congenic mouse strain lacking T-cell receptor beta chain (TCRβ) and p53 (*TCRβ−/−:Trp53−/−*) has been established (Funabashi *et al.*, 2001). Deficient for both genes, the occurrence of adenocarcinomas especially in the cecum is observed with high frequency in these mice. The Min (multiple intestinal neoplasms) mice express the dominant mutation in the *Min* gene in mice, which occurs in mice treated with the chemical ethylnitrosourea (Moser *et al.*, 1990). Min mice develop multiple tumors throughout their intestinal tract resembling those found in patients with familial adenomatous polyposis. Later on, the *APC* gene was identified in these patients as the gene responsible for this particular phenotype and it became clear that *APC* represents the human counterpart for the murine *Min*, which is now identified as APC^{Min} (Su *et al.*, 1992). Whereas animals derived from induction of homozygous APC mutants ($APC^{Min}-/-$) do not survive embryogenesis, heterozygous mutant mice ($APC^{Min}+/-$) develop intestinal adenomas similar to those found in humans (Oshima *et al.*, 1995). Consistent with findings in humans affected from colon carcinoma, adenomatous cells have lost the wt APC allele already at the microadenoma stage, suggestive of the mutation in the *APC gene* being an early event during colon carcinogenesis (Oshima *et al.*, 1997). Newer approaches involving methodologies targeting specific loci of recombination (lox) that at first do not interfere with the expression of the targeted gene result in generation of homozygous knockout mice. Subsequently, an adenoviral vector encoding the enzyme cre recombinase, capable of deleting the sequences between the lox sites, was introduced into the intestine of APC knockout mice, leading to development of intestinal adenomas derived from adenovirally infected cells (Shibata *et al.*, 1997).

Recent studies in SMAD knockout mice revealed that *SMAD* gene products, which are involved in transducing signals from the transforming growth factor β (TGF-β) receptor, are often deleted in human colon carcinoma (Gryfe *et al.*, 1997). Double heterozygotes mutant mice for Smad4 and $APC^{\Delta716}$ develop more malignant tumors than were observed in simple $APC^{\Delta716}$ heterozygotes (Takaku *et al.*, 1998). Targeting another *SMAD* gene, *Smad3*, gives rise to Smad3−/− mice that develop metastatic colorectal disease, representing another interesting mouse model to study potentially existing tumor antigens associated with intestinal neoplasia (Zhu *et al.*, 1998).

The EGP-2 transgenic FVN mouse strain was constructed by McLaughlin *et al.* (2001) to study immunotherapy targeting this pan-carcinoma antigen. Using parental and EGP-2 transfected B16 tumor cell lines, their initial investigations using anti-EGP-2 antibody for anti-tumor therapy showed specific localization of the antibody to the tumor site but not to normal epithelial tissues expressing the transgene. Collectively, these genetic modified mouse strains represent potentially clinically relevant murine models for future studies aimed at antigen discovery, as well as development of cancer vaccines and immunotherapy.

References

Abrams, S. I., Dobrzanski, M. J., Well, D. T., Stanziale, S. F., Zaremba, S., Masuelli, L. *et al.* (1995) Peptide-specific activation of cytolytic $CD4^+$ T lymphocytes against tumor cells bearing mutated epitopes of K-ras p21. *European Journal of Immunology*, **25**, 2588–2597.

Abrams, S. I., Stanziale, S. F., Lunin, S. D., Zaremba, S. and Schlom J. (1996) Identification of overlapping epitopes in mutant ras oncogene peptides that activate $CD4^+$ and $CD8^+$ T cell responses. *European Journal of Immunology*, **26**, 435–443.

Alpen, B., Gure, A., Scanlan, M., Old, L. and Chen, Y. (2002) A new member of the NY-ESO-1 gene family is ubiquitously expressed in somatic tissues and evolutionarily conserved. *Gene*, **297**, 141–149.

Amici, A., Smorlesi, A., Noce, G., Santoni, G., Cappalletti, P., Capparuccia, L. *et al.* (2000) DNA vaccination with full-length or truncated neu induces protective immunity against the development of spontaneous mammary tumors in HER-2/neu transgenic mice. *Gene Therapy*, **7**, 703–706.

Amici, A., Venanzi, F. M. and Concetti, A. (1998) Genetic immunization against neu/erbB2 transgenic breast cancer. *Cancer Immunology and Immunotherapy*, **47**, 183–190.

Beauchemin, N., Turbide, C., Afar, D., Bell, J., Raymond, M., Stanners, C. P. *et al.* (1989) A mouse analogue of the human carcinoembryonic antigen. *Cancer Research*, **49**, 2017–2021.

Bemhard, E. J., Le, A. X., Barbosa, J. A., Lacy, E. and Engelhard, V. H. (1988) Cytotoxic T lymphocytes from HLA-A2 transgenic mice specific for HLA-A2 expressed on human cells. *Journal of Experimental Medicine*, **168**, 1157–1162.

Bloom, M. B., Perry-Lalley, P. F., Robbins, P. F., Li, Y., el-Gamil, M., Rosenberg, S. A. *et al.* (1997) Identification of tyrosinase-related protein 2 as a tumor rejection antigen for the B16 melanoma. *Journal of Experimental Medicine*, **185**, 453–459.

Bos, J. L. (1989) *Ras* oncogenes in human cancer: a review. *Cancer Research*, **49**, 4682–4689.

Bristol, J. A., Orsini, C., Lindinger, P., Thalhamer, J. and Abrams, S. I. (2000) Identification of a ras oncogene peptide that contains both CD4(+) and CD8(+) T cell epitopes in a nested configuration and elicits both T cell subset responses by peptide or DNA immunization. *Cellular Immunology*, **205**, 73–83.

Bronte, V., Apolloni, E., Ronca, R., Zamboni, P., Overwijk, W. W., Surman, D. R. *et al.* (2000) Genetic vaccination with "self" tyrosinase-related protein 2 causes melanoma eradication but not vitiligo. *Cancer Research*, **60**, 253–258.

Bueler, H. and Mulligan, R. C. (1996) Induction of antigen-specific tumor immunity by genetic and cellular vaccines against MAGE: enhanced tumor protection by coexpression of granulocyte-macrophage colony-stimulating factor and B7-1. *Molecular Medicine*, **2**, 545–555.

Cefai, D., Morrison, B. W., Sckell, A., Favre, L., Balli, M., Leunig, M. *et al.* (1999) Targeting HER-2/neu for active-specific immunotherapy in a mouse model of spontaneous breast cancer. *Inernational Journal of Cancer*, **83**, 393–400.

Chen, Y., Hu, D., Eling, D. J., Robbins, J. and Kipps, T. J. (1998) DNA vaccines encoding full-length or truncated Neu induce protective immunity against Neu-expressing mammary tumors. *Cancer Research*, **58**, 1965–1971.

Clarke, P., Mann, J., Simpson, J. F., Rickard-Dickson, K. and Primus, F. J. (1998) Mice transgenic for human carcinoembryonic antigen as a model for immunotherapy. *Cancer Research*, **58**, 1469–1477.

Dahl, A. M., Beverly, P. C. and Stauss, H. J. (1996) A synthetic peptide derived from the tumor-associated protein mdm2 can stimulate autoreactive, high avidity cytotoxic T lymphocytes that recognize naturally processed protein. *Journal of Immunology*, **157**, 239–246.

DeLeo, A. B., Jay, G., Apella, E., DuBois, G. C., Law, L. W. and Old, L. J. (1979) Detection of a transformation-related antigen in chemically induced sarcomas and other transformed cells of the mouse. *Proceedings of the National Academy of Sciences USA*, **76**, 2420–2424.

De Plaen, E., De Backer, O., Arnaud, D., Bonjean, B., Chomez, P., Martelange, V. *et al.* (1999) A new family of mouse genes homologous to the human MAGE genes. *Genomics*, **55**, 176–184.

Disis, M. L. and Cheever, M. A. (1996) Oncogenic proteins as tumor antigens. *Current Opinion in Immunology*, **8**, 637–642.

Donehower, L. A., Harvey, M., Slagle, B. L., McArthur, M. J., Montgomery, C. A., Butel, J. S. *et al.* (1992) Mice deficient for p53 are developmentally normal but susceptible to spontaneous tumors. *Nature*, **356**, 215–221.

Donnellan, R. and Chetty, R. (1998) Cyclin D1 and human neoplasia. *Molecular Pathology*, **51**, 1–7.

Eades-Pemer, A. M., van der Putten, H., Hirth, A., Thompson, J., Neumaier, M., von Kleist, S. *et al.* (1994) Mice transgenic for the human carcinoembryonic antigen gene maintain its spatiotemporal expression pattern. *Cancer Research*, **54**, 4169–4176.

Esserman, L. J., Lopez, T., Montes, R., Bald, L. N., Fendly, B. M. and Campbell, M. J. (1999) Vaccination with the extracellular domain of p185neu prevents mammary tumor development in neu transgenic mice. *Cancer Immunology and Immunotherapy*, **47**, 337–342.

Fenton, R. G., Taub, D. D., Kwak, L. W., Smith, M. R. and Longo, D. L. (1993) Cytotoxic T cell response and *in vivo* protection against tumor cells harboring activated ras proto-oncogenes. *Journal of the National Cancer Institute*, **85**, 1266–1268.

Funabashi, H., Uchida, K., Kado, S., Matsuoka, Y. and Ohwaki, M. (2001) Establishment of a *TCR*β and *Trp53* genes deficient mouse strain as an animal model for spontaneous colorectal cancer. *Experimental animals/Japanese Association for Laboratory Animal Science*, **50**, 41–47.

Gaiger, A., Reese, V., Disis, M. L. and Cheever, M. A. (2000) Immunity to WT1 in the animal model and in patients with acute myeloid leukemia. *Blood*, **96**, 1480–1489.

Gao, L., Bellantuono, I., Elsasser, A., Marley, S. B., Gordon, M. Y., Goldman *et al.* (2000) Selective elimination of leukemic CD34(+) progenitor cells by cytotoxic T lymphocytes specific for WT1. *Blood*, **95**, 2198–2203.

Gilboa, E. (2001) The risk of autoimmunity associated with tumor immunotherapy. *Nature Immunology*, **2**, 789–792.

Gorelik, E., Jay, G., Kim, M., Hearing, V. J., DeLeo, A. and McCoy, J. P. Jr (1991) Effects of *H-2K*[b] gene on expression of melanoma-associated antigen and lectin-binding sites on BL6 melanoma cells. *Cancer Research*, **51**, 5212–5218.

Graham, R. A., Burchell, J. M., Beverley, P. and Taylor-Papadimitriou, J. (1996) Intramuscular immunisation with muc1 DNA can protect C57 mice challenged with muc1-expressing syngeneic mouse tumor cells. *International Journal of Cancer*, **65**, 664–670.

Grand, R. J. A. and Owen, D. (1991) The biochemistry of *ras* p21. *Biochemistry*, **279**, 609–631.

Grimm, C. F., Ortmann, D., Mohr, L., Michalak, S., Krohne, T. U., Meckel, S. *et al.* (2000) Mouse alpha-fetoprotein-specific DNA-based immunotherapy of hepatocellular carcinoma leads to tumor regression in mice. *Gastroenterology*, **119**, 1104–1112.

Gryfe, R., Swallow, C., Bapat, B., Redston, M. Gallinger, S. and Couture, J. (1997) Molecular biology of colorectal cancer. *Current Problems in Cancer*, **21**, 233–300.

Hasegawa, T., Isobe, K., Tsuchiya, Y., Oikawa, S., Nakazato, H., Ikezawa, H. *et al.* (1991) Establishment and characterisation of human carcinoembryonic antigen transgenic mice. *British Journal of Cancer*, **64**, 710–714.

Hernandez, J., Ko, A. and Sherman, L. A. (2001) CTLA-4 blockade enhances the CTL responses to the p53 self-tumor antigen. *Journal of Immunology*, **166**, 3908–3914.

Hollstein, M. D., Sidransky, D., Vogelstein, B. and Harris, C. C. (1991) p53 mutations in human cancers. *Science* (Washington DC), **253**, 49–53.

Houbiers, J. G., Nijman, H. W., van der Burg, S. H., Drijfhout, J. W., Kenemans, P., van de Velde, C. J., *et al.* (1993) *In vitro* induction of human cytotoxic T lymphocyte responses against peptides of mutant and wild-type p53. *European Journal of Immunology*, **23**, 2072–2077.

Ikuta, Y., Okugawa, T., Furugen, R., Nagata, Y., Takahashi, Y., Wang, L. *et al.* (2000) A HER2/NEU-derived peptide, a K(d)-restricted murine tumor rejection antigen, induces HER2-specific HLA-A2402-restricted CD8(+) cytotoxic T lymphocytes. *International Journal of Cancer*, **87**, 553–558.

Imboden, M., Murphy, K. R., Rakhmilevich, A. L., Neal, Z. C., Xiang, R., Reisfeld, R. A. *et al.* (2001) The level of MHC class I expression on murine adenocarcinoma can change the antitumor effector mechanism of immunocytokine therapy. *Cancer Research*, **61**, 1500–1507.

Ishida, T., Chada, S., Stipanov, M., Nadaf, S., Ciernik, F. I., Gabrilovich *et al.* (1999) Dendritic cells transduced with wild-type p53 gene elicit potent anti-tumour immune responses. *Clinical and Experimental Immunology*, **117**, 244–251.

Jacks, T., Remington, L., Williams, B. O., Schmitt, E. M., Halachmi, S., Bronson, R. T. *et al.* (1994) Tumor spectrum analysis in p53-mutant mice. *Current Biology*, **4**, 1–7.

Jurk, M., Kremmer, R., Schwarz, U., Forster, R. and Winnacker, E. L. (1998) MAGE-11 protein is highly conserved in higher organisms and located predominately in the nucleus. *International Journal of Cancer*, **75**, 762–766.

Kantor, J., Irvine, K., Abrams, S., Kaufman, H., Di Pietro, J. and Schlom, J. (1992) Antitumor immunity and immune responses induced by a recombinant carcinoembryonic antigen vaccinia virus vaccine. *Journal of the National Cancer Institute*, **84**, 1059–1061.

Karre, K. (1995) Express yourself or die: peptides, MHC molecules, and NK cells. *Science*, **267**, 978–979.

Kass, E., Schlom, J., Thompson, J., Guadagni, F., Graziano, P. and Greiner, J. W. (1999) Induction of protective host immunity to carcinoembryonic antigen (CEA), a self-antigen in CEA transgenic mice, by immunizing with a recombinant vaccinia-CEA virus. *Cancer Research*, **59**, 676–683.

Klein, G., Sjogren, H. O., Klein, E. and Hellstrom, K. E. (1960) Demonstration of resistance against methylcholanthrene induced sarcomas in the primary autochthonous host. *Cancer Research*, **20**, 1561–1572.

Koido, S., Kashiwaba, M., Chen, D., Gendler, S., Kufe, D. and Gong, J. (2000) Induction of antitumor immunity by vaccination of dendritic cells transfected with MUC1 RNA. *Journal of Immunology*, **165**, 5713–5719.

Lee, J. H., Sung, B. W., Youn, H. J. and Park, J. H. (2000) Identification, expression and nuclear location of murine Mage-b2 protein, a tumor-associated antigen. *Molecules and Cells*, **10**, 647–653.

Lee, T. (1990) Distribution of HLA antigens in North American Caucasians, North American Blacks and Orientals. In *The HLA System*, edited by J. Lee, pp. 141, New York: Springer-Verlag.

Liu, X., Peralta, E. A., Ellenhorn, J. D. and Diamond, D. J. (2000) Targeting of human p53-overexpressing tumor cells by an HLA A*0201-restricted murine T-cell receptor expressed in Jurkat T cells. *Cancer Research*, **60**, 693–701.

McLauglin P. M. J., Harmsen, M. C., Dokter, W. H. A., Kroesen, B.-J., van der Molen, H. *et al.* (2001) The epithelial glycoprotein 2 (EGF-2) promoter-driven epithelial-specific expression of EGP-2 in transgenic mice: a new model to study carcinoma-directed immunotherapy. *Cancer Research*, **61**, 4105–4111.

Malkin, D., Li, F. P., Strong, L. C., Fraumeni, J. F., Nelson, C. E., Kim, D. H. *et al.* (1990) Germ line p53 mutations in a familial syndrome of breast cancer, sarcomas and other neoplasms. *Science*, **250**, 1233–1238.

Mayordomo, J. I., Loftus, D. J., Sakamoto, H., De Cesare, C. M., Appasamy, P. M., Lotze, M. T. *et al.* (1997) Therapy of murine tumors with p53 wild-type and mutant sequence peptide-based vaccines. *Journal of Experimental Medicine*, **183**, 1357–1365.

Moser, A. R., Pitot, H. C. and Dove, W. F. (1990) A dominant mutation that predisposes to multiple intestinal neoplasia in the mouse. *Science*, **247**, 322–324.

Mumberg, D., Wick, M. and Schreiber H. (1996) Unique tumor antigens redefined as mutant tumor-specific antigens. *Seminars in Immunology*, **8**, 289–293.

Nagata, Y., Furugen, R., Hiasa, A., Ikeda, H., Ohta, N., Furukawa, K. *et al.* (1997) Peptides derived from a wild-type murine proto-oncogene *c-erbB-2/HER2/neu* can induce CTL and tumor suppression in syngeneic hosts. *Journal of Immunology*, **159**, 1336–1343.

Nelson, A. J., Dunn, R. J., Peach, R., Aruffo, A. and Farr, A. G. (1996) The murine homolog of human Ep-CAM, a homotypic adhesion molecule, is expressed by thymocytes and thymic epithelial cells. *European Journal of Immunology*, **26**, 401–408.

Noguchi, Y., Chen, Y. T. and Old, L. J. (1994) A mouse mutant p53 product recognized by $CD4^+$ and $CD8^+$ T cells. *Proceedings of the National Academy of Sciences USA*, **91**, 3171–3175.

Noguchi, Y., Richards, E. C., Chen, Y. T. and Old, L. J. (1995) Influence of interleukin 12 on peptide vaccination against established Meth/A sarcoma. *Proceedings of the National Academy of Sciences USA*, **92**, 2219–2223.

Oka, Y., Udaka, K., Tsuboi, A., Elisseeva, O. A., Ogawa, H., Aozasa *et al.* (2000) Cancer immunotherapy targeting Wilms' tumor gene WT1 product. *Journal of Immunology*, **164**, 1873–1880.

Okugawa, T., Ikuta, Y., Takahashi, Y., Obata, H., Tanida, K., Watanabe, M. *et al.* (2000) A novel human HER2-derived peptide homologous to the mouse K(d)-restricted tumor rejection antigen can induce HLA-A24-restricted cytotoxic T lymphocytes in ovarian cancer patients and healthy individuals. *European Journal of Immunology*, **11**, 3338–3346.

Old, L. J. (1981) Cancer immunology: the search for specificity – G. H. A. Clowes Memorial Lecture. *Cancer Research*, **41**, 361–375.

Oshima, M., Oshima, H., Kitagawa, K., Kobayashi, M., Itakura, C. and Taketo, M. (1995) Loss of Apc heterozygosity and abnormal tissue building in nascent intestinal polyps in mice carrying a truncated *APC* gene. *Proceedings of the National Academy of Sciences USA*, **92**, 4482–4486.

Oshima, H., Oshima, M., Kobayashi, M., Tsutsumi, M. and Taketo, M. M. (1997) Morphological and molecular processes of polyp formation in APC (delta716) knockout mice. *Cancer Research*, **57**, 1644–1649.

Osterlund, C., Tohonen, V., Forslund, K. O. and Nordqvist, K. (2000) Mage-b4, a novel melanoma antigen (MAGE) gene specifically expressed during germ cell differentiation. *Cancer Research*, **60**, 1054–1061.

Overwijk, W. W., Lee, D. S., Surman, D. R., Irvine, K. R., Touloukian, C. E., Chan, C. *et al.* (1999) Vaccination with recombinant vaccinia virus encoding a "self" antigen induces autoimmune vitiligo and tumor cell destruction in mice: requirement for $CD4^+$ T lymphocytes. *Proceedings of the National Academy of Sciences USA*, **96**, 2982–2987.

Peace, D. J., Chen, W., Nelson, H. and Cheever, M. A. (1991) T cell recognition of transforming proteins encoded by mutated *ras* proto-oncogenes. *Journal of Immunology*, **146**, 2059–2065.

Peace, D. J., Smith, J. W., Disis, M. L., Chen, W. and Cheever, M. A. (1993) Induction of T cells specific for the mutated segment of oncogenic p21ras protein by immunization *in vivo* with the oncogenic protein. *Journal of Immunotherapy*, **14**, 110–114.

Peace, D. J., Smith, J. W., Chen, W., You, S. G., Cosand, W. L., Blake, J. *et al.* (1994) Lysis of *ras* oncogene-transformed cells by specific cytotoxic T lymphocytes elicited by primary *in vitro* immunization with mutated ras peptide. *Journal of Experimental Medicine*, **179**, 473–479.

Prezioso, J. A., Wang, N., Duty, L., Bloomer, W. D. and Gorelik, E. (1993) Enhancement of pulmonary metastasis formation and gamma-glutamyltranspeptidase activity in B16 melanoma induced by differentiation *in vitro*. *Clinical and Experimental Metastasis*, **11**, 263–274.

Renkvist, N., Castelli, C., Robbins, P. F. and Parmiani, G. (2001) A listing of human tumor antigens recognized by T cells. *Cancer Immunology and Immunotherapy*, **50**, 3–15.

Rovero, S., Amici, A., Carlo, E. D., Bei, R., Nanni, P., Quaglino, E. *et al.* (2000) DNA vaccination against rat her-2/Neu p185 more effectively inhibits carcinogenesis than transplantable carcinomas in transgenic BALB/c mice. *Journal of Immunology*, **65**, 5133–5142.

Rovero, S., Boggio, K., Carlo, E. D., Amici, A., Quaglino, E., Porcedda, P. *et al.* (2001) Insertion of the DNA for the 163–171 peptide of IL1beta enables a DNA vaccine encoding p185(neu) to inhibit mammary carcinogenesis in Her-2/neu transgenic BALB/c mice. *Gene Therapy*, **8**, 447–452.

Sadovnikova, E. and Stauss, H. J. (1996) Peptide-specific cytotoxic T lymphocytes restricted by nonself major histocompatibility complex class I molecules: reagents for tumor immunotherapy. *Proceedings of the National Academy of Sciences USA*, **93**, 13114–13118.

Satoh, T. and Kaziro, Y. (1992) *Ras* in signal transduction. *Seminars in Cancer Biology*, **3**, 169–177.

Shibata, H., Toyama, K., Shioya, H., Ito, M., Hirota, M., Hasegawa, S. *et al.* (1997) Rapid colorectal adenoma formation initiated by conditional targeting of the *APC* gene. *Science*, **278**, 120–123.

Skipper, J. and Stauss, H. J. (1993) Identification of two cytotoxic T lymphocyte-recognized epitopes in the ras protein. *Journal of Experimental Medicine*, **177**, 1493–1498.

Smithies, O., Gregg, R. G., Boggs, S. S., Koralewski, M. A. and Kucherlapati, R. S. (1985) Insertion of DNA sequences into the human chromosomal beta-globin locus by homologous recombination. *Nature*, **317**, 230–234.

Sogn, J. A. (1998) Tumor immunology: the glass is half full. *Immunity*, **9**, 757–763.

Stanislawski, T., Voss, R. H., Lotz, C., Sadovnikova, E., Willemsen, R. A., Kuball, J. *et al.* (2001) Circumventing tolerance to a human MDM2-derived tumor antigen by TCR gene transfer. *Nature Immunology*, **2**, 962–970.

Steitz, J., Bruck, J., Knop, J. and Tuting T. (2001) Adenovirus-transduced dendritic cells stimulate cellular immunity to melanoma via a CD4(+) T cell-dependent mechanism. *Gene Therapy*, **16**, 1255–1263.

Steitz, J., Bruck, J., Steinbrink, K., Enk, A., Knop, J. and Tuting T. (2000) Genetic immunization of mice with human tyrosinase-related protein 2: implications for the immunotherapy of melanoma. *International Journal of Cancer*, **86**, 89–94.

Su, L. K., Kinzler, K. W., Vogelstein, B., Preisinger, A. C., Moser, A. R., Luongo, C. *et al.* (1992) Multiple intestinal neoplasia caused by a mutation in the murine homolog of the *APC* gene. *Science*, **256**, 668–670.

Sypniewska, R. K., Hoflack, L., Bearss, D. J. and Gravekamp, C. (2002) Potential mouse tumor model for pre-clinical testing of mage-specific breast cancer vaccines. *Breast Cancer Research and Treatment*, **74**, 221–233.

Takaku, K., Oshima, M., Miyoshi, H., Matsui, M., Seldin, M. F. and Taketo, M. M. (1998) Intestinal tumorigenesis in compound mutant mice of both *Dpc4* (*Smad4*) and *Apc* genes. *Cell*, **92**, 645–656.

Theobald, M., Biggs, J., Dittmer, D., Levine, A. J. and Sherman, L. A. (1995) Targeting p53 as a general tumor antigen. *Proceedings of the National Academy of Sciences USA*, **92**, 11993–11997.

Theobald, M., Biggs, J., Hernandez, J., Lustgarten, J., Labadie, C. and Sherman, L. A. (1997) Tolerance to p53 by A2.1-restricted cytotoxic T lymphocytes. *Journal of Experimental Medicine*, **185**, 833–841.

Thomas, K. R., Folger, K. R. and Capecchi, M. R. (1986) High frequency targeting of genes to specific sites in the mammalian genome. *Cell*, **14**, 419–428.

Tsuboi, A., Oka, Y., Ogawa, H., Elisseeva, O. A., Li, H., Kawasaki, K. *et al.* (2000) Cytotoxic T-lymphocyte responses elicited to Wilms' tumor gene WT1 product by DNA vaccination. *Journal of Clinical Immunology*, **20**, 195–202.

Tüting, T., DeLeo, A. B., Lotze, M. T. and Storkus, W. J. (1997) Genetically modified bone marrow-derived dendritic cells expressing tumor-associated viral or "self" antigens induce antitumor immunity *in vivo*. *European Journal of Immunology*, **27**, 2702–2707.

Tüting, T., Gambotto, A., Robbins, P. D., Storkus, W. J. and DeLeo, A. B. (1999a) Co-delivery of T helper 1-biasing cytokine genes enhances the efficacy of gene gun immunization of mice: studies with the model tumor antigen beta-galactosidase and the BALB/c Meth A p53 tumor-specific antigen. *Gene Therapy*, **6**, 629–636.

Tüting, T., Gambotto, A., DeLeo, A., Lotze, M. T., Robbins, P. D. and Storkus, W. J. (1999b) Induction of tumor antigen-specific immunity using plasmid DNA immunization in mice. *Cancer Gene Therapy*, **6**, 73–80.

Tüting, T., Steitz, J., Bruck, J., Gambotto, A., Steinbrink, K., DeLeo, A. B. *et al.* (1999c) Dendritic cell-based genetic immunization in mice with a recombinant adenovirus encoding murine TRP2 induces effective anti-melanoma immunity. *Journal of Gene Medicine*, **1**, 400–406.

Van den Eynde, B., Lethe, B., Van Pel, A., De Plaen, E. and Boon, T. (1991) The gene coding for a major tumor rejection antigen of tumor P815 is identical to the normal gene of syngeneic DBA/2 mice. *Journal of Experimental Medicine*, **173**, 1373–1384.

Van Pel, A., De Plaen, E., Duffour, M.-T., Warnier, G., Uyttenhove, C., Perricaudet, M. *et al.* (2001) Induction of cytolytic T lymphocytes by immunization of mice with an adenovirus containing a mouse homolog of the human MAGE-A genes. *Cancer Immunology and Immunotherapy*, **49**, 593–602.

Vierboom, M. P. M., Nijman, H. W., Offringa, R., van der Voort, E. I. H., van Hall, T., van den Broek *et al.* (1997) Tumor eradication by wild-type p53-specific cytotoxic T lymphocytes. *Journal of Experimental Medicine*, **186**, 695–704.

Vierboom, M. P., Zwaveling, S., Bos, G. M. J., Ooms, M., Krietemeijer, G. M., Melief, C. J. *et al.* (2000) High steady-state levels of p53 are not a prerequisite for tumor eradication by wild-type p53-specific cytotoxic T lymphocytes. *Cancer Research*, **60**, 5508–5513.

Vos, H. L., de Vries, Y. and Hilkens, J. (1991) The mouse episialin (*Muc 1*) gene and its promoter: rapid evolution of the repetitive domain in the protein. *Biochemical and Biophysical Research Communications*, **181**, 121–130.

Watanabe, T., Ichikawa, A., Saito, H. and Hotta, T. (1996) Overexpression of the MDM2 oncogene in leukemia and lymphoma. *Leukemia Lymphoma*, **21**, 391–397.

Xiang, R., Lode, H. N., Dolman, C. S., Dreirer, T., Varki, N. M., Qian, X. *et al.* (1997) Elimination of established murine colon carcinoma metastases by antibody-interleukin 2 fusion protein therapy. *Cancer Research*, **57**, 4948–4955.

Xiang, R., Lode, H. N., Chao, T. H., Ruehlmann, J. M., Dolman, C. S., Rodriguez *et al.* (2000) An autologous oral DNA vaccine protects against murine melanoma. *Proceedings of the National Academy of Sciences USA*, **97**, 5492–5497.

Xiang, R., Silletti, S., Lode, H. N., Dolman, C. S., Ruehlmann, J. M., Niethammer, A. G. *et al.* (2001) Protective immunity against human carcinoembryonic antigen (CEA) induced by an oral DNA vaccine in CEA-transgenic mice. *Clinical Cancer Research*, **7**, 856s–864s.

Xing, P. X., Poulos, G. and McKenzie, I. F. (2001) Breast cancer in mice: effect of murine MUC-1 immunization on tumor incidence in C3H/HeOuj mice. *Journal of Immunotherapy*, **24**, 10–18.

Yanuck, M., Carbone, D. P., Pendleton, C. D., Tsuki, T., Winter, S. F., Minna, J. D. *et al.* (1993) A mutant p53 tumor suppressor protein is a target for peptide induced CD8$^+$ cytotoxic T cells. *Cancer Research*, **53**, 3257–3261.

Zeh, H. J. 3rd Perry-Lalley, D., Dudley, M. E., Rosenberg, S. A. and Yang, J. C. (1999) High avidity CTLs for two self-antigens demonstrate superior *in vitro* and *in vivo* antitumor efficacy. *Journal of Immunology*, **162**, 989–994.

Zhu, Y., Richardson, J. A., Parada, L. F. and Graff, J. M. (1998) Smad3 mutant mice develop metastatic colorectal cancer. *Cell*, **94**, 703–714.

Chapter 2

Role of heat shock protein in chaperoning tumor antigens and modulating anti-tumor immunity

Zihai Li

Summary

Searching for tumor-specific transplantation antigens for chemically induced tumors in rodents have led to uncovering the immunological properties of heat shock proteins (HSPs). Known best for their roles in protein folding and chaperoning, HSPs are now found to, (a) chaperone antigenic peptides, (b) modulate the functions of professional antigen presenting cells (APCs) and, (c) mediate presentation and cross-presentation of antigens to MHC molecules for T-cell activations. Thus, the roles of HSPs have extended beyond tumor immunity. This article summarizes the general immunological principles associated with HSPs. Features of specific HSPs including gp96, HSP90, HSP70, calreticulin (CRT), HSP110 and GRP170 are discussed in detail in the context of anti-tumor immune responses.

Introduction: discovery of HSPs in chaperoning anti-tumor immunity

According to the concept of immunosurveillance first proposed by Burnet, one of the major functions of the adaptive immunity is to patrol and protect the host against malignancies due to the constant risk of somatic mutations and transformations (Burnet, 1970). Over the years, there has been a large collection of evidence for and against this theory. Nevertheless, it is increasingly appreciated that the immune system does play a critical role in the interaction between the host and malignancy. This is reinforced by the recent demonstration that an adequate immune system is critical in preventing the onset of clinically detectable tumors induced by carcinogens, or developed spontaneously (Shankaran *et al.*, 2001). Therefore, understanding the mechanism of anti-tumor immune response is essential for generating immunotherapeutic approaches against cancer, and for realizing the dream that one day tumors can be prevented by a simple "shot" of tumor vaccines.

Tumor-specific immunity was convincingly demonstrated by a series of transplantation experiments of syngeneic tumors as early as in the 1950s (Gross, 1943; Baldwin, 1955; Prehn and Main, 1957; Klein *et al.*, 1960; Old *et al.*, 1962). Inactivated tumor cells were shown to immunize syngeneic animals against the subsequent challenge by the same, but live tumors. This phenomenon is not restricted to tumor types or hosts. Serological and biochemical studies have linked the activity of "tumor rejection antigens" to a number of HSPs including HSP90, HSP70, CRT and an endoplasmic reticulum (ER) residential HSP90 paralog gp96 (reviewed by Li, 1997; Srivastava *et al.*, 1998), and most recently HSP110 and GRP170 (Wang *et al.*, 2001). HSPs are a family of proteins that are essential for all the cellular functions involving the folding/unfolding of the polypeptide chains (Lindquist and Craig, 1988;

Hartl, 1996). It was surprising initially, that HSPs were "tumor rejection antigens" since HSPs are among the most conserved molecules in evolution (see a review by Srivastava and Heike, 1991). Indeed, after an extensive cloning and sequencing work, the tumor-associated structural mutations were never found in the case of the *gp96* gene (Srivastava and Maki, 1991).

Close scrutiny has led to the conclusion that HSPs are not immunogenic *per se*, but rather act as carriers for small molecular weight peptides (Li and Srivastava, 1993; Udono and Srivastava, 1993; Isshi *et al.*, 1999; Nair *et al.*, 1999). Thus, HSPs exist as HSP–peptide complexes (HSP–PC), and immunization with HSP–PC induces T-cell immunity against the peptides, but not against HSPs themselves. The basis for specific immunity elicited by tumor-derived HSPs, but not HSPs from normal tissues, is because of the tumor-specific peptides associated with HSPs of the tumor origin. Moreover, purified tumor-derived HSP–PCs are potent vaccines for pre-established cancers or subsequent tumor challenges in a wide variety of tumor models such as lung, melanoma, lymphoma, fibrosarcoma, adenocarcinoma of the colon, sarcoma, prostate or breast cancers (see reviews by Li, 1997; Srivastava *et al.*, 1998).

While the peptide-binding property of HSP is predictable in light of the protein chaperoning function of HSPs in general, the recent discovery of the immunomodulating properties of HSPs was entirely unexpected. It was found that gp96, HSP90, HSP60, CRT, HSP70 can all bind to the surface of APC in a receptor-dependent manner. In the case of gp96, HSP70, HSP90 and CRT, one of the receptors on APCs is CD91 (Binder *et al.*, 2000; Basu *et al.*, 2001). The interaction of HSPs with APCs results in two functional consequences: the activation of APC to increase the expression of antigen presenting and co-stimulatory molecules, and the cross-presentation of HSP-associated peptides to MHC class I molecules on the surface of APCs for the priming of cytotoxic T lymphocytes (CTL).

Therefore, HSPs themselves are not tumor antigens. They bind to peptides, and are involved in modulating the function of APCs. Since APC is important in bridging innate and adaptive immunity, HSPs are now sitting at the intersection between these two arms of immunity. Thus the roles of HSPs are now extending beyond tumor immunity. To keep with the scope and spirit of this book, a detailed description of each HSPs in the context of tumor immunity ensues. A model will be provided to unify what we know about HSPs in immune responses and to perspectively illustrate what lies ahead in the de-coding of this class of mysterious but fascinating immunomodulating molecules.

Roles of major mammalian HSPs in immune responses

gp96

Gp96 stands for glycoprotein of 96 kDa. In humans, only one true gene locus has been mapped and was nomenclatured as *tra-1* (Maki *et al.*, 1993). In literature, gp96 is also referred as GRP94, Erp99, endoplasmin, etc. (Csermely *et al.*, 1998; Argon and Simen, 1999). The first report to link gp96 with tumor immunity came from Srivastava and Das who showed that in a Wistar rat Zajdela ascitic hepatoma model, a homogenous preparation of an ~100 kDa protein (named ZAH-TATA) was able to immunize against the challenge by the parental tumors (Srivastava and Das, 1984). Although ZAH-TATA was not molecularly defined in this report, the biochemical property of it suggests that ZAH-TATA is a rat homolog of gp96.

Using methylcholanthrene (MCA)-induced fibrosarcoma (Meth A) as a model in BALB/c mice, a series of papers in 1980s described the identification of a glycoprotein of 96 kDa (gp96) in its ability to immunize naïve mice against the challenges from the tumor where gp96 was purified (Srivastava *et al.*, 1986, 1987; Palladino Jr *et al.*, 1987). Gp96 was found to be ubiquitously expressed in both normal and tumor cells and turned out to be identical to Erp99, an abundant protein of the ER (Mazzarella and Green, 1987), although surface expression of gp96 was also demonstrated. Subsequently, gp96 was found to be a *bona fide* protein chaperone or HSP because of heat or stress inducibility (Altmeyer *et al.*, 1996), adenosine nucleotide binding and ATPase activity (Li and Srivastava, 1993), and apparent unselective binding to unfolded proteins (Csermely *et al.*, 1998). There are no tumor associated structural mutations of gp96 (Srivastava, personal communication). It became clear that gp96 was not a tumor-rejection antigen *per se* but is a protein carrier for peptides including tumor-specific peptides.

The function of gp96 as carriers for antigenic peptides has now been validated by both structural and immunological studies in multiple systems. This is largely achieved by identifying known peptides such as viral antigens, minor histocompatibility antigens and other model antigens that are present in the gp96–PCs (Srivastava *et al.*, 1998). For example, gp96 purified from vesicular stomatitis virus (VSV)-infected cells, but not uninfected cells, was associated with an 8-mer VSV-derived peptide as revealed by both structural and immunological assays (Nieland *et al.*, 1996). Similarly, it was shown that highly purified gp96 from cells expressing β-galactosidase (β-gal), minor histocompatibility antigens (Arnold *et al.*, 1995), ovalbumin (Breloer *et al.*, 1998; Nair *et al.*, 1999) and murine leukemia Rl male symbol antigens (Ishii *et al.*, 1999), was found to associate with respective peptides derived from these proteins. In all these systems, immunization with gp96–PCs in the absence of any exogenous adjuvant, primed MHC class I-restricted CTLs against the corresponding antigens, but not against gp96 itself. Moreover, compared with peptide alone, gp96–PCs immunization is several orders of magnitudes more efficient, in both sensitization of target cells for CTL recognition *in vitro* (Suto and Srivastava, 1995) and in the priming of naïve CTLs *in vivo* (Blachere *et al.*, 1997). This data implies that gp96 could serve as both a peptide carrier and an adjuvant. Recently, it was found in patients with hepatitis virus B-associated hepatocellular carcinomas that a virus-specific peptide is associated with gp96 in three out of three patients (Meng *et al.*, 2001).

The binding of gp96 to peptides *in vitro* has no apparent bias towards the sequence and length of the peptides (Blachere *et al.*, 1997; Li, unpublished). Therefore, if a tumor-associated peptide is defined, it can be easily complexed with gp96 *in vitro* by a simple heat-dependent refolding assay. The reconstituted gp96–PC is more efficient than the peptide alone to prime CD8+ T cells (Blachere *et al.*, 1997). The peptide-binding property of gp96 has also been supported by a number of elegant biophysical and structural assays (Wearsch *et al.*, 1998; Sastry and Linderoth, 1999). Recently, by protease mapping and cross-linking approaches, the minimal peptide-binding site of gp96 was mapped to amino acid residues 624–630 in a highly conserved region (Linderoth *et al.*, 2000).

The cellular mechanism for gp96-elicited immunity has been studied. It was shown that depletion of APCs during immunization phase abolished the vaccination effect of gp96 (Udono *et al.*, 1994). Moreover, macrophage-like cells in the peritoneal exudates can re-present or cross-present peptides from gp96–PCs to MHC class I for T-cell recognition (Suto and Srivastava, 1995). The unexpected potent efficiency of gp96–PC vaccination and the dependence on APCs led Srivastava *et al.* (1994) to suggest the presence of a receptor for gp96 on APCs. By immunofluorescence and electron microscopy, multiple groups have indeed provided convincing evidence for a receptor-like molecule on APCs for gp96 (Arnold-Schild *et al.*, 1999; Wassenberg *et al.*, 1999; Binder *et al.*, 2000a; Singh-Jasuja *et al.*, 2000a). Furthermore, by using affinity purification with gp96-conjugated column, Binder *et al.* (2000b) have purified the cell surface ligand (receptor) for gp96 from a macrophage cell line, RAW 264.7 to its homogeneity. A polyclonal antibody raised against this protein can block the re-presentation of gp96 chaperoned peptide to MHC class I. Microsequencing by mass spectrometry of this molecule, confirmed it to be CD91, a protein known as α2-macroglobulin receptor or the low-density lipoprotein-related protein. CD91 as a receptor for gp96 was further supported by the evidence that α2-macroglobulin, a previously known CD91 ligand, inhibited re-presentation of gp96 chaperoned antigenic peptides by macrophages, as did antibodies against CD91 (Binder *et al.*, 2000b; Basu *et al.*, 2001). In addition, gp96 can activate immature dendritic cells (DCs) in a peptide-independent manner to secrete cytokines and to induce the surface expression of co-stimulatory molecules such as CD80, CD86 and MHC class II molecules (Basu *et al.*, 2000; Singh-Jasuja *et al.*, 2000b). Thus, the dual properties of gp96 in chaperoning antigenic peptides and modulating the function of APCs have been established.

The therapeutic roles of tumor-derived autologous gp96 against malignancies are currently under clinical testing. Thus far, it was found that gp96 vaccination is well tolerated and shows no or minimal toxicity. In some cases, tumor-specific T-cell responses can be detected in the peripheral blood (Janeztki *et al.*, 2000). Whether or not gp96 is effective in the treatment of cancers awaits further phase III testing (Caudill and Li, 2001).

HSP90

HSP90 is an abundant cytosolic HSP that is essential in the folding, activation, and assembly of proteins that are particularly involved in signal transduction, cell cycle control and transcriptional regulations (Csermely *et al.*, 1998). HSP90 and gp96 are structurally related, and it is thought that *gp96* gene is perhaps duplicated from the *HSP90* gene and has been evolved to perform specific functions related to the ER. Not surprisingly, HSP90 purified from tumor cells was also shown to be a "tumor rejection antigen" (Ullrich *et al.*, 1986). HSP90 has two isoforms: the inducible HSP90α (HSP84) and the constitutively expressed HSP90β (HSP86). Immunization with the mixture of both isoforms purified from Meth A fibrosarcoma induced protection against subsequent Meth A challenge (Ullrich *et al.*, 1986). The two HSP90 isoforms are 76% homologous as a result of gene duplication about 500 million years ago (Moore *et al.*, 1989; Krone and Sass, 1994). They probably perform similar functions, although direct comparison of the immunological properties of these two isoforms on the equal molar basis has never been performed. It is unclear, therefore, which form is associated with more peptides, and therefore which one is more immunogenic. The ability of HSP90 to act as "tumor rejection antigen" was subsequently confirmed by Udono *et al.* (1994) who directly compared the efficiency of HSP90 with gp96 and HSP70 in eliciting immunity against Meth A (Udono and Srivastava, 1993). It was found that gp96 and HSP70 were equally immunogenic, whereas the immunogenidty of HSP90 was approximately 10% of that of gp96 or HSP70 on equal molar basis. However, since HSP90 is abundant (estimated to be 1% of the total cytosolic proteins) and could be released in large quantities upon cell death, the relative contribution of HSP90–PCs in priming T cells *in vivo* is expected to be more significant.

The peptide-binding property of HSP90 was confirmed both functionally and biochemically in an RLmale1 mouse leukemia model, in which the tumor-rejection antigen was known to be the octamer epitope (IPGLPLSL or pRL1a), which was derived from a mutated *akt* gene product (Ishii *et al.*, 1999). This epitope was recognized by a CTL clone in L^d-restricted manner. To demonstrate that HSP90 associated with this epitope, HSP90 was purified from Rlmale1 cells to its homogenicity. Small molecular weight peptides were then stripped by acid extraction, size enriched and resolved by a reverse phase HPLC according to the methodology developed for analyzing peptides associated with MHC molecules (Falk *et al.*, 1990; Van Bleek and Nathenson, 1990). Mass of the peptides was determined by mass spectrometry, and further confirmed by the ability of peptides to pulse an antigen-negative target cell for lysis by a pRL1a-specific CTL clone. It was found that HSP90 not only associated with the final 8-mer CTL epitope, but it also associated with two other longer precursor peptides, one of which was larger than 10-mer (Ishii *et al.*, 1999).

The immunomodulating function of HSP90 is not as well studied compared to gp96, although HSP90 clearly binds to CD91 on the surface of APCs (Basu *et al.*, 2001). The cellular mechanism underlying the immunological property of HSP90 is also unclear. Interestingly, two immunoregulatory drugs namely taxol and geldanomycin seem to elicit their functions on direct binding to HSP90 (Byrd *et al.*, 1999). Taxol is a microtubule-stablizing plant-derived antitumor agent. It appears to induce cell signals in a manner indistinguishable from bacterial lipopolysaccharide (LPS). Using biotin-labeled Taxol, avidin-agarose affinity chromatography and peptide mass fingerprinting, two Taxol targets from mouse macrophages and brain were identified as HSP70 and HSP90 (Byrd *et al.*, 1999). Geldanamycin is a specific inhibitor of the HSP90 family by binding directly to the adenosine nucleotide binding site of HSP90 at its N-terminus (Stebbins *et al.*, 1997). It was found that geldanamycin blocked the nuclear translocation of NF-kappaB and the expression of tumor necrosis factor in macrophages treated with Taxol or LPS.

HSP70

HSP70 is a family of multiple members with high sequence homology to each other. It is one of the best-studied protein chaperones. While the roles of HSP90 and gp96 in chaperoning tumor peptides was discovered serenpediously, the ability of tumor-derived HSP70 to behave operationally like "tumor rejection antigen" was correctly predicted (Udono and Srivastava, 1993) on the basis of well-known properties of HSP70 to bind to peptides (Flynn *et al.*,1989, 1991). HSP70 purified from Meth A, but not from normal livers and spleens could immunize BALB/c mice against challenge with live Meth A cells. It is well known that ATP binding and hydrolysis by HSP70 causes the release of HSP70-associated proteins or peptides. As predicted, treatment of an antigenically active HSP70 preparation with ATP, followed by removal of low molecular weight material, resulted in loss of antigenicity in tumor rejection assays. However, HSP70 purified by ADP affinity chromatography retained activity (Peng *et al.*, 1997), consistent with the notion that HSP70 adopts a more favorable substrate binding conformation when complexed with ADP. Moreover, free peptides can be loaded to HSP70 non-covalently *in vitro* in the presence of ADP, but not ATP. Immunization with HSP70 complexed with viral peptides elicited virus-specific CTLs (Blachere *et al.*, 1997; Ciupitu *et al.*, 1998). In addition, HSP70 isolated from cells that express ovalbumin (Breloer *et al.*, 1998) or leukemia antigen (Isshi *et al.*, 1999) was found to associate with respective antigens. The peptide-binding property of prokaryotic HSP70, DnaK has also been confirmed by the presence of definite peptide-binding pockets within the N-terminal substrate binding domain as revealed by X-ray crystallography (Zhu *et al.*, 1996). The above experiments confirmed unequivocally that HSP70 is a peptide-binding protein, and that the specific immunity elicited by HSP70–PC is towards HSP70-associated peptides, not HSP70 itself.

There is one example, that HSP70 mutation does occur and the mutated HSP70 can serve as a tumor antigen (Gaudin *et al.*, 1999). A CD8+ T-cell clone was obtained from T cells that were infiltrating renal cell carcinoma RCC-7. This clone recognized only the autologous RCC-7 tumor cell line in the context of HLA-A*0201. The antigen was identified by expression cloning and is encoded by a mutated form of the *HSP70* gene found in the tumor cells, but not in autologous lymphocytes, nor in 47 other tumors.

Recombinant mycobacterium tuberculosis HSP70, fused covalently with a model peptide has also been found to be highly immunogenic against the peptide (Suzue and Young, 1996; Suzue *et al.*, 1997; Liu *et al.*, 2000). Using this system, it was found that the immunogenicity associated with HSP70 fusion protein was dependent on a discrete 200-amino acid protein fragment at the N-teriminal ATP-binding domain (Huang *et al.*, 2000). The implication for the function of HSP70 is unclear, since the fusion protein may have adopted a totally different conformation.

Clinical development in using autologous tumor derived HSP70 or HSP70 fused with a tumor antigen for the treatment of human malignancies are ongoing. Studies have indeed shown that HSP70 purified from human melanoma can activate T cells to recognize melanoma differentiation antigens such as MART-1, gp100 and TRP-2 in the MHC restricted manner (Castelli *et al.*, 2001). Moreover, HSP70 isolated from an allogeneic melanoma cell line can pulse target cells for specific recognition by anti-gp100 CTL clones, indicating that peptide binding by HSP70 is not restricted by MHC haptotypes.

The immunomodulating effect of HSP70 has also been confirmed. It was found that expression of an inducible HSP70 in tumor cells led to increased T-cell dependent tumor immunity (Menoret *et al.*, 1995; Melcher *et al.*, 1998). Indeed, when soluble HSP70 is added to APCs such as DCs, surface binding occurs in a saturable, competitive manner, implying that there is a receptor for HSP70 on the surface of APCs (Asea *et al.*, 2000; Binder *et al.*, 2000a). This binding does not seem to be dependent on whether or not HSP70 is associated with peptides (Moroi *et al.*, 2001). This property of HSP70 was referred as chaperokine (Asea *et al.*, 2000) to reflect the dual roles of HSP70 as a chaperone and cytokine. It was found that HSP70 bound with high affinity to the plasma membrane of human monocytes, elicited a rapid intracellular calcium flux, activated nuclear factor (NF)-kappaB and up-regulated the expression of pro-inflammatory cytokines tumor necrosis factor (TNF)-α, IL-1β and IL-6 in human monocytes. Furthermore, two different signal transduction pathways were activated by exogenous HSP70: one dependent on CD14 and intracellular calcium, which resulted in increased IL-1β, IL-6 and TNF-α; and the other independent of CD14 but dependent on intracellular calcium, which resulted in an increase in TNF-α but not IL-1β or IL-6. These findings indicate that CD14 is a co-receptor for HSP70-mediated signaling in human monocytes. The binding of HSP70 with APCs can lead to the presentation of HSP70-associated peptides to MHC class I molecules on the surface of APCs. This process can be inhibited by anti-CD91 antibody, indicating that CD91 is involved in the re-presentation pathway.

Calreticulin

Calreticulin is an abundant lumenal protein of the ER (Michalak *et al.*, 1999). It is known to be a component of MHC class I/transporter associated with antigen presentation (TAP) complex (Cresswell *et al.*, 1999). The ability of tumor-derived CRT to serve as tumor rejection antigen was almost simultaneously and independently reported by two groups (Basu and Srivastava, 1999; Nair *et al.*, 1999). This was stimulated in part by findings that CRT is a major peptide acceptor for peptides translocated through TAP (Spee and Neefjes, 1997), and the fact that CRT is also an HSP. Indeed, when mice are immunized with purified CRT from Meth A tumors, they became resistant to subsequent CRT challenge in a dose-restricted manner. Moreover, CRT can be readily charged with antigenic peptides *in vitro*. Immunization with CRT complexed with a VSV antigen, elicited VSV-specific CD8+ T-cell responses (Basu and Srivastava, 1999). Similarly, immunization with CRT purified from B16/F10.9 melanoma, or E.G7-ova induced tumor-specific CTL responses against the corresponding antigens (Nair *et al.*, 1999).

The relative efficiency of CRT in generating protective immunity was compared with gp96 and HSP70 in A20 murine leukemia/lymphoma tumor model (Graner *et al.*, 2000). BALB/c mice were immunized with 20 μg of HSP70, gp96 or CRT twice at weekly intervals (i.e. day-14 and day-7). Seven days after the last immunization (day 0), mice were challenged via tail vein injection with 1×10^6 viable A20 or BDL-2 B-cell leukemia/lymphoma cells. It was found that HSP70, gp96 and CRT all provided significant improvement in survival over controls, with A20-derived HSP70 being the most effective chaperone protein followed by grp94/gp96 and CRT. It is unclear as to the basis for the differential efficiency of HSPs. In a different system, it was shown that both gp96 and CRT were released after cell lysis induced by lethal infection of cells with rVV ES-OVA(Met258–265), a recombinant, ovalbumin epitope-expressing vaccinia virus or mechanical cell death (freeze/thaw of ovalbumin-expressing cells). For both cell death scenarios, released gp96 contained ovalbumin epitope, as demonstrated by the ability to sensitize for the activation of an ovalbumin-specific T cell-hybridoma (B3Z). In contrast, CRT-derived from rVV ES-OVA (Met258–265)-infected cell extracts did not stimulate B3Z activity, which suggest that the peptide pool associated with each HSP is not identical.

It is unclear if CRT can modulate the function of APCs, however, CRT has been shown to be a cell surface molecule. Pulsing of APCs with CRT–PCs resulted in cross-presentation of CRT-associated peptides to T cells (Nair *et al.*, 1999; Basu *et al.*, 2001). The significance of cell surface expression of CRT is unclear. One report found that a panel of anti-DNA monoclonal antibodies specifically recognized CRT on the surface of multiple cell types, suggesting that CRT may mediate the penetration of anti-DNA antibodies into the cells and play an important role in lupus pathogenesis.

HSP110

HSP110 is perhaps the third most abundant cytosolic HSP in mammanlian cells, next only to HSP90 and HSP70 (Easton *et al.*, 2000). Structurally, it is related to HSP70, although it does not seem to have ATPase activity. Interestingly, even the chaperoning function of HSP110 has only just been appreciated probably due to the fact that it is cloned more recently. The substrate binding property is different between HSP70 and HSP110; the latter seems to bind to the peptide chain more efficiently (Oh *et al.*, 1997, 1999). In examining the native interactions of HSP110, it was observed that HSP110 resided in a large molecular complex that contained the constitutive form of HSP70 and HSP25. When examined *in vitro*, purified HSP25, HSP70 and HSP110 were observed to spontaneously form a large complex and directly interact with one another (Wang *et al.*, 2000).

The ability of HSP110 to potentially serve as a chaperone for tumor antigens was pursued based on the same reasoning that correctly predicted antigen-chaperoning property for HSP70 (Wang *et al.*, 2001). It was found that immunization of naïve mice with highly purified HSP110 from Meth A cells, conferred tumor protection against subsequent challenge with Meth A. Further, HSP110 derived from CT-26 carcinoma, but not from normal liver vaccinated successfully against both CT-26 challenge and established CT-26. Moreover, immunization with tumor-derived HSP110 generated tumor-specific CTLs. Additionally, bone marrow-derived DCs pulsed with HSP110 purified from CT-26 cell led to protection against CT-26 challenge.

Given the functional and structural similarities to HSP70, HSP110 most likely also bind to surface receptors on APCs. It will be interesting to test whether or not HSP110 possesses modulating properties on the function of DCs.

GRP170

GRP170 is another homolog of HSP70 and resides in the lumen of the ER (Easton *et al.*, 2000). The expression of GRP170 is induced less by thermal stress, but more pronounced by metabolic conditions that disrupt the functions of the ER such as glucose starvation and depletion of calcium. The studies of the chaperoning functions and immunological properties of GRP170 parallel with the study of HSP110. GRP170 appears to form a large complex with two other residential ER proteins gp96 and GRP78, which are all induced by glucose starvation (Lin *et al.*, 1993). The peptides that were translocated into the lumen of the microsomes *in vitro* were shown to bind to GRP170 in an *in vitro* translocation assay (Spee *et al.*, 1999). Such peptide-binding property is ATP-independent.

The ability of GRP170 to chaperone tumor peptides to generate tumor-specific immunity has been confirmed in Meth A and CT-26. Similar to HSP110, immunization of mice with tumor derived GRP110 conferred tumor protection as well as induced tumor-specific CTLs (Wang *et al.*, 2001). Tumor protection can also be achieved by vaccination with DCs pulsed with tumor derived GRP170, underscoring again the generality of HSPs to transfer tumor immunity most likely due to their ability to chaperone tumor-specific peptides, and their ability to modulate the function of APCs.

Conclusion: immunological principles associated with HSPs

In summary, four principles of HSPs in anti tumor immune responses have emerged. Each HSP may have just one, or all four properties (Table 2.1).

Principle one: HSPs per se *are rarely tumor antigens.* HSPs are well-conserved molecules in evolution. Polymorphism has not been described between individuals in the same species. Although HSP expression can be down-regulated and up-regulated in relationship to cancers, no cancer specific "hotspots" of mutations have been described. In only one example,

Table 2.1 Four principles of the immunological roles associated with HSPs

Principles	*HSPs*					
	gp96	*HSP90*	*HSP70*	*CRT*	*HSP110*	*GRP170*
1 As mutated antigens	–	–	+[rare]	–	–	–
2 As peptide carriers	+	+	+	+	+	+
3 As immunomodulators	+	+	+	+	?	?
4 In cross-presentation	+	+	+	+	+	+

mutation of HSP70 occurred in RCC, and the mutated HSP70 served as the target for tumor-specific CTL (Gaudin *et al.*, 1999). This is an exception rather than a rule.

Principle two: HSPs are molecular carriers or chaperones for tumor antigens or peptides. This principle has been confirmed structurally and immunologically. However, it is important to realize that it is unclear whether peptide-binding features of HSPs simply reflect the ability of HSPs to bind polypeptides, or whether it represents yet undefined functions of HSPs in regulating important biological functions of peptide biogenesis such as in antigen presentation. There is apparent limited specificity regarding the length or the sequence of the peptides that can associate with HSPs. There is certainly no structural basis for selectivity for tumor-specific peptides. Therefore, the difference of HSPs between normal and cancer cells lies only in the composition of peptides associated with HSPs.

Principle three: HSPs are immunomodulators. HSPs can interact specifically with APCs such as DCs through receptor-dependent mechanisms. This kind of interaction results in DC activation in producing cytokines, and up-regulating surface molecules. Most of the attention is currently focused on the role of HSPs in productive immune responses. It should be kept in mind that HSPs could also play roles in tolerance induction by inducing DCs to produce anti-inflammatory cytokines. For example, it has been reported that small molecular weight HSPs such as HSP27 could stimulate human monocytes to produce IL-10, an anti-inflammatory cytokine (De *et al.*, 2000).

Principle four: HSPs are involved in cross-presentation of tumor antigens. This principle remains speculative, although circumstantial evidence is strong. It has been argued that tumor antigens have to be cross presented from tumor cells to the class I molecules of MHC on APCs for the priming of CD8+ T cells due to lack of co-stimulatory molecules on tumor cells. Since the default pathway for the presentation of exogenous antigens is to MHC II for activation of CD4+ T cells, mechanism might exist for cross-presentation of exogenous antigens to CD8+ T cells. The candidate molecules responsible for such a mechanism are expected to have the following features: universal expression in all somatic cells, binding non-selectively to tumor antigens, specific interaction with APCs, and capacity to target exogenous tumor antigens to MHC class I on APCs. As evident from this chapter, HSPs possess all of these features and are the best candidate molecules for mediating cross-presenting MHC class I-associated antigens.

Perspectives

Although the immunological features of HSPs are discovered in the context of defining tumor antigens, it is now clear that HSPs play far-reaching roles in the immune responses. The story of HSPs is somewhat analogous to that of MHC molecules. MHC was discovered as a major transplant barrier, its "natural" role actually lies in antigen presentation. Different from HSPs, the function of MHC is almost exclusively immunological with a few exceptions (Ober *et al.*, 1997; Huh *et al.*, 2000), which permits the creating of innovative animal models and sophisticated genetic studies. The study of HSPs so far still relies on biochemistry, that is, purification of HSPs and testing the immunological properties of purified materials. This situation must change so that the fundamental roles of HSPs can be addressed more in the physiological context. Many questions remain unanswered. For example, what are the roles of HSPs in antigen presentation? Are HSPs involved in cross-presenting peptides to thymic T cells for positive or negative selections? What are the contributions of HSPs to peripheral tolerance? The recent availability of tools such as the heat shock factor-1 knockout mice which lack the expression of inducible HSPs (Xiao *et al.*, 1999) should facilitate research effort to answer these questions.

The clinical development of HSPs in the treatment of malignancies must continue not only because there are limited arsenals for most human cancers, but also due to the need to understand how human immune responses are modified by HSPs. Detailed knowledge of immune responses initiated by HSPs in humans would allow better design of preventive or therapeutic protocols for human malignancies.

References

Altmeyer, A., Maki, R. G., Feldweg, A. M., Heike, M., Protopopov, V. P., Masur, S. K. *et al.* (1996) Tumor-specific cell surface expression of the-KDEL containing, endoplasmic reticular heat shock protein gp96. *Int. J. Cancer*, 69(4): 340–9.

Argon, Y. and Simen, B. B. (1999) GRP94, an ER chaperone with protein and peptide binding properties. *Semin. Cell Dev. Biol.*, 10(5): 495–505.

Arnold, D., Faath, S., Rammensee, H. G. and Schild, H. (1995) Cross-priming of minor histocompatibility antigen-specific cytotoxic T cells upon immunization with the heat shock protein gp96. *J. Exp. Med.*, 182: 885–9.

Arnold-Schild, D., Hanau, D., Spehner, D., Schmid, C., Rammensee, H. G., de la Salle, H. *et al.* (1999) Cutting edge: receptor-mediated endocytosis of heat shock proteins by professional antigen-presenting cells. *J. Immunol.*, 162(7): 3757–60.

Asea, A., Kraeft, S. K., Kurt-Jones, E. A., Stevenson, M. A., Chen, L. B., Finberg, R. W. *et al.* (2000) HSP70 stimulates cytokine production through a CD14-dependent pathway, demonstrating its dual role as a chaperone and cytokine. *Nat. Med.*, 6(4): 435–42.

Baldwin, R. W. (1955) Immunity to methylchalanthrene-induced tumors in inbred rats following atrophy and regression of the implanted tumors. *Br. J. Cancer*, 9: 652–7.

Basu, S. and Srivastava, P. K. (1999) Calreticulin, a peptide-binding chaperone of the endoplasmic reticulum, elicits tumor- and peptide-specific immunity. *J. Exp. Med.*, 189(5): 797–802.

Basu, S., Binder, R. J., Suto, R., Anderson, K. M. and Srivastava, P. K. (2000) Necrotic but not apoptotic cell death releases heat shock proteins, which deliver a partial maturation signal to dendritic cells and activate the NF-kappaB pathway. *Int. Immunol.*, 12(11): 1539–46.

Basu, S., Binder, R. J., Ramalingam, T. and Srivastava, P. K. (2001) CD91 is a common receptor for heat shock proteins gp96, hsp90, hsp70, and calreticulin. *Immunity*, 14(3): 303–13.

Binder, R. J., Anderson, K. M., Basu, S. and Srivastava, P. K. (2000a) Cutting edge: heat shock protein gp96 induces maturation and migration of CD11c+ cells *in vivo*. *J. Immunol.*, 165(11): 6029–35.

Binder, R. J., Han, D. K. and Srivastava, P. K. (2000b) CD91 is a receptor for heat shock protein gp96. *Nat. Immunol.*, 1(2): 151–5.

Blachere, N. E., Li, Z., Chandawarkar, R. Y., Suto, R., Jaikaira, N. S., Basu, S. *et al.* (1997) Heat shock protein–peptide complexes, reconstituted *in vitro*, elicit peptide-specific cytotoxic T lymphocyte response and tumor immunity. *J. Exp. Med.*, 186(8): 1315–22.

Booth, C. and Koch, G. L. (1989) Perturbation of cellular calcium induces secretion of luminal ER proteins. *Cell*, 59(4): 729–37.

Breloer, M., Marti, T., Fleischer, B. and von Bonin, A. (1988) Isolation of processed, H-2Kb-binding ovalbumin-derived peptides associated with the stress proteins HSP70 and gp96. *Eur. J. Immunol.*, 28(3): 1016–21.

Burnet, F. M. (1970) The concept of immunological surveillance. *Prog. Exp. Tumor Res.*, 13: 1–27.

Byrd, C. A., Bormann, W., Erdjument-Bromage, H., Tempst, P., Pavletich, N., Rosen, N. *et al.* (1999) Heat shock protein 90 mediates macrophage activation by Taxol and bacterial lipopolysaccharide. *Proc. Natl. Acad. Sci. USA.*, 96(10): 5645–50.

Castelli, C., Ciupitu, A. M., Rini, F., Rivoltini, L., Mazzocchi, A., Kiessling, R. *et al.* (2001) Human heat shock protein 70 peptide complexes specifically activate antimelanoma T cells. *Cancer Res.*, 61(1): 222–7.

Caudill, M. and Li, Z. (2001) HSPPC-96: a personalized cancer vaccine. *Exp. Opin. Bio. Ther.*, 1: 539–48.

Ciupitu, A. M., Petersson, M., O'Donnell, C. L., Williams, K., Jindal, S., Kiessling, R. *et al.* (1988) Immunization with a lymphocytic choriomeningitis virus peptide mixed with heat shock protein 70 results in protective antiviral immunity and specific cytotoxic T lymphocytes. *J. Exp. Med.*, 187(5): 685–91.

Cresswell, P., Bangia, N., Dick, T. and Diedrich, G. (1999) The nature of the MHC class I peptide loading complex. *Immunol. Rev.*, 172: 21–8.

Csermely, P., Schnaider, T., Soti, C., Prohaszka, Z. and Nardai, G. (1998) The 90-kDa molecular chaperone family: structure, function and clinical applications. A comprehensive review. *Pharmacol. Ther.*, 79(2): 129–68.

De. A. K., Kodys, K. M., Yeh, B. S. and Miller-Graziano, C. (2000) Exaggerated human monocyte IL-10 concomitant to minimal TNF-alpha induction by heat-shock protein 27 (Hsp27) suggests Hsp27 is primarily an antiinflammatory stimulus. *J. Immunol.*, 165(7): 3951–8.

Easton, D. P., Kaneko, Y. and Subjeck, J. R. (2000) The hsp110 and Grp1 70 stress proteins: newly recognized relatives of the Hsp70s. *Cell Stress Chaperones*, 5(4): 276–90.

Falk, K., Rotzschke, O. and Rammensee, H. G. (1990) Cellular peptide composition governed by major histocompatibility complex class I molecules. *Nature*, 348(6298): 248–51.

Flynn, G. C., Chappell, T. G. and Rothman, J. E. (1989) Peptide binding and release by proteins implicated as catalysts of protein assembly. *Science*, 245(4916): 385–90.

Flynn, G. C., Pohl, J., Flocco, M. T. and Rothman, J. E. (1991) Peptide-binding specificity of the molecular chaperone BiP. *Nature*, 353(6346): 726–30.

Gaudin, C., Kremer, F., Angevin, E., Scott, V. and Triebel, F. (1999) A hsp70–2 mutation recognized by CTL on a human renal cell carcinoma. *J. Immunol.*, 162(3): 1730–8.

Graner, M., Raymond, A., Romney, D., He, L. Whitesell, L. and Katsanis, E. (2000) Immunoprotective activities of multiple chaperone proteins isolated from murine B-cell leukemia/lymphoma. *Clin. Cancer Res.*, 6(3): 909–15.

Gross, L. (1943) Intradernal immunization C3H mice against a sarcoma that originated in an animal of the same line. *Cancer Res.*, 3: 323–6.

Hartl, F. U. (1966) Molecular chaperones in cellular protein folding. *Nature*, 381(6583): 571–9. (Review).

Huh, G. S., Boulanger, L. M., Du, H., Riquelme, P. A., Brotz, T. M. and Shatz, C. J. (2000) Functional requirement for class I MHC in CNS development and plasticity. *Science*, 290(5499): 2155–9.

Huang, Q., Richmond, J. F., Suzue, K., Eisen, H. N. and Young, R. A. (2000) *In vivo* cytotoxic T lymphocyte elicitation by mycobacterial heat shock protein 70 fusion proteins maps to a discrete domain and is CD4(+) T cell independent. *J. Exp. Med.*, 191(2): 403–8.

Ishii, T., Udono, H., Yamano, T., Ohta, H., Uenaka, A., Ono, T., Hizuta, A., Tanaka, N., Srivastava, P. K. and Nakayama, E. (1999) Isolation of MHC class I-restricted tumor antigen peptide and its precursors associated with heat shock proteins Hsp70, Hsp90 and gp96. *J. Immunol.*, 1303–9.

Janetzki, S., Palla, D., Rosenhauer, V., Lochs, H., Lewis, J. J. and Srivastava, P. K. (2000) Immunization of cancer patients with autologous cancer-derived heat shock protein gp96 preparations: a pilot study. *Int. J. Cancer*, 88(2): 232–8.

Klein, G., Sjogren, H. O., Klein, E. and Hellstrom, K. E. (1960) Demonstration of resistance against methylcholanthrene-induced sacromas in the primary autochthonous host. *Cancer Res.*, 20: 1561–72.

Krone, P. H and Sass, J. B. (1994) HSP 90 alpha and HSP 90 beta genes are present in the zebrafish and are differentially regulated in developing embryos. *Biochem. Biophys. Res. Commun.*, 204(2): 746–52.

Li, Z. (1997) Priming of T cells by heat shock protein–peptide complexes as the basis of tumor vaccines. *Semin. Immuno.*, 9(5): 315–22.

Li, Z. and Srivastava, P. K. (1993) Tumor rejection antigen gp96/grp94 is an ATPase: implications for protein folding and antigen presentation. *EMBO J.*, 12: 3143–51.

Lin, H. Y., Masso-Welch, P., Di, Y. P., Cai, J. W., Shen, J. W. and Subjeck, J. R. (1993) The 170-KDa glucose-regulated stress protein is an endoplasmic reticulum protein that binds immunoglobulin. *Mol. Biol. Cell*, 4(11): 1109–19.

Linderoth, N. A., Popowicz, A. and Sastry, S. (2000) Identification of the peptide-binding site in the heat shock chaperone/tumor rejection antigen gp96 (Grp94). *J. Biol. Chem.*, 275(8): 5472–7.

Lindquist, S. and Craig, E. A. (1988) The heat–shock proteins. *Annu. Rev. Genet.*, 22: 631–77.

Liu, D. W., Tsao, Y. P., Kung, J. T., Ding, Y. A., Sytwu, H. K., Xiao, X. *et al.* (2000) Recombinant adeno-associated virus expressing human papillomavirus type 16 E7 peptide DNA fused with heat shock protein DNA as a potential vaccine for cervical cancer. *J. Virol.*, 74(6): 2888–94.

Maki, R. G., Eddy, R. L. Jr, Byers, M., Shows, T. B. and Srivastava, P. K. (1993) Mapping of the genes for human endoplasmic reticular heat shock protein gp96/grp94. *Somat. Cell Mol. Genet.*, 19(1): 73–81.

Mazzarella, R. A. and Green, M. (1987) ERp99, an abundant, conserved glycoprotein of the endoplasmic reticulum, is homologous to the 90-kDa heat shock protein (hsp90) and the 94-kDa glucose regulated protein (GRP94). *J. Biol. Chem.*, 262(18): 8875–83.

Melcher, A., Todryk, S., Hardwick, N., Ford, M., Jacobson, M. and Vile, R. G. (1998) Tumor immunogenicity is determined by the mechanism of cell death via induction of heat shock protein expression. *Nat. Med.*, 4(5): 581–7.

Meng, S. D., Gao, T., Gao, G. F. and Tien, P. (2001) HBV-specific peptide associated with heat-shock protein gp96. *Lancet*, 357(9255): 528–9.

Menoret, A., Patry, Y., Burg, C. and Le Pendu, J. (1955) Co-segregation of tumor immunogenicity with expression of inducible but not constitutive hsp70 in rat colon carcinomas. *J. Immunol.*, 155(2): 740–7.

Michalak, M., Corbett, E. F., Mesaeli, N., Nakamura, K. and Opas, M. (1999) Calreticulin: one protein, one gene, many functions. *Biochem. J.*, 344 (Pt2): 281–92.

Moore, S. K., Kozak, C., Robinson, E. A., Ullrich, S. J. and Appella, E. (1989) Murine 86- and 84-kDa heat shock proteins, cDNA sequences, chromosome assignments, and evolutionary origins. *J. Biol. Chem.*, 264(10): 5343–51.

Moroi, Y., Mayhew, M., Trcka, J., Hoe, M. H., Takechi, Y., Hartl, F. U. *et al.* (2001) Induction of cellular immunity by immunization with novel hybrid peptides complexed to heat shock protein 70. *Proc. Natl. Acad. Sci. USA*, 97(7): 3485–90.

Nair, S., Wearsch, P. A., Mitchell, D. A., Wassenberg, J. J., Gilboa, E. and Nicchitta, C. V. (1999) Calreticulin displays *in vivo* peptide-binding activity and can elicit CTL responses against bound peptides. *J. Immunol.*, 162(11): 6426–32.

Nieland, T. J., Tan, M. C., Monne-van Muijen, M., Koning, F., Kruisbeek, A. M. and van Bleek, G. M. (1996) Isolation of an immunodominant viral peptide that is endogenously bound to the stress protein gp96/grp94. *Proc. Natl. Acad. Med. Sci. USA*, 96: 6135–9.

Ober, C., Weitkamp, L. R., Cox, N., Dytch, H., Kostyu, D. and Elias, S. (1997) HLA and mate choice in humans. *Am. J. Hum. Genet.*, 61(3): 497–504.

Oh, H. J., Chen, X. and Subjeck, J. R. (1997) Hsp110 protects heat-denatured proteins and confers cellular thermoresistance. *J. Biol. Chem.*, 272(50): 31636–40.

Oh, H. J., Easton, D., Murawski, M., Kaneko, Y. and Subjeck, J. R. (1999) The chaperoning activity of hsp110. Identification of functional domains by use of targeted deletions. *J. Biol. Chem.*, 274(22): 15712–8.

Old, L. J., Boyse, E. A., Clarke, D. A. and Carswell, E. A. (1962) Antigenic properties of chemically induced tumors. *Ann. NY Acad. Sci.*, 101: 80–106.

Palladino, M. A. Jr, Srivastava, P. K., Oettgen, H. F. and DeLeo, A. B. (1987) Expression of a shared tumor-specific antigen by two chemically induced BALB/c sarcomas. *Cancer Res.*, 47(19): 5074–9.

Peng, P., Menoret, A. and Srivastava, P. K. (1997) Purification of immunogenic heat shock protein 70–peptide complexes by ADP-affinity chromatography. *J. Immunol. Meth.*, 204(1): 13–21.

Prehn, R. T. and Main, J. M. (1957) Immunity to methylcholanthrene-induced sarcomas. *J. Natl Cancer Inst.*, 18: 769.

Reilly, R. T., Gottleib, M. B., Ercolini, A. M., Machiels, J. P., Kane, C. E., Okoye, F. I. *et al.* (2000) HER-2/neu is a tumor rejection target in tolerized HER-2/neu transgenic mice. *Cancer Res.*, 60(13): 3569–76.

Robert, J., Menoret, A. and Cohen, N. (1999) Cell surface expression of the endoplasmic reticular heat shock protein gp96 is phylogenetically conserved. *J. Immunol.*, 163(8): 4133–9.

Sastry, S. and Linderoth, N. (1999) Molecular mechanisms of peptide loading by the tumor rejection antigen/heat shock chaperone gp96 (GRP94). *J. Biol. Chem.*, 274(17): 12023–35.

Seddiki, N., Nato, F., Lafaye, P., Amoura, Z., Piette, J. C. and Mazie, J. C. (2001) Calreticulin, a potential cell surface receptor involved in cell penetration of anti-DNA antibodies. *J. Immunol.*, 166(10): 6423–9.

Shankaran, V., Ikeda, H., Bruce, A. T., White, J. M., Swanson, P. E., Old, L. J. *et al.* (2001) IFNgamma and lymphocytes prevent primary tumour development and shape tumour immunogenicity. *Nature*, 410(6832): 1107–11.

Singh-Jasuja, H., Toes, R. E. M., Spee, P., Munz, C., Hilf, N., Schoenberger, S. P. *et al.* (2000a) Cross-presentation of glycoprotein 96-associated antigens on major histocompatibility complex class I molecules requires receptor-mediated endocytosis. *J. Exp. Med.*, 191(11): 1965–74.

Singh-Jasuja, H., Scherer, H. U., Hilf, N., Arnold-Schild, D., Rammensee, H. G., Toes, R. E. *et al.* (2000b) The heat shock protein gp96 induces maturation of dendritic cells and down-regulation of its receptor. *Eur. J. Immunol.*, 30(18): 2211–15.

Spee, P. and Neefjes, J. (1997) TAP-translocated peptides specifically bind proteins in the endoplasmic reticulum, including gp96, protein disulfide isomerase and calreticulin. *Eur. J. Immunol.*, 27(9): 2441–9.

Spee, P., Subjeck, J. and Neefjes, J. (1999) Identification of novel peptide binding proteins in the endoplasmic reticulum, ERp72, calnexin, and grp170. *Biochemistry*, 38(32): 10559–66.

Srivastava, P. K. and Das, M. R. (1984) The serologically unique cell surface antigen of Zajdela ascitic hepatoma is also its tumor-associated transplantation antigen. *Int. J. Cancer*, 33(3): 417–22.

Srivastava, P. K. and Heike, M. (1991) Tumor-specific immunogenicity of stress-induced proteins: convergence of two evolutionary pathways of antigens presentation? *Semin. Immunol.*, 3(1): 57–64 (review).

Srivastava, P. K. and Maki, R. G. (1991) Stress-induced proteins in immune response to cancer. *Curr. Top. Microbiol. Immunol.*, 167: 109–23.

Srivastava, P. K., Chen, Y. T. and Old, L. J. (1987) 5′-structural analysis of genes encoding polymorphic antigens of chemically induced tumors. *Proc. Natl. Acad. Sci. USA*, 84(11): 3807–11.

Srivastava, P. K., Deleo, A. B. and Old, L. J. (1986) Tumor rejection antigens of chemically induced sarcomas of inbred mice. *Proc. Natl. Acad. Sci. USA*, 83(10): 3407–11.

Srivastava, P. K., Udono, H., Blachere, N. E. and Li, Z. (1994) Heat shock proteins transfer peptides during antigen processing and CTL priming. *Immunogenetics*, 39(2): 93–8.

Srivastava, P. K., Menoret, A., Basu, S., Binder, R. J. and McQuade, K. L. (1998) Heat shock protein, come of age: primitive functions acquire new roles in an adaptive world. *Immunity*, 8: 656–65.

Stebbins, C. E., Russo, A. A., Schneider, C., Rosen, N., Hartl, F. U. and Pavletich, N. P. (1997) Crystal structure of an Hsp90–geldanamycin complex: targeting of a protein chaperone by an antitumor agent. *Cell*, 89(2): 239–50.

Suto, R. and Srivastava, P. K. (1995) A mechanism for the specific immunogenicity of heat shock protein-chaperoned peptides. *Science*, 269(5230): 1585–8.

Suzue, K. and Young, R. A. (1996) Adjuvant-free hsp70 fusion protein system elicits humoral and cellular immune responses to HIV-1 p24. *J. Immunol.*, 156(2): 873–9.

Suzue, K., Zhou, X., Eisen, H. N. and Young, R. A. (1997) Heat shock fusion proteins as vehicles for antigen delivery into the major histocompatibility complex class I presentation pathway. *Proc. Natl. Acad. Sci. USA*, 94(24): 13146–51.

Todryk, S., Melcher, A. A., Hardwick, N., Linardakis, E., Bateman, A., Colombo, M. P. *et al.* (1999) Heat shock protein 70 induced during tumor cell killing induces Th1 cytokines and targets immature dendritic cell precursors to enhance antigen uptake. *J. Immunol.*, 163: 1398–1408.

Udono, H. and Srivastava, P. K. (1993) Heat shock protein 70-associated peptides elicit specific cancer immunity. *J. Exp. Med.*, 178: 1391–6.

Udono, H., Levey, D. L. and Srivastava, P. K. (1994) Cellular requirements for tumor-specific immunity elicited by heat shock proteins: tumor rejection antigen gp96 primes CD8+ T cells *in vivo*. *Proc. Natl. Acad. Sci. USA*, 91(8): 3077–81.

Ullrich, S. J., Robinson, E. A., Law, L. W., Willingham, M. and Appella, E. (1986) A mouse tumor-specific transplantation antigen is a heat shock-related protein. *Proc. Natl. Acad. Sci. USA*, 83(10): 3121–5.

Van Bleek, G. M. and Nathenson, S. G. (1990) Isolation of an endogenously processed immunodominant viral peptide from the class I H-2Kb molecule. *Nature*, 348(6298): 213–6.

Wang, X. Y., Chen, X., Oh, H. J., Repasky, E., Kazim, L. and Subjeck, J. (2000) Characterization of native interaction of hsp110 with hsp25 and hsc70. *FEBS Lett*, 465(2–3): 98–102.

Wang, X. Y., Kazim, L., Repasky, E. A. and Subjeck, J. R. (2001) Characterization of heat shock protein 110 and glucose-regulated protein 170 as cancer vaccines and the effect of fever-range hyperthermia on vaccine activity. *J. Immunol.*, 166(1): 490–7.

Wassenberg, J. J., Dezfulian, C. and Nicchitta, C. V. (1999) Receptor mediated and fluid phase pathways for internalization of the ER hsp90 chaperone GRP94 in murine macrophages. *J. Cell Sci.*, 112(Pt 13): 2167–75.

Wearsch, P. A., Voglino, L. and Nicchitta, C. V. (1998) Structural transitions accompanying the activation of peptide binding to the endoplasmic reticulum Hsp90 chaperone GRP94. *Biochemistry*, 37(16): 5709–19.

Wiest, D. L., Bhandoola, A., Punt, J., Kreibich, G., McKean, D. and Singer, A. (1997) Incomplete endoplasmic reticulum (ER) retention in immature thymocytes as revealed by surface expression of "ER-resident" molecular chaperones. *Proc. Natl. Acad. Sci. USA*, 94(5): 1884–9.

Xiao, X., Zuo, X., Davis, A. A., McMillan, D. R., Curry, B. B., Richardson, J. A. and Benjamin, I. J. (1999) HSF1 is required for extra-embryonic development, postnatal growth and protection during inflammatory responses in mice. *EMBO J.*, 18(21): 5943–52.

Yamazaki, K., Nguyen, T. and Podack, E. R. (1999) Tumor secreted heat shock-fusion protein elicits CD8 cells for rejection. *J. Immunol.*, 163(10): 5178–82.

Zhu, X., Zhao, X., Burkholder, W. F., Gragerov, A., Ogata, C. M., Gottesman, M. E. *et al.* (1996) Structural analysis of substrate binding by the molecular chaperone DnaK. *Science*, 272(5268): 1606–14.

Part 2

Human tumor antigens recognized by class I HLA-restricted T cells

Chapter 3

WT1 as target for tumor immunotherapy

Hans J. Stauss, Shao-an Xue, Liquan Gao, Gavin Bendle, Angelika Holler, Roopinder Gillmore and Francisco Ramirez

Summary

The Wilms tumor antigen 1 (WT1) is expressed at elevated levels in most leukemias compared with normal hematopoietic cells. Furthermore, WT1 is also activated in a variety of solid cancers, while the corresponding normal cells do not express this protein. Hence, WT1 is an attractive target molecule for tumor immunotherapy. However, low levels of expression in normal tissues is likely to cause immunological tolerance, which may lead to the inactivation of high avidity T lymphocytes. The allo-restricted strategy was developed to avoid tolerance and to isolate high avidity cytotoxic T lymphocytes (CTL) specific for self-proteins, such as WT1. *In vitro* studies have shown that allo-restricted CTL kill WT1 expressing tumor cell lines and leukemic progenitor cells isolated from patients. The expression level in normal hematopoietic cells is insufficient to trigger killing by WT1-specific CTL. However, a major drawback of immunotherapy with allo-restricted CTL is the HLA mismatch between CTL and tumor patients, leading to rejection of infused CTL in immunocompetent recipients. To overcome this limitation, the transfer of cloned T cell receptors (TCR) into patients CD8+ T cells provides a strategy to equip autologous CTL with high avidity receptors specific for defined tumor-associated proteins.

WT1 structure

WT1 is an intracellular protein involved in activation and repression of gene expression, and in the regulation of post-transcriptional RNA processing (Hastie, 2001). Up to 24 different isoforms of the WT1 protein can be generated by alternative splicing, alternative translation initiation and RNA editing (Hastie, 2001; Scharnhorst *et al.*, 2001) (Figure 3.1). To date, most of the WT1 research has focused on four isoforms referred to as WT1+/+, WT1+/−, WT1−/+ and WT1−/−. This nomenclature indicates the presence or absence of two alternative splice sequences. The first is a 17-amino acid sequence encoded by exon 5 of the *WT1* gene, and the second is the three-amino acid KTS sequence generated by alternative splice donor site used at the end of exon 9. All WT1 isoforms have four zinc finger domains at the carboxyl terminus of the protein, which are involved in binding to DNA motifs in the promoter region of genes that are regulated by WT1 (Nakagama *et al.*, 1995). Insertion of three amino acids (KTS) in the zinc finger domains by alternative splicing decreases the DNA binding properties, and produces a WT1 isoform that is primarily involved in RNA processing. In addition to DNA and RNA binding properties, WT1 can also bind to itself and undergo homodimerization, which is mediated by the first 180 amino acids of the protein (Reddy *et al.*, 1995). Assuming that all WT1 isoforms can associate with each other to form dimers, there is scope for up to 300 dimeric combinations, which may

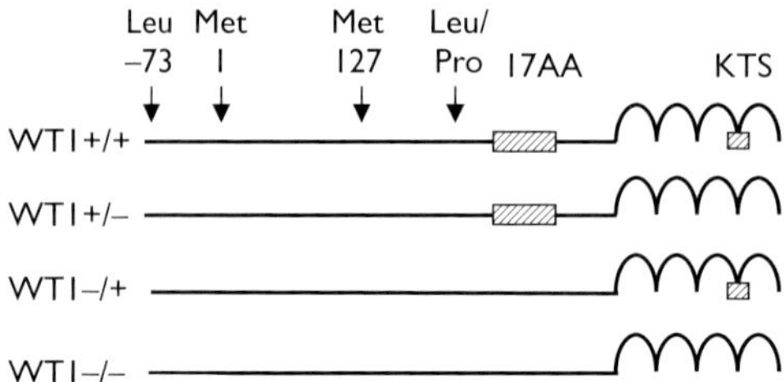

Figure 3.1 Different WT1 isoforms are generated by alternative splicing of 17 amino acids encoded by exon 5 of the *WT1* gene and three amino acids (KTS) generated by alternative splice site usage in exon 9. Additional isoform are produced by alternative translation initiation at position −73 and +127, and by RNA editing changing leucine to proline.

have overlapping but also distinct functions. It is clear from these considerations that WT1 function is not only determined by its expression levels, but also by the relative ratios of the isoforms produced. This is most clearly demonstrated in patients with Frasier syndrome, where a mutation in one allele prevents production of the KTS-containing isoform, whereas the other WT1 allele is normal and capable of producing all isoforms (Barbaux *et al.*, 1997). The resulting imbalance of WT1 isoform production leads to a glomerulopathy, characterized by focal and segmental glomerular sclerosis, and male to female sex reversal in these patients.

WT1 function during embryogenesis

During embryogenesis, WT1 expression is required for undisturbed organogenesis of the kidney and spleen, as revealed by the analysis of the embryonic lethal phenotype of WT1-knockout mice (Kreidberg *et al.*, 1993; Herzer *et al.*, 1999). In kidney development WT1 is postulated to trigger the differentiation of mesenchymal stem cells to become epithelial cells required for nephron formation (Davies *et al.*, 1999). In addition, WT1 also exerts anti-apoptotic functions as revealed by the enhanced apoptosis in the metanephric mesenchyme of WT1-knockout mice (Kreidberg *et al.*, 1993; Donovan *et al.*, 1999). Recently, transgenic mice expressing only the KTS-positive or KTS-negative WT1 isoform have been created (Hammes *et al.*, 2001). Unlike WT1-deficient mice, these mice are born to term, demonstrating some functional overlap of the two WT1 isoforms during embryogenesis. However, functional complementation by each isoform is incomplete, since the mice display severely impaired kidney function leading to death soon after birth. Interestingly, the phenotype of the KTS-negative-WT1 mice is similar to that of Frasier patients with renal dysfunction and male to female sex reversal. Together, the phenotype of mice with WT1 deletions or selective expression of KTS-positive/negative isoforms, illustrate that WT1 has anti-apoptotic functions and plays a central role in the regulation of renal development and female–male sex differentiation.

WT1 function in post-natal life

After birth WT1 expression is switched off in most cells except renal podocytes, Sertoli cells of the testis, granulosa cells of the ovary, myoepithelial progenitor cells and CD34+

hematopoietic cells (Scharnhorst *et al.*, 2001). Most of the information about WT1 function in adult tissues comes from studies of normal CD34+ hematopoietic progenitor/stem cells. Within the CD34+ population WT1 is expressed in early, uncommitted CD34+CD38− and in committed CD34+CD38+ cells (Baird and Simmons, 1997). Studies with 5-fluorouracil treated CD34+CD38− cell populations have shown that WT1 expression is particularly prominent in a small subset of non-cycling quiescent cells, which are highly effective in producing myeloid and lymphoid lineage progeny (Ellisen *et al.*, 2001). This has suggested that WT1 plays a role in the maintenance of the uncommitted, quiescent phenotype of these hematopoietic stem cells. In the committed CD34+CD38+ progenitor cells WT1 expression is switched off when these cells differentiate to become mature hematopoietic cells (Menssen *et al.*, 1997). Retroviral over-expression of WT1 in CD34+CD38+ cells can accelerate their maturation, indicating that WT1 can trigger the differentiation of normal hematopoietic progenitor cells (Ellisen *et al.*, 2001). Hence, WT1 appears to play a dual role in normal hematopoiesis: (i) maintenance of uncommitted stem cells and (ii) induction of differentiation toward more mature cells.

WT1 function in tumors

Over-expression of WT1 has been described in various hematological malignancies, including chronic and acute myeloid leukemia (CML, AML), acute lymphocytic leukemia and myelodysplastic syndrome (Inoue *et al.*, 1994; Patmasiriwat *et al.*, 1999; Tamaki *et al.*, 1999). A high level of over-expression has been correlated with poor disease prognosis in AML patients. The mechanisms by which WT1 over-expression contributes to leukemogenesis are not fully understood, although experimental evidence indicates that WT1 can contribute to the enhanced proliferation and defective differentiation of leukemia cells. WT1 antisense constructs can inhibit the growth of clonogenic progenitor cells isolated from AML and CML patients (Yamagami *et al.*, 1996), and they can also inhibit proliferation of human leukemia cell lines (Algar *et al.*, 1996). In the murine hematopoietic cell line 32D it was found that introduction of WT1 by retroviral transfer inhibited differentiation and resulted in enhanced proliferation in response to granulocyte colony stimulating factor (G-CSF) (Inoue *et al.*, 1998). G-CSF stimulated activation of stat-3 molecules was prolonged in the presence of WT1, suggesting that alteration in the stat-3 signaling pathway may be involved in the defective differentiation and enhanced proliferation in leukemia cells.

The analysis of solid tumors has revealed activation of WT1 expression in cell lines established from glioblastoma, cancer of the lung, colon, stomach, breast, ovary, uterus, liver and thyroid gland (Oji *et al.*, 1999; Menssen *et al.*, 2000). However, in one of these studies the incidence of WT1 expression in fresh tumor tissue was lower compared with cell lines (Menssen *et al.*, 2000). This led to the concern that activation of WT1 expression might be an effect of the *in vitro* adaptation of cell lines derived from tumor specimens, but may not necessarily reflect WT1 expression in patients. To address this concern, a recent study examined WT1 expression in fresh tumor tissue of patients with carcinoma of the breast (Loeb *et al.*, 2001). The analysis of RNA and protein revealed WT1 expression in 27/31 patients with breast cancer. In contrast, WT1 was detected in only 1/20 normal breast samples obtained from women undergoing reduction mammoplasty. These data are similar to studies in our laboratory, showing that WT1 RNA is detectable in a large proportion of freshly isolated breast cancer tissues, but not in adjacent normal tissues isolated from the same patients (unpublished). It is likely that WT1 is also involved in other solid cancers, although a systematic analysis using fresh tumor material and normal tissue is required to determine the incidence and level of WT1 expression.

The mechanisms by which activation of WT1 expression may contribute to tumorigenesis are not clear. Detailed analyses are complicated by the large number of WT1 isoforms that can be produced within one cell, and by the observation that disturbance of the isoform balance can change WT1 function (e.g. Frasier syndrome). Thus, transfection studies with individual isoforms are limited, although they illustrate isoform-specific functional properties. For example, stable transfection of an adenovirus-transformed baby-rat kidney cell line with the WT1−/− isoform increased the tumorigenicity in nude mice (Menke *et al.*, 1996). In contrast, transfection of the same cells with the WT1−/+ isoform strongly inhibits tumor formation (Menke *et al.*, 1995). An additional complexity is that the function of WT1 is dependent upon the cell type in which it is expressed. It appears that even closely related cells provide a distinct set of intracellular molecules capable of modifying WT1 function. For example, the described tumor enhancing effect of WT1−/− described above is not observed in a closely related adenovirus-transformed baby-rat kidney cell line (Scharnhorst *et al.*, 2000). One possible explanation is that proteins that associate with WT1 to enhance its tumorigenic potential are present in one but not the other baby-rat kidney cell line.

A number of intracellular molecules have been shown to associate with WT1. These include proteins involved in RNA splicing, HSP70 and others (Scharnhorst *et al.*, 2001). The most interesting partners of WT1 are p53 and its relatives p63 and p73. Binding of WT1 to p53 can inhibit p53-mediated apoptosis in response to ultraviolet irradiation (Maheswaran *et al.*, 1995). The functional consequences of WT1 binding to the other p53-like molecules have not yet been determined. The question which target genes are regulated by WT1 and its partners has been primarily explored using transient transfection assays. The predominant observation in these assays was that WT1 co-transfection resulted in inhibition of reporter gene activity. In most cases, the effect of WT1 on the regulation of endogenous gene expression has not yet been examined. Using DNA microchip technology, a recent report demonstrated activation of endogenous gene expression in human osteosarcoma cells containing a tetracylin inducible WT1−/− construct (Lee *et al.*, 1999). The microchip analysis did not reveal any suppression of endogenous gene expression, suggesting that it might be an effect of un-physiological conditions of transient transfection assays. Amphiregulin, a member of the epidermal growth factor receptor (EGF-R) family, was strongly upregulated in response to WT1. In addition, enhanced expression of the acidic fibroblast growth factor was detectable in this study. In an independent study, WT1-induced expression of endogenous EGF-R has been demonstrated (Liu *et al.*, 2001), although previous studies showed suppression of EGF-R expression in different cells (Englert *et al.*, 1995), once again demonstrating cell type dependent WT1 function. *Bcl-2* is another endogenous target gene that is up-regulated in response to WT1 (Mayo *et al.*, 1999). Together, the interaction of WT1 with the p53 family of molecules and the activation of expression of genes encoding growth factors, their receptors and anti-apoptotic molecules may account for the ability of WT1 to display transforming activity in certain cell types.

WT1-based immunotherapy

WT1 is an attractive target for tumor immunotherapy because its normal expression pattern in adult life is restricted to a few tissues, and its expression is activated in most leukemias and in a variety of solid tumors. Hence, WT1-based immunotherapy may provide a treatment option for many human malignancies. The necessity to overcome immunological tolerance is particularly important when stimulating CTL responses against antigens such as WT1 that are expressed at low level in normal tissue. HLA class I MHC/self-peptide complexes are likely to represent thymic and peripheral tolerogens preventing immune responses. Using RT-PCR assays in mice, it has been shown that WT1 is *not* expressed in the thymus (unpublished observation). Therefore, immunological tolerance would be expected to be the result of "peripheral" not "central" (i.e. intrathymic deletion) mechanisms. In general, the level of peripheral tolerance is affected by the cell type expressing the antigen and by the amount produced. If expressed in few cells outside the lymphoid system, the antigen may be ignored (Ohashi *et al.*, 1991). If expressed at higher levels, antigen may leak into the lymphoid organs and trigger functional inactivation or physical deletion of high avidity CTL (Kurts *et al.*, 1998). Hematopoietic stem cells (including WT1-expressing CD34+ cells) do recirculate at low frequency in the peripheral blood and are also found in secondary lymphoid tissues such as spleen (Wright *et al.*, 2001), a suitable site for peripheral tolerance induction. Peripheral tolerance of high avidity CTL has been documented in a transgenic model where the CTL recognized antigen, which was expressed by pancreatic islet cells, drained into local lymph nodes (Lo *et al.*, 1992; Morgan *et al.*, 1999). Similarly, tolerance due to tyrosinase expression in normal melanocytes has been shown to blunt CTL responses against this tumor-associated melanoma antigen (Colella *et al.*, 2000). In this model high avidity tyrosinase-specific CTL responses were absent in normal mice, but easily detectable in tyrosinase-deficient mice. Residual intermediate avidity CTL were detectable in normal mice after vaccination with a peptide analog that differed by a point mutation from the native tyrosinase sequence. Together, these murine studies demonstrate that extrathymic antigen expression in a relatively small number of cells in peripheral tissues is sufficient to cause partial immunological tolerance by inactivating high avidity CTL, resulting in residual immune responses by intermediate/low avidity CTL. Hence, it is likely that the described WT1 expression in peripheral tissues and hematopoietic CD34+ cells leads to the purging of high avidity CTL, and that residual intermediate/low avidity CTL may mount an ineffective attack against patient leukemia cells.

Avoiding immunological tolerance

In the past few years the allo-restricted CTL approach has been developed to circumvent immunological tolerance to self-proteins (Sadovnikova and Stauss, 1996; Stauss, 1999). It is based on the finding that tolerance is self-MHC-restricted. Therefore, individuals expressing HLA-A2 are tolerant to peptide epitopes presented by the A2 molecule, but not other HLA alleles. Similarly, the T cells of individuals who do not express A2 have never been exposed to A2-presented peptides and are consequently not tolerant to such epitopes (Figure 3.2). *In vitro* conditions have been developed whereby responder T cells from A2-negative, healthy donors are stimulated with A2-positive stimulator cells. The stimulator cells have a defect in the genes encoding the transporter for antigen presentation (TAP) molecules, and can be conveniently loaded with synthetic A2-binding peptides. These peptides have been identified in proteins that are over-expressed in tumors, such as WT1, using *in vitro* binding assays. After several rounds of stimulation, bulk CTL are plated under limiting dilution conditions to identify CTL that specifically recognize the immunizing peptide epitope presented by A2 class I molecules. Such allo-restricted CTL are then used to determine their avidity and ability to recognize tumor cells expressing the target antigen endogenously. In the past, this

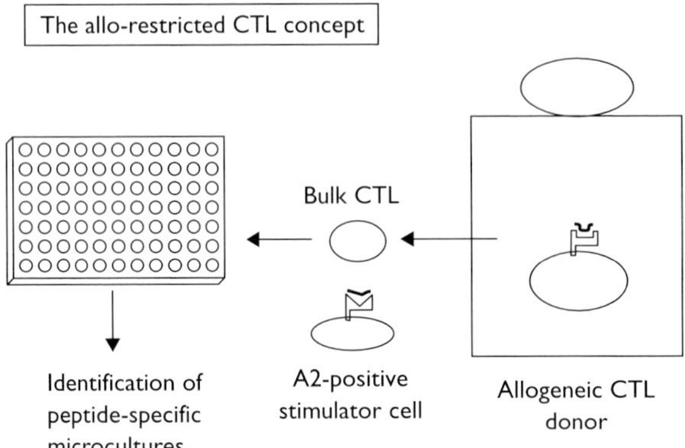

Figure 3.2 The allo-restricted approach can be used to generate CTL against peptides to which autologous CTL are tolerant. Allogeneic HLA molecules have distinctly shaped peptide binding grooves. This leads to the presentation of distinct peptide epitopes from the same self-molecules. As a consequence, the CTL of an allogeneic donor are not tolerant to the self-peptides presented by HLA-A2 class I molecules. Bulk stimulation of T cells from A2-negative allogeneic donors followed by limiting dilution plating, allows identification of rare CTL that are specific for the immunizing peptide presented by A2 molecules.

approach has been used to isolate high avidity CTL against the human tumor associated antigens cyclinD1, mdm2 and WT1 (Sadovnikova *et al.*, 1998; Gao *et al.*, 2000; Stanislawski *et al.*, 2001).

Anti-WT1 CTL in leukemia

To date most of the WT1 studies have been performed with samples from CML patients, and clinical trials with WT1-specific CTL populations will be performed in these patients in the next few years. The analysis of subpopulations of normal and leukemia CD34+ cells showed higher levels of WT1 expression in uncommitted (CD38−, DR−) and committed (CD38+DR+) populations of leukemia cells compared to normal counterparts (Inoue *et al.*, 1997). We have used high avidity, HLA-A2-allo-restricted CTL with specificity for the peptide epitope WT126 (RMFPNAPYL) derived from the WT1 protein to test if such CTL can discriminate between normal and leukemic CD34+ cell populations. The study of a panel of cell lines indicated that the WT126-specific CTL killed WT1-expressing leukemia cell lines in an A2-restricted fashion (Gao *et al.*, 2000). The colony forming units (CFU) were measured to assess the ability of WT126-specific CTL to kill committed CD34+ clonogenic progenitor cells freshly isolated from CML patients. The results indicated that the CTL eliminated 80–100% of CFU progenitors of A2-positive leukemia patients, whilst CFU progenitors of A2-negative control patients were unaffected. Most importantly, the CTL did not inhibit the CFU progenitors of normal A2-positive individuals (Gao *et al.*, 2000). These data indicated that the WT1-specific CTL were able to distinguish between leukemia and normal progenitor cells. The described WT1-specific CTL are similar to CTL against an A2-presented peptide of proteinase-3 (Molldrem *et al.*, 1997). These CTL inhibited the CFU activity of bone marrow cells from CML patients by 34–98%, without affecting CFU of normal bone marrow. The selective killing of CML bone marrow correlated with increased levels of proteinase-3 expression compared with normal bone marrow. Proteinase-3 is a differentiation antigen that is expressed in hematopoietic cells of the myeloid lineage. Hence, proteinase-3-specific CTL would be expected to be particularly effective in the killing of mature myeloid cells and committed CFU progenitors, but less effective against more immature cells. In contrast, unpublished experiments have shown that the WT126-specific CTL can also discriminate between leukemic and normal stem cells (unpublished). The selective killing of leukemic progenitor and stem cells by WT1-specific CTL was superior to that observed with STI572 (Deininger *et al.*, 1997).

Can self-restricted CTL protect against WT1 expressing tumors?

Two reports of human CTL and two reports of murine CTL indicate that tolerance to WT1 is incomplete. In the human experiments, self-HLA-restricted CTL against the WT235 and WT126 peptides presented by HLA-A24 and A2, respectively, were generated by repeated *in vitro* peptide stimulation (Ohminami *et al.*, 2000; Oka *et al.*, 2000a). The obtained CTL lines required nanomolar peptide concentrations for target cell recognition. In the murine experiments, CTL lines were isolated after repeated *in vivo* immunizations with the WT126 peptide (Gaiger *et al.*, 2000; Oka *et al.*, 2000b). Again, the isolated CTL lines required nanomolar peptide concentrations for target cell recognition. In contrast, the allo-HLA-restricted CTL specific for WT126 were capable of recognizing target cells pulsed with picomolar peptide concentrations (Gao *et al.*, 2000), suggesting that they were of higher avidity than the self-restricted CTL. It will be important to determine whether the decreased avidity of self-restricted CTL is sufficient for recognition of CD34+ progenitor/stem cells, which is an essential requirement for their use in immunotherapy in leukemia patients.

Immunotherapy via TCR transfer

The HLA-mismatch between CTL donor and recipient is a major drawback of immunotherapy with allo-restricted CTL, since alloantigens expressed by CTL are likely to stimulate immune rejection by recipient T cells. In order to overcome these limitations, TCR-based gene transfer provides an excellent opportunity to take advantage of the unique specificity and high affinity of the TCRs of allo-restricted CTL. We have recently shown that retroviral transduction can be used to transfer TCRs into human CD8+ T cells, and that the transduced cells display the same specificity and avidity as the CTL from which the TCR was isolated (Stanislawski *et al.*, 2001). This strategy should allow us to equip autologous human CD8+ T cells with the specificity of TCRs derived from allo-restricted CTL. This new technology is not free of problems. For example, transfer of TCR α and β chains into T cells that already express endogenous chains may create novel specificities with a potential autoimmune risk. Transfer of modified TCR constructs that do not pair with endogenous TCR chains may overcome this problem. At present, retrovirus-based TCR transfer requires *in vitro* activation of T cells, which may change their functional properties and their ability to develop into long-lived memory cells. New viral vectors or transfection protocols capable of introducing TCR genes into resting T cells needs to be explored. An attractive option avoiding concerns of unwanted pairing and ineffective T-cell memory is the introduction of TCR chains into CD34+ hematopoietic stem cells. In these cells, which do not express CD3 molecules, the introduced TCR chains would not be expressed on the cell surface and would therefore not interfere with the normal function of the CD34+ cells. Once these cells arrive in the thymus, differentiation towards the T-cell lineage leads to the expression of CD3 molecules, permitting surface expression of the introduced TCR chains while suppressing rearrangement of endogenous chains. This is expected to lead to the maturation of single positive naïve T cells expressing only the introduced TCR. Thus, conceptually, TCR transduced hematopoietic stem cells are an attractive proposition, since it may serve as a permanent supply of high frequency naïve CTL with defined specificities for tumor antigens, such as WT1, or for virus antigens, depending upon the TCR construct used for stem cell manipulation.

References

Algar, E. M., Khromykh, T., Smith, S. I., Blackburn, D. M., Bryson, G. J. and Smith, P. J. (1996) A WT1 antisense oligonucleotide inhibits proliferation and induces apoptosis in myeloid leukaemia cell lines. *Oncogene*, **12**, 1005–14.

Baird, P. N. and Simmons, P. J. (1997) Expression of the Wilms' tumor gene (WT1) in normal hemopoiesis. *Exp. Hematol.*, **25**, 312–20.

Barbaux, S., Niaudet, P., Gubler, M. C., Grunfeld, J. P., Jaubert, F., Kuttenn, F., Fekete, C. N., Souleyreau-Therville, N., Thibaud, E., Fellous, M. and McElreavey, K. (1997) Donor splice-site mutations in WT1 are responsible for Frasier syndrome. *Nat. Genet.*, **17**, 467–70.

Colella, T. A., Bullock, T. N., Russell, L. B., Mullins, D. W., Overwijk, W. W., Luckey, C. J., Pierce, R. A., Restifo, N. P. and Engelhard, V. H. (2000) Self-tolerance to the murine homologue of a tyrosinase-derived melanoma antigen: implications for tumor immunotherapy. *J. Exp. Med.*, **191**, 1221–32.

Davies, R., Moore, A., Schedl, A., Bratt, E., Miyahawa, K., Ladomery, M., Miles, C., Menke, A., van Heyningen, V. and Hastie, N. (1999) Multiple roles for the Wilms' tumor suppressor, WT1. *Cancer Res.*, **59**, 1747s–50s; discussion 1751s.

Deininger, M. W., Goldman, J. M., Lydon, N. and Melo, J. V. (1997) The tyrosine kinase inhibitor CGP57148B selectively inhibits the growth of BCR-ABL-positive cells. *Blood*, **90**, 3691–8.

Donovan, M. J., Natoli, T. A., Sainio, K., Amstutz, A., Jaenisch, R., Sariola, H. and Kreidberg, J. A. (1999) Initial differentiation of the metanephric mesenchyme is independent of WT1 and the ureteric bud. *Dev. Genet.*, **24**, 252–62.

Ellisen, L. W., Carlesso, N., Cheng, T., Scadden, D. T. and Haber, D. A. (2001) The Wilms tumor suppressor WT1 directs stage-specific quiescence and differentiation of human hematopoietic progenitor cells. *EMBO J.*, **20**, 1897–909.

Englert, C., Hou, X., Maheswaran, S., Bennett, P., Ngwu, C., Re, G. G., Garvin, A. J., Rosner, M. R. and Haber, D. A. (1995) WT1 suppresses synthesis of the epidermal growth factor receptor and induces apoptosis. *EMBO J.*, **14**, 4662–75.

Gaiger, A., Reese, V., Disis, M. L. and Cheever, M. A. (2000) Immunity to WT1 in the animal model and in patients with acute myeloid leukemia. *Blood*, **96**, 1480–9.

Gao, L., Bellantuono, I., Elsasser, A., Marley, S. B., Gordon, M. Y., Goldman, J. M. and Stauss, H. J. (2000) Selective elimination of leukemic CD34(+) progenitor cells by cytotoxic T lymphocytes specific for WT1. *Blood*, **95**, 2198–203.

Hammes, A., Guo, J. K., Lutsch, G., Leheste, J. R., Landrock, D., Ziegler, U., Gubler, M. C. and Schedl, A. (2001) Two splice variants of the Wilms' tumor 1 gene have distinct functions during sex determination and nephron formation. *Cell*, **106**, 319–29.

Hastie, N. D. (2001) Life, sex, and WT1 isoforms – three amino acids can make all the difference. *Cell*, **106**, 391–4.

Herzer, U., Crocoll, A., Barton, D., Howells, N. and Englert, C. (1999) The Wilms tumor suppressor gene wt1 is required for development of the spleen. *Curr. Biol.*, **9**, 837–40.

Inoue, K., Tamaki, H., Ogawa, H., Oka, Y., Soma, T., Tatekawa, T., Oji, Y., Tsuboi, A., Kim, E. H., Kawakami, M., Akiyama, T., Kishimoto, T. and Sugiyama, H. (1998) Wilms' tumor gene (WT1) competes with differentiation-inducing signal in hematopoietic progenitor cells. *Blood*, **91**, 2969–76, Issn: 0006-4971.

Inoue, K., Ogawa, H., Sonoda, Y., Kimura, T., Sakabe, H., Oka, Y., Miyake, S., Tamaki, H., Oji, Y., Yamagami, T., Tatekawa, T., Soma, T., Kishimoto, T. and Sugiyama, H. (1997) Aberrant overexpression of the Wilms tumor gene (WT1) in human leukemia. *Blood*, **89**, 1405–12, Issn: 0006-4971.

Inoue, K., Sugiyama, H., Ogawa, H., Nakagawa, M., Yamagami, T., Miwa, H., Kita, K., Hiraoka, A., Masaoka, T., Nasu, K. *et al.* (1994) WT1 as a new prognostic factor and a new marker for the detection of minimal residual disease in acute leukemia. *Blood*, **84**, 3071–9, Issn: 0006-4971.

Kreidberg, J. A., Sariola, H., Loring, J. M., Maeda, M., Pelletier, J., Housman, D. and Jaenisch, R. (1993) WT-1 is required for early kidney development. *Cell*, **74**, 679–91.

Kurts, C., Heath, W. R., Kosaka, H., Miller, J. F. and Carbone, F. R. (1998) The peripheral deletion of autoreactive CD8+ T cells induced by cross-presentation of self-antigens involves signaling through CD95 (Fas, Apo-1). *J. Exp. Med.*, **188**, 415–20.

Lee, S. B., Huang, K., Palmer, R., Truong, V. B., Herzlinger, D., Kolquist, K. A., Wong, J., Paulding, C., Yoon, S. K., Gerald, W., Oliner, J. D. and Haber, D. A. (1999) The Wilms tumor suppressor WT1 encodes a transcriptional activator of amphiregulin. *Cell*, **98**, 663–73.

Liu, X. W., Gong, L. J., Guo, L. Y., Katagiri, Y., Jiang, H., Wang, Z. Y., Johnson, A. C. and Guroff, G. (2001) The Wilms' tumor gene product WT1 mediates the down-regulation of the rat epidermal growth factor receptor by nerve growth factor in PC12 cells. *J. Biol. Chem.*, **276**, 5068–73.

Lo, D., Freedman, J., Hesse, S., Palmiter, R. D., Brinster, R. L. and Sherman, L. A. (1992) Peripheral tolerance to an islet cell-specific hemagglutinin transgene affects both CD4+ and CD8+ T cells. *Eur. J. Immunol.*, **22**, 1013–22.

Loeb, D. M., Evron, E., Patel, C. B., Sharma, P. M., Niranjan, B., Buluwela, L., Weitzman, S. A., Korz, D. and Sukumar, S. (2001) Wilms' tumor suppressor gene (WT1) is expressed in primary breast tumors despite tumor-specific promoter methylation. *Cancer Res.*, **61**, 921–5.

Maheswaran, S., Englert, C., Bennett, P., Heinrich, G. and Haber, D. A. (1995) The WT1 gene product stabilizes p53 and inhibits p53-mediated apoptosis. *Genes Dev.*, **9**, 2143–56.

Mayo, M. W., Wang, C. Y., Drouin, S. S., Madrid, L. V., Marshall, A. F., Reed, J. C., Weissman, B. E. and Baldwin, A. S. (1999) WT1 modulates apoptosis by transcriptionally upregulating the *bcl-2* proto-oncogene. *EMBO J.*, **18**, 3990–4003.

Menke, A. L., Riteco, N., van Ham, R. C., de Bruyne, C., Rauscher, F. J., 3rd, van der Eb, A. J. and Jochemsen, A. G. (1996) Wilms' tumor 1 splice variants have opposite effects on the tumorigenicity of adenovirus-transformed baby-rat kidney cells. *Oncogene*, **12**, 537–46.

Menke, A. L., van Ham, R. C., Sonneveld, E., Shvarts, A., Stanbridge, E. J., Miyagawa, K., van der Eb, A. J. and Jochemsen, A. G. (1995) Human chromosome 11 suppresses the tumorigenicity of adenovirus transformed baby rat kidney cells: involvement of the Wilms' tumor 1 gene. *Int. J. Cancer*, **63**, 76–85.

Menssen, H. D., Bertelmann, E., Bartelt, S., Schmidt, R. A., Pecher, G., Schramm, K. and Thiel, E. (2000) Wilms' tumor gene (WT1) expression in lung cancer, colon cancer and glioblastoma cell lines compared to freshly isolated tumor specimens. *J. Cancer. Res. Clin. Oncol.*, **126**, 226–32.

Menssen, H. D., Renkl, H. J., Entezami, M. and Thiel, E. (1997) Wilms' tumor gene expression in human CD34+ hematopoietic progenitors during fetal development and early clonogenic growth. *Blood*, **89**, 3486–7.

Molldrem, J. J., Clave, E., Jiang, Y. Z., Mavroudis, D., Raptis, A., Hensel, N., Agarwala, V. and Barrett, A. J. (1997) Cytotoxic T-lymphocytes specific for a nonpolymorphic proteinase-3 peptide preferentially inhibit chronic myeloid-leukemia colony-forming-units. *Blood*, **90**, 2529–34.

Morgan, D. J., Kurts, C., Kreuwel, H. T., Holst, K. L., Heath, W. R. and Sherman, L. A. (1999) Ontogeny of T cell tolerance to peripherally expressed antigens. *Proc. Natl Acad. Sci. USA*, **96**, 3854–8.

Nakagama, H., Heinrich, G., Pelletier, J. and Housman, D. E. (1995) Sequence and structural requirements for high-affinity DNA binding by the WT1 gene product. *Mol. Cell Biol.*, **15**, 1489–98.

Ohashi, P. S., Oehen, S., Buerki, K., Pircher, H., Ohashi, C. T., Odermatt, B., Malissen, B., Zinkernagel, R. M. and Hengartner, H. (1991) Ablation of "tolerance" and induction of diabetes by virus infection in viral antigen transgenic mice. *Cell*, **65**, 305–17.

Ohminami, H., Yasukawa, M. and Fujita, S. (2000) HLA class I-restricted lysis of leukemia cells by a CD8(+) cytotoxic T-lymphocyte clone specific for WT1 peptide. *Blood*, **95**, 286–93.

Oji, Y., Ogawa, H., Tamaki, H., Oka, Y., Tsuboi, A., Kim, E. H., Soma, T., Tatekawa, T., Kawakami, M., Asada, M., Kishimoto, T. and Sugiyama, H. (1999) Expression of the Wilms' tumor gene WT1 in solid tumors and its involvement in tumor cell growth. *Jpn. J. Cancer Res.*, **90**, 194–204.

Oka, Y., Elisseeva, O. A., Tsuboi, A., Ogawa, H., Tamaki, H., Li, H., Oji, Y., Kim, E. H., Soma, T., Asada, M., Ueda, K., Maruya, E., Saji, H., Kishimoto, T., Udaka, K. and Sugiyama, H. (2000a) Human cytotoxic T-lymphocyte responses specific for peptides of the wild-type Wilms' tumor gene (WT1) product. *Immunogenetics*, **51**, 99–107.

Oka, Y., Udaka, K., Tsuboi, A., Elisseeva, O. A., Ogawa, H., Aozasa, K., Kishimoto, T. and Sugiyama, H. (2000b) Cancer immunotherapy targeting Wilms' tumor gene WT1 product. *J. Immunol.*, **164**, 1873–80.

Patmasiriwat, P., Fraizer, G., Kantarjian, H. and Saunders, G. F. (1999) WT1 and GATA1 expression in myelodysplastic syndrome and acute leukemia. *Leukemia*, **13**, 891–900.

Reddy, J. C., Morris, J. C., Wang, J., English, M. A., Haber, D. A., Shi, Y. and Licht, J. D. (1995) WT1-mediated transcriptional activation is inhibited by dominant negative mutant proteins. *J. Biol. Chem.*, **270**, 10878–84.

Sadovnikova, E. and Stauss, H. J. (1996) Peptide-specific cytotoxic T lymphocytes restricted by nonself major histocompatibility complex class I molecules: reagents for tumor immunotherapy. *Proc. Natl. Acad. Sci USA*, **93**, 13114–18.

Sadovnikova, E., Jopling, L. A., Soo, K. S. and Stauss, H. J. (1998) Generation of human tumor-reactive cytotoxic T cells against peptides presented by non-self HLA class I molecules. *Eur. J. Immunol.*, **28**, 193–200.

Scharnhorst, V., Menke, A. L., Attema, J., Haneveld, J. K., Riteco, N., van Steenbrugge, G. J., van der Eb, A. J. and Jochemsen, A. G. (2000) EGR-1 enhances tumor growth and modulates the effect of the Wilms' tumor 1 gene products on tumorigenicity. *Oncogene*, **19**, 791–800.

Scharnhorst, V., van der Eb, A. J. and Jochemsen, A. G. (2001) WT1 proteins: functions in growth and differentiation. *Gene*, **273**, 141–61.

Stanislawski, T., Voss, R. H., Lotz, C., Sadovnikova, E., Willemsen, R. A., Kuball, J., Ruppert, T., Bolhuis, R. L., Melief, C. J., Huber, C., Stauss, H. J. and Theobald, M. (2001) Circumventing tolerance to a human MDM2-derived tumor antigen by TCR gene transfer. *Nat. Immunol.*, **2**, 962–70.

Stauss, H. J. (1999) Immunotherapy with CTLs restricted by nonself MHC. *Immunol. Today*, **20**, 180–3.

Tamaki, H., Ogawa, H., Ohyashiki, K., Ohyashiki, J. H., Iwama, H., Inoue, K., Soma, T., Oka, Y., Tatekawa, T., Oji, Y., Tsuboi, A., Kim, E. H., Kawakami, M., Fuchigami, K., Tomonaga, M., Toyama, K., Aozasa, K., Kishimoto, T. and Sugiyama, H. (1999) The Wilms' tumor gene WT1 is a good marker for diagnosis of disease progression of myelodysplastic syndromes. *Leukemia*, **13**, 393–9.

Wright, D. E., Wagers, A. J., Gulati, A. P., Johnson, F. L. and Weissman, I. L. (2001) Physiological migration of hematopoietic stem and progenitor cells. *Science*, **294**, 1933–6.

Yamagami, T., Sugiyama, H., Inoue, K., Ogawa, H., Tatekawa, T., Hirata, M., Kudoh, T., Akiyama, T., Murakami, A. and Maekawa, T. (1996) Growth inhibition of human leukemic cells by WT1 (Wilms tumor gene) antisense oligodeoxynucleotides: implications for the involvement of WT1 in leukemogenesis. *Blood*, **87**, 2878–84, Issn: 0006-4971.

Chapter 4

Human melanoma antigens recognized by CD8+ T cells

Yutaka Kawakami

Summary

Melanoma is a relatively immunogenic cancer and responds to various immunotherapies. Clinical and immunological observations obtained from patients received various immunotherapy protocols indicate that CD8+ T cells are involved in *in vivo* tumor rejection, and melanoma reactive CD8+ T cells with various antigen specificity were generated from the patients. Using these T cells and cDNA expression cloning techniques, various melanoma antigens recognized by CD8+ T cells were identified. The representative antigens are tissue (melanocyte)-specific proteins (e.g. gp100), cancer–testis antigens that are preferentially expressed in various cancers and normal testis (e.g. MAGE) and tumor specific mutated peptides (e.g. β-catenin) and others. These antigens can also be isolated using various methods including cDNA expression cloning with patients' sera (SEREX) and cDNA subtraction with RDA, SAGE, DNAChip and EST databases. Analysis of T-cell epitopes in these antigens revealed the mechanisms for generation of tumor antigens recognized by CD8+ T cells and the nature of anti-tumor T-cell responses. The identification of antigenic peptides also allowed us to analyze anti-tumor T-cell responses in more detail particularly using the histocompatibility leukocyte antigen (HLA) tetramer technique as well as to develop new types of immunotherapy. In addition, modified antigens with higher immunogenicity can be generated for effective immunotherapy. These results obtained from the melanoma research may be useful for development of immunotherapy for patients with various cancers.

Introduction

Melanoma is a highly metastatic cancer and resistant to chemotherapy and radiotherapy, however, it responds relatively well to various immunotherapies. Administration of high dose IL2 resulted in about 15% response rate (PR + CR) for patients with metastatic melanoma (Rosenberg *et al.*, 1994). Immunohistochemical study in biopsied tissues from regressing tumor after the treatment demonstrated massive infiltrates of T cells and macrophages (Rubin *et al.*, 1989). In some cases, dominant CD8+ T-cell infiltrates were observed. Adoptive transfer of cultured tumor infiltrating T lymphocytes (TIL) along with IL2 resulted in about 35% response rate (Rosenberg *et al.*, 1995). In this clinical protocol, accumulation of the injected T cells in tumor tissues and melanoma recognition by these T cells appeared to associate with tumor regression. In many cases, the cultured TIL were dominantly CD8+ T cells and contained cytotoxic T cells (CTL) against autologous melanoma cells. Since most melanoma cells express MHC class I, but not always express MHC class II, tumor reactive CD8+ CTL appear to play an important role in *in vivo* immunological rejection of melanoma. These

Table 4.1 Antigen specificity of melanoma reactive CTL

Specificity	*Autologous melanoma specific T cells*	*Melanocyte specific shared Ag T cells*	*Cancer specific shared Ag T cells*	*Melanoma specific shared Ag T cells*
Autologous melanoma cells	+	+	+	+
Allogeneic melanoma cells	−	+	+	+
Other types of cancer cells	−	−	+	−
Cultured melanocytes	−	+	−	−

CTL demonstrate various antigen specificity. Some recognize only autologous melanoma cells, some also recognize allogeneic melanoma and cultured melanocytes, some recognize allogeneic melanoma and other types of cancers, and some recognize only melanoma cells (Table 4.1). Identification of melanoma antigens recognized by these CD8+ CTL enabled us to evaluate T-cell immune responses to melanoma in more detail as well as to develop more effective immunotherapies. In this chapter, identification and characterization of human melanoma antigens recognized by CD8+ T cells will be discussed particularly in the aspect of anti-tumor immune responses and development of immunotherapy.

Methods for the identification of antigens recognized by CD8+ T cells

DNA expression cloning using melanoma reactive T cells

Functional DNA expression cloning methods using tumor reactive CD8+ T cells was first developed by Boon and his colleagues (De Plaen *et al.*, 1988; Van der Bruggen *et al.*, 1991), then, various modifications were added to improve the cloning efficacy (Brichard *et al.*, 1993; Kawakami *et al.*, 1994). Most of the identified melanoma antigens were isolated by this way (Table 4.2). cDNA cloning methods that use transient cDNA expression system with highly transfectable cell lines such as COS cells, 293 cells, and VO cells, are frequently used (Brichard *et al.*, 1993; Robbins *et al.*, 1994). However, isolated antigens should be carefully evaluated whether they were truly tumor antigens, since cross-reactive unrelated antigens can be isolated in this high expression system. Retroviral cDNA library has recently been used to isolate NY-ESO-I (Wang *et al.*, 1998c). It allows us to utilize autologous antigen presenting cells (APCs) such as fibroblasts, without determining antigen presenting MHC for the cloning, although it appears to be less sensitive for screening antigens than the conventional methods using COS cells. It is particularly valuable for the isolation of unique antigens that are only expressed in autologous tumor cells. It is sometimes difficult to determine MHC restriction for unique antigens unless MHC blocking mAb or various MHC loss tumor variants from the same tumor are available.

Direct identification of antigenic peptides on melanoma cells

Direct identification of epitope peptides bound to HLA on tumor cells is an invaluable method to identify naturally presented epitopes (Skipper *et al.*, 1999). It is sometimes difficult to identify natural epitopes using synthetic peptides, particularly in the case of epitopes with posttranslational modification (Skipper *et al.*, 1996a; Kittlesen *et al.*, 1998). One of the

Table 4.2 Isolation methods for melanoma antigens and candidates recognized by CD8+ T cells

Criteria for isolation	*Methods*
Immunogenicity	cDNA expression cloning with patient's serum (SEREX database) cDNA expression cloning with tumor reactive T cells
Specific expression (tissue specific, tumor specific, tumor overexpressed)	
mRNA	cDNA subtraction (RDA) cDNA profile comparison (DNA chip, SAGE, EST database)
Protein	Protein expression profile comparison (2D EP, MS, protein database)
HLA bound peptide	HLA binding peptide isolation by HPLC and MS

Notes
SEREX: Serological analysis of autologous tumor antigens by recombinant cDNA expression cloning; SAGE: Serial Analysis of Gene Expression; RDA: representational differential analysis; EP: electrophoresis; MS: mass spectrometry; HPLC: high pressure liquid chromatography.

gp100 epitopes was isolated by this way from HPLC fractions of HLA bound peptides using electrospray ionization triple quadrupole tandem mass spectrometer (Cox *et al.*, 1994).

Isolation of melanoma antigens using sera from patients (SEREX)

A number of melanoma antigens recognized by CD8+ T cells, including tyrosinase, TRP-1, gp100, MAGE-1 and NY-ESO-I, were also isolated using cDNA expression cloning using IgG in patients' sera, a technique called SEREX (serological analysis of recombinant cDNA expression libraries) (Sahin *et al.*, 1995). Presence of IgG responses indicates activation of CD4+ helper T cells specific for the same antigens. In addition, some antigens were found to be also recognized by CD8+ CTL (Tureci *et al.*, 1997; Old and Chen, 1998). It has recently been reported that IgG and CD8+ T-cell responses correlate positively in the immune response to NY-ESO-1 (Jager *et al.*, 2000b).

Systematic isolation of candidates for melanoma antigens by cDNA subtraction and DNA databases

Using various methods described above, melanocyte specific proteins, testis specific proteins and overexpressed proteins in tumor cells were found to be representative melanoma antigens recognized by CD8+ CTL. These antigens can be systematically isolated by various mRNA/cDNA subtraction methods among various tissues and cancer cells. The classical cDNA subtraction and differential hybridization methods that actually hybridize cDNA or mRNA from different tissues are relatively difficult to optimize experimental conditions. PCR differential display is a relatively easy technique, but it may miss many differentially expressed genes. Representational differential analysis (RDA) is a superior technique to isolate genes expressed at low frequency. MAGE-C1, a candidate cancer–testis antigen, was isolated using RDA by subtracting cDNAs among testis, melanoma and other normal tissues (Lucas *et al.*, 1998).

Alternatively, cDNA databases that have recently been increasing under the human genome projects may be utilized for the identification of candidates. XAGE was identified from the cDNA databases by searching homology with known cancer testis antigens

Table 4.3 Comparison of cDNA tag expression of known melanoma antigens among various tissues using SAGE

Antigens	*TAG sequence*	*Melanoma*	*Melanocytes*	*Other tissues*
Melanocyte-specific antigens				
gp100	CCTGGTCAAG	+	+	−
Tyrosinase	GAGAAAGAGG	+	+	+/−
TRP1	AAATATATTT	+	+	−
TRP2	CCTTACCTAA	+	+	−
MART-1	TGAGGAAATG	+	+	−
KU-MEL-1	ACCAACACGG	+	+	+/−
Cancer testis antigens				
MAGE-1	AGGCCCATTC	+	−	+
MAGE-3	AGATAACTCA	+	−	+
Others				
PRAME	TAGGAGTTAA	+	−	+
SART-1	AACGCGAACA	+	−	+

Note
Other tissues include brain, colon and testis tissues, and brain and colon cancer tissues.

(Brinkmann *et al.*, 1999). However, cDNA databases containing sufficient number of genes have not yet been available for melanoma. We have recently analyzed cDNA profiles of various cancer cells and normal tissues, including melanoma and melanocytes using serial analysis of gene expression (SAGE) and DNAChip, and have identified many candidate genes for melanoma antigens. One of the newly identified melanocyte specific proteins by SAGE and DNAChip analysis, KU-MEL-1, was also identified as a melanoma antigen by SEREX (Table 4.3) (Kiniwa *et al.*, 2001).

Induction of melanoma reactive CTL using candidate molecules

Effective methods for the evaluation of candidate molecules identified by the methods described above have not yet been developed. Unless established tumor reactive T cells recognize the candidates, tumor reactive T cells need to be induced by stimulation with the candidate molecules. One of the methods is *in vitro* CTL induction by stimulation with many peptides synthesized based on HLA allele specific binding motifs and proteasome cleavage prediction from PBMC or TIL of cancer patients or healthy individuals (Celis *et al.*, 1994; Rivoltini *et al.*, 1995; Kessler *et al.*, 2001). *In vitro* induced CTL that recognize only target cells incubated with high concentration of peptides (more than 1 μM) has a tendency not to recognize tumor cells that present endogenously processed tumor peptides at relatively low density on the cell surface possibly due to low avidity of TCR binding to the peptide/MHC complex (Yee *et al.*, 1999). CTL induced by stimulation with HLA-A2 binding MAGE-3 peptide (FLWGPRALV) recognized target cells incubated with nanomolar concentration of the peptide and recognized COS cells transfected with a minigene encoding the MAGE-3 peptide. However, this CTL did not lyse COS cells transfected with full length MAGE-3 because of ineffective cleavage of COOH terminus of the peptide by proteasomes (Valmori *et al.*, 1999). Thus, it is important to confirm that the peptide induced T cells are actually able to recognize tumor cells.

It is cumbersome to set up *in vitro* CTL induction for many peptides from many candidate molecules. Alternatively, APC transfected with cDNA or RNA of candidate proteins can be used as stimulator cells. Recombinant virus including adenovirus, retrovirus and vaccinia virus were used for the efficient transduction of tumor antigen genes into APC including dendritic cells (DCs) and B cells (Kim *et al.*, 1997; Butterfield *et al.*, 1998; Perez-Diez *et al.*, 1998). HLA transgenic mice can also be used to identify immunogenic peptides presented by HLA. One of the problems on the use of transgenic mice is different sequence of the corresponding antigens in mice. If amino acid difference exists in epitopes between human and mouse, human peptides are highly immunogenic in the HLA transgenic mice. However, if the sequence is identical, it is relatively difficult to induce T cells against those peptides (Overwijk *et al.*, 1998; Irvine *et al.*, 1999; Bullock *et al.*, 2000).

Characteristics of the identified melanoma antigens recognized by CD8+T cells and their implications to development of immunotherapy

Using various techniques described above, many human melanoma antigens recognized by CD8+ T cells have been identified and their biological and immunological characteristics were investigated (Table 4.4).

Melanocyte specific antigens

Some melanoma reactive CTL recognize autologous and allogeneic melanoma cells and cultured melanocytes sharing antigen presenting MHC class I, but do not recognize other types of cells, suggesting that these CTL recognize non-mutated peptides derived from melanocyte specific proteins. Melanosomal proteins including tyrosinase, TRP-1, TRP-2, gp100, MART-1/Melan-A (MART-1), and AIM-1 were isolated as melanoma antigens by cDNA expression cloning with melanoma reactive CTL. MSH receptor M1CR was identified as an antigen using *in vitro* CTL induction by stimulation with HLA-A2 binding synthetic peptides. The reasons for the relatively high immunogenicity of melanosomal proteins are not well understood. Since melanin pigments and melanosomal proteins are transferred to keratinocytes from melanocytes in skin, they may also be taken by antigen presenting DC in the skin, Langerhans cells, around melanoma and these DC move to draining lymph nodes and efficiently sensitize specific T cells. Trafficking of melanin to draining lymph nodes by TGFβ dependent cells, possibly Langerhans cells, has been demonstrated using transgenic mice with melanocytosis (Hemmi *et al.*, 2001).

Tyrosinase, an enzyme that has tyrosine hydroxylase activity involved in the melanin synthesis, was identified as a melanoma antigen recognized by HLA-A2 restricted CTL by cDNA expression cloning (Brichard *et al.*, 1996). Numbers of epitopes presented by various HLA alleles including HLA-A1, -A2, -A24, and -B44 have been identified (Robbins *et al.*, 1994; Wolfel *et al.*, 1994; Brichard *et al.*, 1996). The HLA-A2 binding epitope, YMDGTMSQV, was found to have posttranslational deamidation of asparagine (Skipper *et al.*, 1996a). The HLA-A1 binding epitope, DAEKCDICTDEY, contains two cysteines and the sulfohydryl group of the second one appears to be oxidized (Kittlesen *et al.*, 1998). Adoptive transfer of one of the HLA-A24 restricted, tyrosinase reactive TIL along with IL2 into the autologous patient resulted in complete regression of tumor (Robbins *et al.*, 1994), and immunization with HLA-A2 binding tyrosinase epitopes, MLLAVLYCL and MDGTMSQV, with GM-CSF

Table 4.4 Human melanoma antigens recognized by CD8+ T cells

Antigen	*Ag presenting MHC*	*Epitope*
Possibly autoreactive antigens		
Melanosomal proteins		
gp100	HLA-A2	KTWGQYWQV
	HLA-A2	AMLGTHTMEV
	HLA-A2	MLGTHTMEV
	HLA-A2	ITDQVPFSV
	HLA-A2	YLEPGPVTA
	HLA-A2	LLDGTATLRL
	HLA-A2	VLYRYGSFSV
	HLA-A2	SLADTNSLAV
	HLA-A2	RLMKQDFSV
	HLA-A2	RLPRIFCSC
	HLA-A3	ALLAVGATK
	HLA-A3	LIYRRRLMK
	HLA-A3	ALNFPGSQK
	HLA-A11	ALNFPGSQK
	HLA-A24	VYFFLPDHL[a]
	HLA-Cw8	SNDGPTLI
MART-1/Melan-A	HLA-A2	AAGIGILTV
	HLA-A2	EAAGIGILTV
	HLA-A2	ILTVILGVL
	HLA-B45	AEEAAGIGIL
	HLA-B45	AEEAAGIGILT
TRP1 (gp75)	HLA-A31	MSLQRQFLR[b]
TRP2	HLA-A2	SVYDFFVWL
	HLA-A2	SLDDYNHLV
	HLA-A2	YAIDLPVSV
	HLA-A31	LLGPGRPYR
	HLA-A33	LLGPGRPYR
	HLA-A33	EVISCKLIKR[a]
	HLA-A68	EVISCKLIKR[a]
	HLA-Cw8	ANDPIFVVL
Tyrosinase	HLA-A1	DAEKCDKTDEY
	HLA-A1	SSDYVIPIGTY
	HLA-A2	MLLAVLYCL
	HLA-A2	YMDGTMSQV
	HLA-A24	FLPWHRLF
	HLA-B44	SEIWRDIDF
AIM-1	HLA-A2	AMFGREFCYA
MC1-R	HLA-A2	TILLGIFFL
	HLA-A2	FLALIICNA
	HLA-A2	AIIDPLIYA
Tumor specific shared antigens		
Cancer–testis antigens		
MAGE-1	HLA-A1	EADPTGHSY
	HLA-A3	SLFRAVITK
	HLA-A24	NYKHCFPEI
	HLA-A28	EVYDGREHSA
	HLA-B37	REPVTKAEML
	HLA-B53	DPARYEFLW
	HLA-Cw2	SAFPTTINF
	HLA-Cw3	SAYGEPRKL
	HLA-Cw16	SAYGEPRKL

(Continued)

Table 4.4 (Continued)

Antigen	*Ag presenting MHC*	*Epitope*
MAGE-2	HLA-A2	YLQLVFGIEV
	HLA-A24	EYLQLVFGI
	HLA-B37	REPVTKAEML
MAGE-3	HLA-A1	EVDPIGHLY
	HLA-A2	FLWGPRALV
	HLA-A2	KVAELVHFL
	HLA-A24	IMPKAGLLI
	HLA-A24	TFPDLESEF
	HLA-B37	REPVTKAEML
	HLA-B44	MEVDPIGHLY
	HLA-B52	WQYFFPVIF
MAGE-4	HLA-A2	GVYDGREHTV
MAGE-6	HLA-B34	MVKISGGPR
	HLA-B37	REPVTKAEML
MAGE-10	HLA-A2	GLYDGMEHL
MAGE-12	HLA-CW7	VRIGHLYIL
MAGE-B2	HLA-A2	FLWGPRAYA
BAGE	HLA-Cw16	AARAVFLAL
GAGE-1,2,8	HLA-Cw6	YRPRPRRY
GAGE3,4,5,6,7B	HLA-A29	YYWPRPRRY
NY-ESO-1	HLA-A2	SLLMWITQC
	HLA-A2	SLLMWITQCFL
	HLA-A2	QLSLLMWIT
	HLA-A31	ASGPGGGAPR
	HLA-A31	LAAQERRVPR[b]
GnT-V	HLA-A2	VLPDVFIRC[a]
p15	HLA-A24	AYGLDFYIL
PRAME	HLA-A24	LYVDSLFFL
Tumor specific unique peptides		
β-catenin	HLA-A24	SYLDSGIHF*
CDK4	HLA-A2	AC*DPHSGHFV
MUM-1	HLA-B44	EEKL*IVVLF[a]
MUM-2	HLA-B44	SELFRSG*LDSY
	HLA-Cw6	FRSG*LDSYV
MUM-3	HLA-A28	EAF*IQPITR
MART-2	HLA-A1	FLE*GNEVGKTY
Myosin class I	HLA-A3	K*INKNPKYK

Notes
a Peptides from intron sequences.
b Peptides from alternative ORFs.
* Mutation.

administration resulted in tumor regression in some patients, suggesting that tyrosinase may function as a tumor rejection antigen (Jager *et al.*, 1998).

Other tyrosinase family proteins, TRP-1 and TRP-2, enzymes involved in melanin synthesis as DHI-2-carboxylic acid oxidase and DOPA chrome tautomerase, were recognized by CD8+ TIL (Wang *et al.*, 1995). A TRP-1 epitope, MSLQRQFLR, recognized by HLA-A31 restricted CTL, was derived from a short 24 amino acid peptide encoded by an alternative open reading frame (ORF) different from the ORF encoding the functional TRP-1 (Wang *et al.*, 1995, 1996b). TRP-2 was isolated by cDNA cloning with HLA-A31 restricted TIL (Wang *et al.*, 1996a). It has been shown to be recognized by HLA-A2, -A33, -A68 and -Cw8 restricted CTL (Castelli *et al.*, 1999; Sun *et al.*, 2000; Harada *et al.*, 2001). One of the TRP-2 epitopes, LLPGGRPYR, was recognized by CTL restricted by both HLA-A31 and HLA-A33, members of the HLA-A3 superfamily (Wang *et al.*, 1998a). The other epitope, EVISCKLIKR, that was derived from the intron 2 sequence in the incompletely spliced mRNA, was also recognized by both HLA-A*68011 and HLA-A33 restricted CTL (Lupetti *et al.*, 1998). In contrast to the other intron derived gp100 epitope, VYFFLPDHL, that was expressed in both melanoma and melanocytes (Robbins *et al.*, 1997), this TRP-2 intron 2 peptide was present in about 50% of melanoma cell lines and tissues, but not in melanocytes. HLA-A2 restricted epitope, SVYDEFVWL, was identified by *in vitro* CTL induction with peptides synthesized based on the HLA-A2 binding motif (Parkhurst *et al.*, 1998). This epitope was found to have the same sequence as the mouse TRP-2 epitope that was isolated using H-2Kb restricted, B16 melanoma reactive TIL (Bloom *et al.*, 1997). Administration of TIL that was used for the gene cloning of TRP-1 and TRP-2, along with IL2 into the autologous patient resulted in tumor regression, suggesting possible involvement of TRP-1 and TRP-2 in anti-tumor immune responses (Topalian *et al.*, 1988).

MART-1 (Melanoma Antigen Recognized by T-cells-1)/Melan-A (MART-1) was isolated by cDNA expression cloning with HLA-A2 restricted melanoma reactive CTL from TIL and PBMC (Coulie *et al.*, 1994; Kawakami *et al.*, 1994c) and is a membrane protein in melanosomes (Kawakami *et al.*, 1994c). Its biological function has not yet been identified. MART-1 was found to be an immunodominant antigen recognized by the majority of melanoma reactive TIL in HLA-A*0201 patients (Kawakami *et al.*, 1994b, 1995, 2000b). Two peptides, AAGIGILTV and EAAGIGILTV, were found to be responsible for this immunodominance (Kawakami *et al.*, 1994c; Romero *et al.*, 1997; Valmori *et al.*, 1998). Many MART-1 specific CTL recognize both peptides, and some recognized either of them. The 9 mer peptide has been proved to be a naturally presented epitope on melanoma cells by mass spectrometry analysis on HPLC fractions from HLA-A2 bound melanoma peptides (Skipper *et al.*, 1999). This peptide could induce CTL only in HLA-A*0201 individuals among many HLA-A2 subtypes tested (Bettinotti *et al.*, 1998). Although preferential usage of *TCRVA2* and *TCRVB14* genes in TIL containing HLA-A2 restricted MART-1 specific CTL was reported in some patients (Sensi *et al.*, 1995), a variety of Vβ were utilized in a single MART-1 peptide stimulated CTL and in the MART-1 peptide/HLA-A2 tetramer sorted CTL (Cole *et al.*, 1994; Valmori *et al.*, 2000a; Dietrich *et al.*, 2001).

These peptides have relatively low (intermediate) binding affinity to HLA-A*0201, likely because of the presence of non-optimal amino acid, alanine, at the primary anchor position (P2) for HLA-A*0201 binding (Kawakami *et al.*, 1995). Characteristics of the identified HLA-A2 binding epitopes in various melanocyte specific antigens are shown in Table 4.5 (Kawakami *et al.*, 1995). It should be noted that epitopes identified using CTL generated by stimulation with tumor cells, not with APC pulsed with synthetic peptides, have tendency to contain non-optimal amino acids (Table 4.5). Despite of their low HLA affinity, these peptides are immunogenic in *in vitro* induction of melanoma reactive CTL. MART-1 specific CTL could be induced even from PBMC of healthy individuals, although it is easier to induce T cells from patients (Marincola *et al.*, 1996). Analysis using HLA tetramers revealed that MART-1 specific T cells were frequently detected in HLA-A2+ individuals, with naïve phenotype (CD45RA+, CD45RO−, CCR7+) in healthy individuals and with the increased

Table 4.5 HLA-A*0201 binding affinity of T-cell epitopes derived from melanosomal proteins

Antigen	*Length (aa)*	*Sequence*	*Natural peptide*[a]	*HLA-A2binding affinity*[b]
gp100	9	K<u>T</u>WGQYWQV[c]	Yes	High
	10	AMLGTHTMEV	n.e.	Intermediate
	9	MLGTHTMEV	n.e.	High
	9	I<u>T</u>DQVPFSV[c]	Yes	Intermediate
	9	YLEPGPVT<u>A</u>[c]	Yes	Intermediate
	10	LLDGTATLRL[c]	n.d.	Intermediate
	10	VLYRYGSFSV[c]	yes	High
	10	SLADTNSLAV	n.e.	Intermediate
	9	RLMKQDFSV[c]	n.e.	High
MART-1	9	A<u>A</u>GIGILTV[c]	Yes	Intermediate
	10	E<u>A</u>AGIGILTV[c]	n.d.	Low
	9	ILTVILGVL[c]	n.d.	Intermediate
Tyrosinase	9	MLLAVCYLL	n.e.	Intermediate
	9	YMDGTMSQV[c]	Yes	High
TRP-2	9	S<u>V</u>YDFFVWL	n.e.	High
	9	SLDDYNHLV	n.e.	High
	9	Y<u>A</u>IDLPVSV[c]	n.e.	n.e.
AIM-1	10	AMFGREFCY<u>A</u>[c]	n.e.	n.e.

Notes

n.d.: not detected; n.e.: not examined; underline: non-optimal primary anchor amino acids.

a Peptides that were proved to be naturally presented by HLA-A2 on melanoma cells by mass spectrometry.

b High affinity <50 nM (IC50), intermediate affinity <50 nm <500 nM, low affinity <500 nM in competitive inhibition assay.

c Epitopes that were identified using CTL generated by stimulation with melanoma, not with peptide pulsed APC.

memory phenotype (CD45RA−, CD45RO+, CCR7−) in patients with melanoma (D'Souza *et al.*, 1998; Pittet *et al.*, 1999). Upon *in vitro* stimulation with the antigen, MART-1/HLA-A2 tetramer positive T cells with native phenotype did not secrete IFNγ, but those with memory phenotype secreted IFNγ (Pittet *et al.*, 2001b). One possible explanation for this high precursor frequency of the MART-1 specific CTL in healthy individuals was that these T cells were primed with cross-reactive antigens derived from self or microbial proteins, because these MART-1 specific CTL were found to recognize homologous peptides derived from other proteins including microbial proteins (Loftus *et al.*, 1996). However, the fact that MART-1 specific T cells have naïve phenotype in healthy individuals suggest the other possibility that immunological tolerance was not induced in these T cells. In HLA-A2 transgenic mouse study, incomplete tolerance to tyrosinase specific CTL with high avidity TCR and induction of depigmentation after transfer of these activated CTL were demonstrated (Colella *et al.*, 2000). One of the mechanisms for the incomplete tolerance may be explained by relatively cryptic nature of the melanoma epitopes due to low HLA binding or low antigen processing. These epitopes may not be presented at high density on the cell surface of melanocytes and APC. Other possible mechanism is the inefficient processing of the

MART-1 epitopes by immunoproteasomes in professional APC such as DC, resulting in inefficient priming of T cells *in vivo*. It was shown that some epitopes from MART-1 and gp100 were not efficiently processed by immunoproteasomes induced by IFNγ, while they were processed by constitutively expressed proteasomes in melanoma cells (Morel *et al.*, 2000). These may be the reasons for the high frequency of naïve MART-1 CTL precursors in healthy individuals.

In patients with melanoma, increase of MART-1 specific CD8+ CTL with memory phenotype (CD45RA−, CD45RO+) was observed in peripheral blood, metastatic lymph nodes and tumor tissues (Romero *et al.*, 1998; Anichini *et al.*, 1999; Pittet *et al.*, 1999), suggesting that T cells were sensitized with melanosomal proteins *in vivo* in melanoma patients and accumulated in tumor tissues. These T cells are likely involved in melanoma regression occurred spontaneously or after treatments. Various changes occurred in melanoma patients, including increase of HLA expression, increase of peptide loading into MHC class I antigen processing pathway through aberrant transport of melanosomal proteins (Halaban *et al.*, 1997), increase of total antigen supply, and inflammatory conditions, may lead to priming of T cells specific for the melanoma epitopes. *In vitro* culture of these T cells with IL2 may further expand and activate melanoma reactive T cells.

Although MART-1 specific CTL are accumulated in tumor tissues and lyse melanoma cells after *in vitro* culture with IL2, tumor rejection does not usually occur without treatment, suggesting the presence of inhibitory mechanisms against anti-tumor immune responses. These inhibitory mechanisms remain to be investigated. The recent HLA tetramer study demonstrated possible development of T cells with an unusual phenotype (CD8β low, CD45RA+, CD45RO−, CD16+), which were specifically unresponsive to tyrosinase (Lee *et al.*, 1999). Although tumor regression was observed in a small number of patients who received MART-1 reactive CTL or who were immunized with the MART-1 peptide in incomplete Freund's adjuvant (IFA) in the clinical trials in the NCI Surgery Branch, further investigation is necessary to clarify the role of MART-1 in *in vivo* melanoma rejection (Cormier *et al.*, 1997). It was recently reported that VB16+ oligoclonal MART-1 specific T cells were detected in PBMC, vitiligo site and DTH site in patients who responded to i.d. immunization with MART-1, tyrosinase, and gp100, suggesting the involvement of MART-1 in the vitiligo development and tumor regression (Jager *et al.*, 2000a).

Gp100 was identified as a melanoma antigen recognized by T cells using three different methods, cDNA expression cloning, direct epitope identification and screening candidate molecules using melanoma reactive CTL (Bakker *et al.*, 1994; Cox *et al.*, 1994; Kawakami *et al.*, 1994a). It was reported to have a melanin polymerase activity in melanosomes (Chakraborty *et al.*, 1996). Three immunodominant peptides, gp100–154 (KTWGQYWQV), gp100–209 (ITDQVPFSV) and gp100–280 (YLEPGPVTA), that were recognized by many HLA-A2 restricted TIL were identified (Cox *et al.*, 1994; Kawakami *et al.*, 1995). Adoptive transfer of TIL that responded strongly to these gp100 epitopes into the autologous patients, correlated significantly with tumor regression (Kawakami *et al.*, 1995, 2000b). Many other HLA-A2 binding peptides were identified by *in vitro* CTL induction with synthetic peptides predicted by the HLA-A*0201 binding motif (Tsai *et al.*, 1997). One of the HLA-A2 binding epitope, RLPRIFCSC, contained two cysteines. Substitution of either cysteine by α-amino butyric acid that has a similar size to the side chain of cysteine, but cannot be oxidized, led to enhancement of the CTL recognition, suggesting that this CTL recognized unoxidized peptide on melanoma cells (Kawakami *et al.*, 1998). Since cysteines in synthetic peptides may easily be oxidized in medium, modification of cysteine residues may be effective for use of these peptides in clinical trials. Gp100 epitopes presented by various HLA including HLA-A1, -A3, -A11, -A24 and -Cw8 have been identified (Skipper *et al.*, 1996b; Robbins *et al.*, 1997; Kawakami *et al.*, 1998; Castelli *et al.*, 1999). The HLA-A24 binding epitope, VYFFLPDHL, was encoded by an intron sequence of an incompletely spliced mRNA present in both melanoma cells and cultured normal melanocytes (Robbins *et al.*, 1997).

Immunization with gp100–209 and –280 in IFA resulted in augmentation of melanoma reactive CTL precursors in PBMC of patients, but immunization with gp100–154 that had high HLA-A2 binding affinity did not augment immune response. It may be explained by different degree of tolerance induction. The augmentation obtained with gp100–209 and –280, however, was relatively weak. CTL precursor frequency measured by a limiting dilution analysis was only increased to less than 1/30,000 even after the immunization. gp100–209 and –280 have relatively low (intermediate) HLA binding affinity possibly because of non-optimal amino acids in primary anchor positions (Table 4.5) (Kawakami *et al.*, 1995). Replacement of thereonine at the second position or alanine at the C-terminus, to optimal anchor amino acids, methionine or valine respectively, increased 10-fold their HLA-A2 binding affinity. These modified peptides were demonstrated to be more immunogenic in both *in vitro* and *in vivo* induction of melanoma reactive CTL (Table 4.6) (Parkhurst *et al.*, 1996; Rosenberg *et al.*, 1998). The immunization with gp100–209(210M) along with IFA increased CTL precursors specific for the native gp100–209 peptide in PBMC in 10 of 11 patients, whereas the immunization with gp100–209 augmented immune response in two of eight patients. The immunization with gp100–209(210M) could increase CTL precursor frequency up to about 1/3,000 in PBMC. Although the immunization with the modified peptide alone lead to only mixed response in some patients, co-administration of high dose IL2 resulted in either CR or PR in 13 of 33 (42% response rate) patients with metastatic melanoma. IL2 might be effective through various mechanisms, including changing endothelial cell structure allowing T-cell migration into tumor tissues, *in vivo* activation and expansion of T cells, and cascade production of various cytokines. However, increase of the gp100 specific CTL precursors in PBMC was not observed when the peptide was administered simultaneously with high dose IL2 at the same time (Rosenberg *et al.*, 1998, 1999). gp100 specific CTL might not always be detected in biopsied tumor tissues after the immunization (Lee *et al.*, 1998). This anti-tumor effect needs to be confirmed in further

Table 4.6 Immunogenicity and anti-tumor activity of the modified gp100 peptide with high HLA binding affinity

Peptide[a]	*Epitope sequence*	*HLA-A*0201 binding affinity (IC50)*	*CTL precursor frequency after immunization*[b]	In vivo *immune augmentation*[c] *(No. of patients)*	*Clinical response (CR + PR)*[d] *(No. of patients (response rate)) (IL2 co-administration)*
gp100-209	ITDQVPFSV	172 nM	<1/30,000	2/8 (25%)	1/9 (11%) (−IL2)
gp100-209(210 M)	IMDQVPFSV	19 nM	1/3,000	10/11 (91%)	0/11 (0%) (−IL2) 13/31 (42%) (+IL2)

Notes

a gp100–209; native peptide with intermediate HLA-A*0201 binding affinity, gp100–209(210 M); modified peptides with high HLA-A*0201 binding affinity.

b CTL precurssor frequency in PBMC measured by limiting dilution analysis.

c Peptides were injected subcutenously with incomplete Freund's adjuvant.

d Response rate with IL2 alone in the other trial was about 15%.

clinical trials and the mechanisms for the tumor regression should be investigated by analyzing T cells infiltrating in regressing tumors. Nevertheless, gp100 is an attractive antigen for development of immunotherapies.

Antigen isolated from immunoselected melanoma-1 (AIM-1) was isolated by HLA-A2 restricted CTL generated from PBMC by stimulation with MART-1 negative, gp100 negative autologous melanoma cell line that was immunoselected with CTL specific for these immunodominant antigens (Harada *et al.*, 2001). The use of immunoselected tumor cells may facilitate isolation of subdominant antigens. *AIM-1* has 12 transmembrane domains with sycrose transporter signature sequence and is homologous to plant sucrose transporters. It is preferentially expressed in melanocytes and melanoma cells. It has recently been reported that *AIM-1* homologous gene is involved in melanin synthesis and their mutations cause oculocutaneous albinism in medaka fish (*b*-locus mutant), mouse (*underwhite*) and human (*OCA4*) (Fukamachi *et al.*, 2001; Newton *et al.*, 2001).

Role of melanocyte specific antigens in the immunotherapy

The melanocyte specific antigens can be applied in the immunotherapy for many patients, since melanoma reactive T cells could be induced from patients with diverse HLA types. The significant correlation between vitiligo development and tumor regression in patients who received the IL2-based immunotherapies (Rosenberg and White, 1996), and the induction of melanocyte reactive T cells from the patients, suggest the involvement of melanocyte specific antigens in *in vivo* melanoma regression. Progressive growth of metastases that lost expression of melanocyte specific antigens has been observed, while other multiple metastases regressed (Jager *et al.*, 1996), and decrease of gp100 expression after the immunization of the gp100 peptide was reported (Riker *et al.*, 1999) in clinical trials. These observations also suggest that melanocyte specific antigens may be useful as tumor rejection antigens in the immunotherapy, although tumor rejection ability may not be so potent because of relatively low immunogenicity and low expression on tumor cell surface.

Melanocyte specific antigens including MART-1, gp100 and tyrosinase, express heterogeneously even in a single metastasis (Cormier *et al.*, 1998; de Vries *et al.*, 2001), and emergence of antigen loss variants appear to increase in metastatic melanoma, compared to primary melanoma lesions (Kageshita *et al.*, 1997). Since these melanocyte specific antigens are not necessary for growth of tumor cells, tumor cells may easily escape from T-cell recognition through loss of the antigens. Metastasis that lost expression of MART-1 or gp100 were found in 5–20% of patients with metastatic melanoma particularly in patients after the immunotherapy (Cormier *et al.*, 1998). Various other mechanisms for tumor escape from T-cell recognition, including loss of HLA, or molecules necessary for antigen processing, have been reported (Marincola *et al.*, 2000). Use of multiple antigens may be effective against antigen loss variants. Other types of treatments including chemotherapy are probably necessary for the eradication of the MHC loss variants, although they may be eliminated by other immune mechanisms including NK cells, NKT cells and macrophages. Immunological tolerance specific for melanoma antigens, including anergy and deletion may be induced in patients as described above (Lee *et al.*, 1999). It is important to extensively analyze these tumor escape mechanisms for further improvement of immunotherapies.

Immune responses against melanocyte specific antigens may cause autoimmune destruction of melanocytes. Development of vitiligo was observed in some patients who received IL2-based immunotherapies in the Surgery Branch, NCI (Rosenberg *et al.*, 1996). However, no ophthalmic problem due to destruction of melanocytes in uvea or pigmented epithelial cells in retina has been observed. Immune responses against self-peptides, particularly cryptic and subdominant epitopes, may not develop autoimmune adverse effects, since the susceptibility to T cells may be different between normal cells and tumor cells because of difference in tissue structure, inflammatory status and epitope density on the cell surface. However, these antigens may also be involved in autoimmune diseases against melanocytes, including autoimmune vitiligo, sympathetic ophthalmia and Vogt–Koyanagi–Harada (VKH) disease in certain conditions (Kawakami *et al.*, 2000a). Latter two diseases highly correlate with HLA-DR4 type, particularly HLA-DRB1*0405. Tyrosinase specific CD4+ T cells and IgG antibody specific for KU-MEL-1 that have recently been isolated from a melanoma patient with vitiligo, were frequently detected in patients with VKH disease (Gocho *et al.*, 2001; Kiniwa *et al.*, 2001). MART-1 specific CTL was induced from anterior chamber of eye in VKH disease (Sugita *et al.*, 1996). MART-1 specific CTL expressing skin homing receptor cutaneous lymphocyte associated antigen (CLA) were frequently present in autoimmune vitiligo (Ogg *et al.*, 1998) Thus, immunization with melanocyte specific antigens of patients having autoimmune prone background including HLA-DRB1*0405 type, should be carefully performed, although HLA-DRB1*0405 is expressed in only 1.5% of Caucasian population.

Cancer–testis antigens

MAGE-1, one of the large MAGE family, was first identified as a melanoma antigen recognized by CTL using cDNA expression cloning (Van der Bruggen *et al.*, 1991). MAGE-1 and MAGE-3 are expressed in approximately 40% and 70% of melanoma, respectively (Gaugler *et al.*, 1994). NY-ESO-1 that expresses approximately 34% of melanoma, was isolated by cDNA expression cloning using tumor reactive T cells as well as patient's serum (Chen *et al.*, 1997; Wang *et al.*, 1998c). These antigens expressed in various cancers, including melanoma, adenocancers, squamous cell cancers and sarcoma, but do not express in normal tissues in the exception of testis. Numbers of tumor antigens with the similar expression pattern were identified and named as "cancer–testis antigens" (Chen *et al.*, 1999) (Table 4.4). Many of them were isolated using melanoma reactive CD8+ CTL. They were also isolated by SEREX using sera from patients with various cancer and cDNA subtraction methods (Tureci *et al.*, 1997; Chen *et al.*, 1998; Lucas *et al.*, 1998; Old *et al.*, 1998; Brinkmann *et al.*, 1999). Many of the identified cancer testis antigens were located in X chromosomes and their biological function has not been known except synaptonemal complex protein 1 (SCP1) that is involved in the pairing of homologous chromosomes in meiosis (Tureci *et al.*, 1998). The mechanism for their expression in tumor cells has not been well understood, although hypomethylation appears to be associated with their expression (de Smet *et al.*, 1996).

In testis, MAGEs express in spermatogonia and spermatocytes that do not express MHC class I, so that MAGE specific T cells do not recognize normal testis cells, indicating that MAGEs are tumor specific shared antigens (Takahashi *et al.*, 1995). Although induction of MAGE specific CTL from PBMC of patients was rather difficult (Salgaller *et al.*, 1994), the second dominant T-cell clones infiltrated in primary melanoma lesion that spontaneously regressed, was found to recognize MAGE-6 (Zorn and Hercenda, 1999a). MAGE-12 was also isolated from a melanoma metastasis that lost gp100 expression after the immunization with the modified gp100 peptide (Panelli *et al.*, 2000a). IgG response to NY-ESO-1 positively correlated to CTL responses in patients with NY-ESO-1 positive tumor (Jager *et al.*, 2000b). Using HLA tetramer, NY-ESO-I specific T cells with memory phenotype were detected in cancer patients (Valmori *et al.*, 2000b). These observations suggest that immune responses to cancer testis antigens occur *in vivo*, and may be involved in tumor regression in some cases. Thus, cancer–testis antigens may be useful targets for immunotherapy for a broad population of patients with various cancers. Tumor regression has been observed in some patients immunized with the MAGE-3 HLA-A1 binding peptide, EVDPIGHLY, by either administration of the peptide alone or the peptide pulsed DC, suggesting that MAGE-3 might be a tumor regression antigen (Marchand *et al.*, 1999; Thurner *et al.*, 1999).

Tumor specific antigens with mutated amino acids derived from genetic alteration in tumor cells

Molecules derived from genetic alterations in tumor cells have long been expected to be tumor specific antigens recognized by immune system as foreign molecules. However, most melanoma antigens isolated with CD8+ CTL were rather self-peptides derived from melanosomal proteins and cancer–testis antigens, although presence of CTLs specific for unique tumor antigens were indicated. Attempts to demonstrate that the known mutated molecules, including products of an oncogene, *ras*, and a tumor suppressor gene, *p53*, could be targets for T cells, was not so successful. Only several unique antigens derived from abnormal sequences in tumor cells have so far been identified using autologous melanoma specific CTL.

β-catenin with a mutation was isolated as a melanoma antigen recognized by HLA-A24 restricted TIL (Robbins *et al.*, 1996). A single C to T transition that may be the result from UV-induced DNA damage generated an HLA-A24 binding epitope, SYLDSGIHF, by replacing S to F at the N-terminus. Administration of cultured TIL containing CTL specific for this mutated β-catenin to autologous patient resulted in complete tumor regression. β-catenin interacts with the adhesion molecule E-cadherin and the APC tumor suppressor gene product. The β-catenin protein was increased in 7 of 25 melanoma cell lines tested due to either mutations or unusual splicing of β-catenin or inactivation of APC (Rubinfeld *et al.*, 1997). These mutations are located in the phosphorylation sites by glycogen synthase kinase-3β (GSK-3β). The same mutation, S to F at residue 37, was found in three melanoma cell lines. β-catenin and APC are involved in the Wnt signaling pathway important in embryonic development and their alterations resulted in tumorigenesis. Thus, similar to colon cancer, abnormal Wnt signaling appears to be involved in development of melanoma.

A mutated peptide in the cyclin dependent kinase 4 (CDK4) was identified as an antigen recognized by HLA-A2 restricted CTL (Wolfel *et al.*, 1995). This mutation was also a C to T transition, and generated an HLA-A2 binding epitope, ACDPHSGHFV. The mutated CDK4 had decreased binding activity to CDK4 inhibitor $p16^{INK4a}$. Mutations in CDK4 were found in familial melanoma (Zuo *et al.*, 1996). Alterations of CDK4/$p16^{INK4a}$ regulation have been observed in various cancers including melanoma. These observations suggest that the mutation of CDK4 is involved in melanoma development in some patients.

Three mutated antigens, melanoma ubiquitous mutated (MUM)-1, -2 and -3, were isolated from patient LB33 who was disease-free more than 10 years after the treatment. MUM-1 was isolated using HLA-B44 restricted CTL. The *MUM-1* gene was expressed in various normal tissues and its biological function is not known (Coulie *et al.*, 1995). The isolated cDNA was derived from an incompletely spliced mRNA. An epitope, EEKLIVVLF, was derived from a region spanning the exon–intron boundary that contained a point mutation within the intron. Since the mutation in the *MUM-1* gene was not found in 300 tumor samples tested, it dose not appear to generally associate with melanoma phenotype.

MUM-2 was isolated using both HLA-B44 and HLA-C6 restricted CTL (Chiari *et al.*, 1999). Two overlapped peptides, SELFRSGLDSY and FRSGLDSYV, containing the same mutated residue were identified as epitopes presented by HLA-B44 and -C6. The mutation in the *MUM-2* gene was not found among 150 tumor samples tested. MUM-2 is ubiquitously expressed and homologous to yeast gene *bet5* that is involved in vesicular transport of protein from endoplasmic reticulum to Golgi apparatus. Since both wild type and the mutated MUM-2 could complement for the bet 5 function, this mutated MUM-2 may not contribute to particular phenotype of LB33 melanoma cells.

A mutated peptide, EAFIQPITR, derived from MUM-3 that is homologous to RNA helicase, was isolated with HLA-A28 restricted CTL (Baurain *et al.*, 2000). A mutated helicase p68 was previously identified as an antigen recognized by CTL in an UV-induced murine sarcoma (Dubey *et al.*, 1997). High precursor frequency (1.2% of blood CD8+ T cells) of CTL specific for this mutated MUM-3 peptide was demonstrated using the HLA tetramer, suggesting the role of the MUM-3 specific CTL in the unusual favorable prognosis of this patient. Since mutations in helicases that are involved in DNA repair mechanisms were found in cancer-prone syndromes such as xeroderma pigmentosum, Bloom's syndrome, Werner's disease, X-linked mental retardation associated with α-thalassemia and Cockayne's syndrome, the mutated MUM-3 may be involved in tumorigenesis of the melanoma.

MART-2 containing a mutated residue was isolated by HLA-A1 restricted, autologous melanoma specific TIL from patients in whom some metastasis regressed after administration of TIL in combination with chemotherapy (Kawakami *et al.*, 2001). The mutation substituted G to E at the third position of the 11 mer peptide and generated HLA-A1 binding epitope, FLEGNEVGKT. The negative charged amino acid at the third position is an important anchor residue for the HLA-A1 peptide binding. MART-2 has recently been found to be a human homolog of Skinny Hedgehog (Ski) acyltransferase that add palmitate to amino-terminus of Hedgehog protein in Drosophila (Chamoun *et al.*, 2001). Hedgehog protein is α secreted protein with both palmitate and cholesterol modification, and involved in Hedgehog signaling that is important in the embryonic and post-embryonic patterning. Abnormality of Hedgehog signaling, such as a mutation in the Hh receptor, Patched, was associated with tumorigenesis in basal cell carcinoma and medulloblastoma. Acyltransferase activity of the mutated MART-2 has not yet been evaluated. However, the mutation resulted in loss of important reside G in the motif for phosphate binding loop (P-loop: GXXXGKT), and the isolated MART-2 with the mutation had actually lost GTP binding ability. Since oncogenic activity of *ras* with P-loop mutations was reported, ability of the mutated MART-2 to transform NIH3T3 was evaluated. Although transforming activity on NIH3T3 cells was not detected by transfection of the mutated MART-2 alone, the mutated MART-2 that lost GTP binding activity may still be involved in some of the melanoma phenotype. Possible mutations in MART-2 were also found in other cancers including lung cancer and teratocarcinoma.

The peptide, KINKNPKYK, derived from mutated myosin class I gene was identified to be recognized by dominant T-cell clones expressing TCRVB16 infiltrated in primary melanoma lesion that was spontaneously regressing, suggesting the involvement of T-cell response to this tumor specific antigen in the melanoma rejection (Zorn and Hercend, 1999b). This HLA-A*0301 restricted CTL did not recognize 17 allogeneic melanoma cell lines transfected with HLA-A3, suggesting that this was a relatively unique mutation.

Role of tumor specific mutated antigens in the immunotherapy

Mutated antigens are tumor-specific and antigen loss variants may not be easily developed if mutations are important for tumor cells to survive and proliferate. Mutated tumor antigens appear to be potent rejection antigens in murine tumor models (Srivastava, 1996). In human melanoma, mutated antigens were isolated using T cells from patients who had relatively good prognosis after the treatment, suggesting that immune responses to these mutated peptides might be involved in the tumor regression and maintenance of tumor free status. Thus, these antigens appear to be ideal targets for immunotherapy.

However, immunogenic mutated peptides may not always be presented on melanoma cells, since various requirements, including appropriate antigen processing and MHC binding, should be fulfilled for these mutated peptides to be presented by MHC on tumor cell surface. In addition, tumor cells that expressed highly immunogenic peptides may have already been eliminated before tumors were clinically detected. Observations that memory T-cell response was detected against mutated ras peptide that was not present in the autologous tumor cells (Gedde-Dahl *et al.*, 1992), and that p53 mutations were located significantly outside the region encoding HLA-A2 binding peptides in HLA-A2 lung cancers (Wiedenfeld *et al.*, 1994), may support this immunosurveillance theory. While most CD8+ CTL established from TIL in the Surgery Branch, NCI recognized normal self-peptides, most melanoma reactive CD4+ TIL appeared to recognize unique mutated peptides. These observations may suggest that CD8+ T cells are capable of eliminating tumor cells that expressed immunogenic mutated peptides, and that CD4+ T cells alone may not be sufficient to reject tumor cells.

It may be difficult to apply unique mutated epitopes for immunotherapy unless mutations are common, or more rapid antigen identification techniques can be developed. However, mutated peptides may practically be used for relapsed melanoma, since patients with immunogenic mutated peptides have relatively good prognosis. Alternatively, methods that do not require identified antigens, including immunization with DC pulsed with tumor cell derived products and immunization with immunogenic modified tumor cells, may be useful for the induction of immunity to unique antigens. Immunization with melanocyte specific antigens may also trigger immune responses to unidentified unique antigens expressed on the same tumor through antigenic spreading. This may explain the difficulty to detect specific CTL for the immunized peptides in regressing metastasis in some patients, and different clinical outcome of the immunization with shared tumor antigens among patients.

Other melanoma antigens

PRAME was isolated by cDNA cloning using an HLA-A24 restricted CTL clone that was generated from PBMC by stimulation with an HLA-A24+ melanoma cell line (Ikeda *et al.*, 1997). This cell line established from a patient after immunotherapy, lost most HLA molecules including HLA-Cw7 possibly by *in vivo* immunoselection. PRAME is expressed in various cancers and some normal tissues including testis, ovary and endometrium. Since this CTL clone expressed killer inhibitory receptors (KIR) that bind to HLA-Cw7 on the patient's cells, it could lyse only HLA-Cw7 lost melanoma cell variants. It has not been successful to induce PRAME specific CTL without expressing KIR that could recognize the original HLA-Cw7 positive melanoma cells. Although various trials to induce PRAME specific CTL was unsuccessful, a recent report demonstrated the induction of PRAME specific melanoma reactive CTL from PBMC of healthy donors by *in vitro* stimulation with HLA-A2 binding peptides that were synthesized based on HLA-A2 binding motifs and evaluation of proteasome cleavage sites (Kessler *et al.*, 2001). However, only allogeneic melanoma cells were tested in this study, thus, these CTL may still express KIR and not recognize autologous melanoma cells. Nevertheless, PRAME may be a useful target for immunotherapy of various cancers including melanoma.

A non-mutated transcript of the ubiquitously expressed N acetylglucosaminyl-transferase-V (*GnT-V*) gene was isolated using HLA-A2 restricted, melanoma specific CTL (Guilloux *et al.*, 1996). This transcript was initiated from a cryptic promoter in the intron, which was activated in about a half of melanoma. An epitope was encoded by this intron, so that the antigen was melanoma specific. A non-mutated peptide derived from p15, the transcript of which was expressed in various normal tissues, was identified using HLA-A24 restricted TIL (Robbins *et al.*, 1995). Since this CTL did not recognize normal cells expressing the p15 mRNA, post-transcriptional regulation may be responsible for the differential presentation of this epitope on melanoma cells. Induction of melanoma specific T cells from additional patients should be evaluated for GnT-V and p15.

Survivin, a new member of the inhibitor of apoptosis proteins (IAP) gene family, was identified as melanoma antigens recognized by HLA-A2 restricted CTL (Andersen *et al.*, 2001a,b). Two peptides, LTLGEFLKL and ELTLGEFLKL, among 10 peptides synthesized based on the HLA–A*0201 binding motif were able to induce the peptide specific CTL from PBMC of melanoma patients. T cells reactive to these peptides were detected by ELISPOT assay in PBMC from 7 of 14 HLA-A2 patients, but not from healthy individuals. These peptides had relatively low affinity to HLA–A*0201. One of the peptides, LTLGEFLKL, was modified to increase MHC binding affinity by replacing T to M at position 2, and this modified peptide was used to generate multimeric peptide/MHC complexes. Using the FITC conjugated multimeric peptide/MHC complexes, survivin specific T cells were visualized in frozen sections of primary melanoma and sentinel lymph nodes. T cells isolated from metastatic lymph nodes with the magnetic beads coated survivin-peptide/HLA-A2 complex could lyse HLA-A2+ surviving positive melanoma cells, suggesting that this peptide was presented on melanoma cells. Since survivin preferentially expresses in a wide variety of tumors, it is an attractive target for immunotherapy.

Implication of the identification of MHC class I restricted melanoma antigens for the understanding of immune responses to melanoma cells and the development of immunotherapy

Molecular nature of human melanoma antigens recognized by CD8+ T cells have been revealed in the past 10 years as described above. Mechanisms for the generation of T-cell epitopes on melanoma cells are summarized in Table 4.7. The identification of these T-cell epitopes enabled us to perform a more detailed analysis of anti-tumor T cell responses in humans. Following investigations were mostly performed with the identified MHC class I restricted melanoma antigens. *In vivo* immunological status of melanoma specific T cells was evaluated using HLA tetramers, not only quantitatively, but also qualitatively such as naïve or memory phenotype, expression of adhesion/co-stimulatory molecules, cytokine production and cytotoxic machinery (Romero *et al.*, 1998; Lee *et al.*, 1999). The use of peptides, instead of tumor cells, enabled us to reliably measure anti-tumor T cells quantitatively. Immunization effects can be easily monitored using various methods, including DTH skin

Table 4.7 Mechanisms for generating T-cell epitopes on melanoma cells

Mechanism	*Antigen*
Translation of functional genes	
Tissue specific proteins	Melanosomal proteins (e.g. gp100)
Cancer–testis antigens	MAGEs, NY-ESO-1, etc.
Mutation	β-catenin, CDK4, MUM-1, -2, -3, MART-2
Translation of alternative ORF	TRP-1, NY-ESO-1
Translation of intron	
Incomplete splicing	gp100, TRP-2, MUM-1
Transcription from cryptic promoter	N-acetyl glucosaminyl transferase-V
Post translational modification	
Deamidation	Tyrosinase
Cysteine oxidation	Tyrosinase
Different processing by immuno-proteasomes	MART1, gp100, MAGE-1, MAGE-3,

reaction, *in vitro* CTL induction, limiting dilution, ELISPOT assay and HLA tetramer (Salgaller *et al.*, 1995; Jager *et al.*, 1996; Scheibenbogen, 1997; Pass *et al.*, 1998; Romero *et al.*, 1998; Pittet *et al.*, 2001a). Even in the immunotherapies that do not use the identified antigens, mechanism for the tumor regression may be analyzed with the identified peptides. Modified peptides with higher immunogenicity may be useful for efficient measurement in these assays (Salgaller *et al.*, 1996; Pass *et al.*, 1998). However, results should be carefully evaluated with adequate controls including native peptides with titration as well as tumor cells, since T cells that are only reactive to the modified peptides or the native peptides at high concentration may arise in the culture for the assay. T-cell recognition of peptides at low concentration usually indicates the ability to recognize tumor cells (Yee *et al.*, 1999). Although immunomonitoring during immunotherapy has mainly been performed with PBMC, no clear correlation has been observed between the immune response in peripheral blood and tumor rejection (Cormier *et al.*, 1997). It is important to analyze immune responses in tumor tissues (Panelli *et al.*, 2000). The HLA tetramer or other multimeric peptide/HLA complex may be useful for the detection of tumor reactive T cells in tumor tissues (Skinner *et al.*, 2000; Andersen *et al.*, 2001a).

To induce strong immune responses, it is important to administer antigens in optimal conditions. The identification of tumor antigens allowed us to administer sufficient amounts of antigens in various forms in appropriate sites with controlled timing for immunotherapy. A variety of immunotherapies using the identified melanoma antigens have been developed and clinical trials have been conducted (Table 4.8). The immunization with peptides from MART-1, gp100 and tyrosinase, along with IL2 or GM-CSF resulted in tumor regression in some patients (Jager *et al.*, 1996; Rosenberg *et al.*, 1998). Since tumor antigens have relatively low immunogenic nature due to various reasons caused by host–tumor relationship. Strategy for effective induction of immune responses against tumor antigens is important as shown in Table 4.9. One of them, the modification of tumor antigens, could be performed through the identification and characterization of T-cell epitopes. Immunization with the modified gp100 peptide with higher MHC binding and immunogenicty along with IL2 administration resulted in tumor regression in more patients as described above (Parkhurst *et al.*, 1996; Rosenberg *et al.*, 1998), although T-cell clones that only recognize the modified peptide arose by repetitive stimulation with the modified peptide (Clay *et al.*, 1999; Bullock *et al.*, 2000; Dudley *et al.*, 2000). Immunization protocols using DCs pulsed with melanoma lysates or synthetic peptides including MAGE-1, MAGE-3, MART-1, gp100 and tyrosinase, were reported to result in tumor regression in some patients (Nestle *et al.*, 1998; Thurner *et al.*, 1999; Banchereau *et al.*, 2001). Direct administration of recombinant viruses or plasmids containing melanoma antigen genes has also been performed. In the clinical trials in the Surgery branch, NCI, recombinant vaccinia virus, fowlpox virus and adenovirus containing MART-1 or gp100 have been used (Rosenberg *et al.*, 1998). However, high titer neutralizing antibodies against viruses that were detected in patients appeared to reduce the efficacy of immunization against tumor antigens. Immunization with plasmids containing the modified gp100 cDNA that produced two high HLA-A2 binding modified epitopes did not induce efficient anti-tumor effects. Melanoma reactive CTL could be generated from PBL of patients by *in vitro* stimulation with the identified melanoma peptides (Celis *et al.*, 1994; Rivoltini *et al.*, 1995). These T cells could be generated more efficiently from the patients pre-immunized with tumor antigens (Salgaller *et al.*, 1996). Tumor reactive T cells can be

Table 4.8 Reported immunotherapy protocols for patients with melanoma

Treatment	*Investigator (institution)*	*PR + CR / total (response rate)*
High dose IL2 alone	Rosenberg (NCI)	27/182 (15%)
TIL + IL2	Rosenberg (NCI)	29/86 (34%)
MAGE3 peptide alone	Marchant (Ludwig Inst. Cancer Res)	5/25 (20%)
Tyrosinase, gp100, MART-1 peptides + GM-CSF	Jager (Nordostrand West)	
MART-1 peptide + IFA	Rosenberg (NCI)	
Modified gp100 peptide + IFA + IL2	Rosenberg (NCI)	13/31 (42%)
Adenovirus MART1 + IL2	Rosenberg (NCI)	4/20 (20%)
DC + peptides/tumor lysates	Nestle (Univ. Zurich)	5/16 (31%)
DC + peptides (MAGE-3)	Schuler (Univ. Erlangen-Nuremberg)	
BM-DC + peptides (MART-1, gp100, MAGE-3)	Banchereau (Baylor Inst for Immunol Res)	3/17 (18%)

Table 4.9 Methods to improve immunization efficacy against melanoma antigens

Modification of epitopes
High MHC binding peptides, superagonists, increased localization, increased stability
Association with other molecules
Liposomes, lipoproteins, cholesterol-polysaccharides
Conjugation with multiple peptides, helper epitope, stress protein, leader sequence
Use of dendritic cells
Adjuvants
IFA, synthetic adjuvants
Cytokines
IL2, GM-CSF, IL12, IFN-γ

selectively expanded using HLA tetramers (Dunbar *et al.*, 1999; Yee *et al.*, 1999), and used in adoptive immunotherapy.

Concluding remarks

Melanoma has been the most advanced system to analyze anti-tumor immune responses in human and to evaluate the possibility of immunotherapy against cancer. The identification of many melanoma antigens extended our ability to analyze anti-tumor T-cell responses in further detail and to develop new types of immunotherapy. Although tumor regression has already been observed in some patients in various clinical trials that utilized the identified melanoma antigens, anti-tumor effect is still limited, and immune response leading to tumor regression has not been completely understood. The role of the most identified antigens in *in vivo* tumor rejection remains to be evaluated through clinical trials. Many subjects, including identification of additional antigens, further analysis of immune responses particularly at tumor sites, understanding of systemic and local tumor escape mechanisms and development of more effective immunization methods, should be addressed for the development of more effective immunotherapy.

References

Andersen, M. H., Pedersen, L., Capeller, B., Brocker, E. B., Becker, J. C. and thor Straten, P. (2001a) Spontaneous cytotoxic T-cell responses against survivin-derived MHC class I-restricted T-cell epitopes *in situ* as well as *ex Vivo* in cancer patients. *Cancer Res.*, **61**, 5964–5498.

Andersen, M. H., Pedersen, L. O., Becker, J. C. and Straten, P. T. (2001b) Identification of a cytotoxic T lymphocyte response to the apoptosis inhibitor protein survivin in cancer patients. *Cancer Res.*, **61**, 869–872.

Anichini, A., Molla, A., Mortarini, R., Tragni, G., Bersani, I., Di Nicola, M. *et al.* (1999) An expanded peripheral T cell population to a cytotoxic T lymphocyte (CTL) defined, melanocyte-specific antigen in metastatic melanoma patients impacts on generation of peptide-specific CTLs, but does not overcome tumor escape from immune surveillance in metastatic lesions. *J. Exp. Med.*, **190**, 651–657.

Bakker, A. B. H., Schreurs, M. W. J., de Boer, A. J., Kawakami, Y., Rosenberg, S. A., Adema, G. J. *et al.* (1994) Melanocyte lineage-specific antigen gp100 is recognized by melanocyte-derived tumor-infiltrating lymphocytes. *J. Exp. Med.*, **179**, 1005–1009.

Banchereau, J., Palucka, A. K., Dhodapkar, M., Burkeholder, S., Taquet, N., Rolland, A. *et al.* (2001) Immune and clinical responses in patients with metastatic melanoma to cd34(+) progenitor-derived dendritic cell vaccine. *Cancer Res.*, **61**, 6451–6458.

Baurain, J. F., Colau, D., van Baren, N., Landry, C., Martelange, V., Vikkula, M. *et al.* (2000) High frequency of autologous anti-melanoma CTL directed against an antigen generated by a point mutation in a new helicase gene. *J. Immunol.*, **164**, 6057–6066.

Bettinotti, M. P., Kim, C. J., Lee, K. H., Roden, M., Cormier, J. N., Panelli, M. *et al.* (1998) Stringent allele/epitope requirements for MART-1/melan A immunodominance: implications for peptide-based immunorotherapy. *J. Immunol.*, **161**, 877–889.

Bloom, M. B., Lalley, D. P., Robbins, P. F., Li, Y., El-Gamil, M., Rosenberg, S. A. *et al.* (1997) Identification of TRP2 as a tumor rejection antigen for the B16 melanoma. *J. Exp. Med.*, in press.

Brichard, V. G., Herman, J., Van Pel, A., Wildmann, C., Gaugler, B., Wolfel, T. *et al.* (1996) A tyrosinase nonapeptide presented by HLA-B44 is recognized on a human melanoma by autologous cytolytic T lymphocytes. *Eur. J. Immunol.*, **26**, 224–230.

Brichard, V., Van Pel, A., Wolfel, T., Wolfel, C., De Plaen, E., Lethe, B. *et al.* (1993) The tyrosinase gene codes for an antigen recognized by autologous cytolytic T lymphocytes on HLA-A2 melanomas. *J. Exp. Med.*, **178**, 489–495.

Brinkmann, U., Vasmatzis, G., Lee, B. and Pastan, I. (1999) Novel genes in th PAGE and GAGE family of tumor antigens found by homology walking in the dbEST database. *Cancer Res.*, **59**, 1445–1448.

Bullock, T. N., Colella, T. A. and Engelhard, V. H. (2000) The density of peptides displayed by dendritic cells affects immune responses to human tyrosinase and gp100 in HLA-A2 transgenic mice. *J. Immunol.*, **164**, 2354–2361.

Butterfield, L. H., Jilani, S. M., Chakraborty, N. G., Bui, L. A., Ribas, A., Dissette, V. B. *et al.* (1998) Generation of melanoma-specific cytotoxic T lymphcytes by dendritic cells transduced with a MART-1 adenovirus. *J. Immunol.*, **161**, 5607–5613.

Castelli, C., Tarsini, P., Mazzocchi, A., Rini, F., Rivoltini, L., Ravagnani, F. *et al.* (1999) Novel HLA-Cw8-restricted T cell epitopes derived from tyrosinase-related protein-2 and gp100 melanoma antigens. *J. Immunol.*, **162**, 1739–1748.

Celis, E., Tsai, V., Crimi, C., DeMars, R., Wentworth, P. A., Chestnut, R. W. *et al.* (1994) Induction of anti-tumor cytotoxic T lymphocytes in normal humans using primary cultures and synthetic peptide epitopes. *Proc. Natl. Acad. Sci. USA*, **91**, 2105–2109.

Chakraborty, A., Platt, J., Kim, K., Kwon, B., Bennett, D. and Pawelek, J. (1996) Polymerization of 5,6-dihydroxyindole-2-carboxylic acid to melanin by the pmel17/silver locus protein. *Eur. J. Biochem.*, **236**, 180–188.

Chamoun, Z., Mann, R., Nellen, D., von Kessler, D. P., Belloto, M., Beachy, P. A. *et al.* (2001) Skinny Hedgehog, an acyltransferase required for palmitoylation and activity of the Hedgehog signal. *Science*, **293**, 2080–2084.

Chen, Y.-T. and Old, L. J. (1999) Cancer–testis antigens: targets for cancer immunotherapy. *Cancer J. Sci. Am.*, 16–17.

Chen, Y. T., Gure, A. O., Tsang, S., Stockert, E., Jager, E., Knuth, A. *et al.* (1998) Identification of multiple cancer/testis antigens by allogeneic antibody screening of a melanoma cell line library. *Proc. Natl. Acad. Sci. USA*, **95**, 6919–6923.

Chen, Y. T., Scanlan, M. J., Sahin, U., Tureci, O., Gure, A. O., Tsang, S. *et al.* (1997) A testicular antigen aberrantly expressed in human cancers detected by autologous antibody screening. *Proc. Natl. Acad. Sci. USA*, **94**, 1914–1918.

Chiari, R., Foury, F., De Plaen, E., Baurain, J. F., Thonnard, J. and Coulie, P. G. (1999) Two antigens recognized by autologous cytolytic T lymphocytes on a melanoma result from a single point mutation in an essential housekeeping gene. *Cancer Res.*, **59**, 5785–5792.

Clay, T. M., Custer, M. C., McKee, M. D., Parkhurst, M., Robbins, P. F., Kerstann, K. *et al.* (1999) Changes in the fine specificity of gp100(209–217)-reactive T cells in patients following vaccination with a peptide modified at an HLA-A2.1 anchor residue. *J. Immunol.*, **162**, 1749–1755.

Cole, D. J., Weil, D. P., Shamamian, P., Rivoltini, L., Kawakami, Y., Topalian, S. *et al.* (1994) Identification of MART-1-specific T-cell receptors: T cells utilizing distinct T-cell receptor variable and joining regions recognize the same tumor epitope. *Cancer Res.*, **54**, 5265–5268.

Colella, T. A., Bullock, T. N., Russell, L. B., Mullins, D. W., Overwijk, W. W., Luckey, C. J. *et al.* (2000) Self-tolerance to the murine homologue of a tyrosinase-derived melanoma antigen: implications for tumor immunotherapy. *J. Exp. Med.*, **191**, 1221–1232.

Cormier, J., Hijazi, Y., Abati, A., Fetsch, P., Bettinotti, M., Steinberg, S. *et al.* (1998) Heterogeneous expression of melanoma-associated antigens and HLA-A2 in metastatic melanoma in vivo. *Int. J. Cancer*, **75**, 517–524.

Cormier, J., Salgaller, M., Prevette, T., Barracchini, K., Rivoltini, L., Restifo, N. *et al.* (1997) Enhancement of cellular immunity in melanoma patients immunized with a peptide from MART-1/Melan A. *Cancer J. Sci. Am.*, **3**, 37–44.

Coulie, P. G., Brichard, V., Van Pel, A., Wolfel, T., Schneider, J., Traversari, C. *et al.* (1994) A new gene coding for a differentiation antigen recognized by autologous cytolytic T lymphocytes on HLA-A2 melanomas. *J. Exp. Med.*, **180**, 35–42.

Coulie, P. G., Lehmann, F., Lethe, B., Herman, J., Lurquin, C., Andrawiss, M. *et al.* (1995) A mutated intron sequence codes for an antigenic peptide recognized by cytolytic T lymphocytes on a human melanoma. *Proc. Natl. Acad. Sci. USA*, **92**, 7976–7980.

Cox, A. L., Skipper, J., Cehn, Y., Henderson, R. A., Darrow, T. L., Shabanowitz, J. *et al.* (1994) Identification of a peptide recognized by five melanoma-specific human cytotoxic T cell lines. *Science*, **264**, 716–719.

Dietrich, P. Y., Walker, P. R., Quiquerez, A. L., Perrin, G., Dutoit, V., Lienard, D. *et al.* (2001) Melanoma patients respond to a cytotoxic T lymphocyte-defined self-peptide with diverse and nonoverlapping T-cell receptor repertoires. *Cancer Res.*, **61**, 2047–2054.

De Vries, T. J., Smeets, M., de Graaf, R., Hou-Jensen, K., Brocker, E. B., Renard, N. *et al.* (2001) Expression of gp100, MART-1, tyrosinase, and S100 in paraffin-embedded primary melanomas and locoregional, lymph node, and visceral metastases: implications for diagnosis and immunotherapy. A study conducted by the EORTC Melanoma Cooperative Group. *J. Pathol.*, **193**, 13–20.

De Plaen, E., Lurquin, C., Van Pel, A., Mariame', B., Szikora, J.-P., Wolfel, T. *et al.* (1988) Immunogenic (tum-) variants of mouse tumor P815: cloning of the gene of tum- antigen P91A and identification of the tum- mutation. *Proc. Natl. Acad. Sci. USA*, **85**, 2274–2278.

De Smet, C., De Backer, O., Faraoni, I., Lurquin, C., Brasseur, F. and Boon, T. (1996) The activation of human gene *MAGE-1* in tumor cells is correlated with genome-wide demethlation. *Proc. Natl. Acad. Sci. USA*, **93**, 7149–7153.

D'Souza, S., Rimoldi, D., Lienard, D., Lejeune, F., Cerottini, J. and Romero, P. (1998) Circulating melan-A/MART-1 specific cytolytic T lymphocyte precursors in HLA-A2(+)melanoma patients have a memory phenotype. *Int. J. Cancer*, **78**, 699–706.

Dubey, P., Hendrickson, R. C., Meredith, S. C., Siegel, C. T., Shabanowitz, J., Skipper, J. C. *et al.* (1997) The immunodominant antigen of an ultraviolet-induced regressor tumor is generated by a somatic point mutation in the DEAD box helicase p68. *J. Exp. Med.*, **185**, 695–705.

Dudley, M. E., Ngo, L. T., Westwood, J., Wunderlich, J. R. and Rosenberg, S. A. (2000) T-cell clones from melanoma patients immunized against an anchor-modified gp100 peptide display discordant effector phenotypes. *Cancer J.*, **6**, 69–77.

Dunbar, P., Chen, J., Chao, D., Rust, N., Teisserenc, H., Ogg, G. *et al.* (1999) Cutting edge: rapid cloning of tumor-specific CTL suitable for adoptive immunotherapy of melanoma. *J. Immunol.*, **162**, 6959–6962.

Fukamachi, S., Shimada, A. and Shima, A. (2001) Mutations in the gene encoding B, a novel transporter protein, reduce melanin content in medaka. *Nat. Genet.*, **28**, 381–385.

Gaugler, B., Van Den Eynde, B., Van der Bruggen, P., Romero, P., Gaforio, J. J., De Plaen, E. *et al.* (1994) Human gene *MAGE-3* codes for an antigen recognized on a melanoma by autologous cytolytic T lymphocytes. *J. Exp. Med.*, **179**, 921–930.

Gedde-Dahl, T., Spurkland, A., Eriksen, J., Thorsby, E. and Gaudernack, G. (1992) Memory T cells of a patient with follicular thyroid carcinoma recognize peptides derived from mutated p21 ras (Gln-Leu61). *Int. Immunol.*, **4**, 1331–1337.

Gocho, K., Kondo, I. and Yamaki, K. (2001) Identification of autoreactive T cells in Vogt–Koyanagi–Harada disease. *Invest. Ophthalmol. Vis. Sci.*, **42**, 2004–2009.

Guilloux, Y., Lucas, S., Brichard, V. G., VanPel, A., Viret, C., De Plaen, E. *et al.* (1996) A peptide recognized by human cytolytic T lymphocytes on HLA-A2 melanoma is encoded by an intron sequence of the N-acetylglucosaminyltransferase V gene. *J. Exp. Med.*, **183**, 1173–1183.

Halaban, R., Cheng, E., Zhang, Y., Moellmann, G., Hanlon, D., Michalak, M. *et al.* (1997) Aberrant retention of tyrosinase in the endoplasmic reticulum mediates accelerated degradation of the enzyme and contributes to the differentiated phenotype of amelanotic melanoma cells. *Proc. Natl. Acad. Sci. USA*, **94**, 6210–6215.

Harada, M., Li, Y. F., El-Gamil, M., Rosenberg, S. A. and Robbins, P. F. (2001) Use of an *in vitro* immunoselected tumor line to identify shared melanoma antigens recognized by HLA-A*0201-restricted T cells. *Cancer Res.*, **61**, 1089–1094.

Hemmi, H., Yoshino, M., Yamazaki, H., Naito, M., Iyoda, T., Omatsu, Y. *et al.* (2001) Skin antigens in the steady state are trafficked to regional lymph nodes by transforming growth factor-beta1-dependent cells. *Int. Immunol.*, **13**, 695–704.

Ikeda, H., Lethe, B., Lehmann, F., Van Baren, N., Chambost, H., Vitale, M. *et al.* (1997) Characterization of an antigen that is recognized on a melanoma showing partial HLA loss by CTL expressing an NK inhibitory receptor. *Immunity*, **6**, 199–208.

Irvine, K., Parkhurst, M., Shulman, E., Tupesis, J., Custer, M., Touloukina, C. *et al.* (1999) Recombinant virus vaccination against "self" antigens using anchor-fixed immunogens. *Cancer Res.*, **59**, 2536–2540.

Jager, E., Chen, Y. T., Drijfhout, J. W., Karbach, J., Ringhoffer, M., Jager, D. *et al.* (1998) Simultaneous humoral and cellular immune response against cancer–testis antigen NY-ESO-1: definition of human histocompatibility leukocyte antigen (HLA)-A2-binding peptide epitopes. *J. Exp. Med.*, **187**, 265–270.

Jager, E., Maeurer, M., Hohn, H., Karbach, J., Jager, D., Zidianakis, Z. *et al.* (2000a) Clonal expansion of Melan A-specific cytotoxic T lymphocytes in a melanoma patient responding to continued immunization with melanoma-associated peptides. *Int. J. Cancer*, **86**, 538–547.

Jager, E., Nagata, Y., Gnjatic, S., Wada, H., Stockert, E., Karbach, J. *et al.* (2000b) Monitoring CD8 T cell responses to NY-ESO-1: correlation of humoral and cellular immune responses. *Proc. Natl. Acad. Sci. USA*, **97**, 4760–4765.

Jager, E., Ringhoffer, M., Karbach, J., Arand, M., Oesch, F. and Knuth, A. (1996) Inverse relationship of melanocyte differentiation antigen expression in melanoma tissues and CD8+ cytotoxic-T-cell responses: evidence for immunoselection of antigen-loss variants *in vivo*. *Int. J. Cancer*, **66**, 470–476.

Kageshita, T., Kawakami, Y., Hirai, S. and Ono, T. (1997) Differential expression of MART-1 in primary and metastatic melanoma lesions. *J. Immunother.*, **20**, 460–465.

Kawakami, Y. and Rosenberg, S. (1995) T-cell recognition of self peptides as tumor rejection antigens. *Immunol. Res.*, **15**, 179–190.

Kawashima, I., Tsai, V., Southwood, S., Takesako, K., Celis, E. and Sette, A. (1998) Identification of gp100-derived, melanoma-specific cytotoxic T-lymphocyte epitopes restricted by HLA-A3 supertype molecules by primary *in vitro* immunization with peptide-pulsed dendritic cells. *Int. J. Cancer*, **78**, 518–524.

Kawakami Y., Suzuki Y., Shofuda T., Kiniwa Y., Inozume T., Dan, K. *et al.* (2000a) T cell immune responses against melanoma and melanocytes in cancer and autoimmunity. *Pigment Cell Res.*, **13**, 163–169.

Kawakami, Y., Dang, N., Wang, X., Tupesis, J., Robbins, P., Wunderlich, J. *et al.* (2000b) Recognition of shared melanoma antigens in association with major HLA-A alleles by tumor infiltrating T lymphocytes from 123 patients with melanoma. *J. Immunother.*, **23**, 17–27.

Kawakami, Y., Battles, J. K., Kobayashi, T., Wang, X., Tupesis, J. P., Marincola, F. M. *et al.* (1997) Production of recombinant MART-1 proteins and specific anti-MART-1 polyclonal and monoclonal antibodies: use in the characterization of the human melanoma antigen MART-1. *J. Immunol. Methods*, **202**, 13–25.

Kawakami, Y., Eliyahu, S., Delgado, C. H., Robbins, P. F., Sakaguchi, K., Appella, E. *et al.* (1994a) Identification of a human melanoma antigen recognized by tumor infiltrating lymphocytes associated with *in vivo* tumor rejection. *Proc. Natl. Acad. Sci. USA*, **91**, 6458–6462.

Kawakami, Y., Eliyahu, S., Delgaldo, C. H., Robbins, P. F., Rivoltini, L., Topalian, S. L. *et al.* (1994b) Cloning of the gene coding for a shared human melanoma antigen recognized by autologous T cells infiltrating into tumor. *Proc. Natl. Acad. Sci. USA*, **91**, 3515–3519.

Kawakami, Y., Robbins, P., Wang, X., Tupesis, J., Fitzgerald, E., Li, Y. *et al.* (1998) Identification of new melanoma epitopes on melanosomal proteins recognized by tumor infiltrating T lymphocytes restricted by HLA-A1, -A2, and -A3 alleles *J. Immunol.*, **161**, 6985–6992.

Kawakami, Y., Eliyahu, S., Jennings, C., Sakaguchi, K., Kang, X.-Q., Southwood, S. *et al.* (1995) Recognition of multiple epitopes in the human melanoma antigen gp100 associated with *in vivo* tumor regression. *J. Immunol.*, **154**, 3961–3968.

Kawakami, Y., Eliyahu, S., Sakaguchi, K., Robbins, P. F., Rivoltini, L., Yannelli, J. B. *et al.* (1994c) Identification of the immunodominant peptides of the MART-1 human melanoma antigen recognized by the majority of HLA-A2 restricted tumor infiltrating lymphocytes. *J. Exp. Med.*, **180**, 347–352.

Kawakami, Y., Wang, X., Shofuda, T., Sumimoto, H., Tupesis, J., Fitzgerald, E. *et al.* (2001) Isolation of a new melanoma antigen, MART-2, containing a mutated epitope recognized by autologous tumor-infiltrating T lymphocytes. *J. Immunol.*, **166**, 2871–2877.

Kessler, J. H., Beekman, N. J., Bres-Vloemans, S. A., Verdijk, P., van Veelen, P. A., Kloosterman-Joosten, A. M. *et al.* (2001) Efficient identification of novel HLA-A(*)0201-presented cytotoxic T lymphocyte epitopes in the widely expressed tumor antigen PRAME by proteasome-mediated digestion analysis. *J. Exp. Med.*, **193**, 73–88.

Kim, C. J., Prevette, T., Cormier, J., Overwijk, W., Roden, M., Restifo, N. P. *et al.* (1997) Dendritic cells infected with pozviruses encoding MART-1/melan a sensitize T lymphocytes *in vitro*. *J. Immunother.*, **20**, 276–286.

Kiniwa, Y., Fujita, T., Akada, M., Ito, K., Shofuda, T., Suzuki, Y. *et al.* (2001) Tumor antigens isolated from a patient with vitiligo and T cell infiltrated melanoma. *Cancer Res.*, **61**, 7900–7907.

Kittlesen, D., Thompson, L. W., Gulden, P. H., Skipper, J. C. A., Colella, T. A., Shabanowitz, J. A. *et al.* (1998) Human melanoma patients recognize an HLA-A1-restricted CTL epitope from tyrosinase containing two cysteine residues: implications for vaccine development. *J. Immunol.*, **160**, 2099–2106.

Lee, K. H., Panelli, M. C., Kim, C. J., Riker, A. I., Bettinotti, M. P., Roden, M. M. *et al.* (1998) Functional dissociation between local and systemic immune response during anti-melanoma peptide vaccination. *J. Immunol.*, **161**, 4183–4194.

Lee, P. P., Yee, C., Savage, P. A., Fong, L., Brockstedt, D., Weber, J. S. *et al.* (1999) Characterization of circulating T cells specific for tumor associated antigens in melanoma patients. *Nat. Med.*, **5**, 677–685.

Loftus, D. J., Castelli, C., Clay, T. M., Squarcina, P., Marincola, F. M., Nichimura, M. I. *et al.* (1996) Identification of epitope mimics recognized by CTL reactive to the melanoma/melanocyte-derived peptide MART-1_{27-35}. *J. Exp. Med.*, **184**, 647–657.

Lucas, S., Smet, C. D., Arden, K. C., Viars, C. S., Lethe, B., Lurquin, C. *et al.* (1998) Identification of a new *MAGE* gene with tumor-specific expression by representational difference analysis. *Cancer Res.*, **58**, 743–752.

Lupetti, R., Pisarra, P., Verrecchia, A., Farina, C., Nicolini, G., Anichini, A. *et al.* (1998) Translation of a retained intron tyrosinase-related protein (TRP) 2 mRNA generates a new cytotoxic T lymphocyte (CTL)-defined and shared human melanoma antigen not expressed in normal cells of the melanocytic lineage. *J. Exp. Med.*, **188**, 1005–1016.

Marchand, M., Baren, N., Weynants, P., Brichard, D., Dreno, B., Tessier, M. H. *et al.* (1999) Tumor regressions observed in patients with metastatic melanoma treated with an antigenic peptide encoded by MAGE-3 and presented by HLA-A1. *Int. J. Cancer*, **80**, 219–230.

Marincola, F. M., Jaffee, E. M., Hicklin, D. J. and Ferrone, S. (2000) Escape of human solid tumors from T-cell recognition: molecular mechanisms and functional significance. *Adv. Immunol.*, **74**, 181–273.

Marincola, F. M., Rivoltini, L., Salgaller, M. L., Player, M. and Rosenberg, S. A. (1996) Differential anti-MART-1/MelanA CTL activity in peripheral blood of HLA-A2 melanoma patients in comparison to healthy donors: evidence for *in vivo* priming by tumor cells. *J. Immunother.* **19**, 266–277.

Morel, S., Levey, F., Burlet-Schiltz, O., Peitrequin, A., Monsarrat, B., Van Velthoven, R. *et al.* (2000) Processing of some antigens by the standard proteasome but not by the immunoproteasome results in poor presentation by dendritic cells. *Immunity*, **12**, 107–117.

Nestle, F., Alijagic, S., Gilliet, M., Sun, Y., Grabbe, S., Dummer, R. *et al.* (1998) Vaccination of melanoma patients with peptide-or tumor lysate-pulsed dendritic cells. *Nat. Med.*, **4**, 328–332.

Newton, J. M., Cohen-Brak, O., Hagiwara, N., Gardner, J. M., Davisson, M. T., King, R. A. *et al.* (2001) Mutations in the human orthologue of the mouse *underwhite* gene (UW) underlie a new form of oculocutaneous albinism, OCA4. *Am. J. Hum. Genet.*, **69**, 981–988.

Old, L. and Chen, Y. (1998) New paths in human cancer serology. *J. Exp. Med.*, **187**, 1163–1167.

Ogg, G. S., Dunbar, P. R., Romero, P., Chen, J. L. and Cerundolo, V. (1998) High frequency of skin homing melanocyte specific cytotoxic T lymphocytes in autoimmune vitiligo. *J. Exp. Med.*, **188**, 1203–1208.

Overwijk, W. W., Tsung, A., Irvine, K. R., Parkhurst, M. R., Goletz, T. J., Tsung, K. *et al.* (1998) gp100/pmel 17 is a murine tumor rejction antigen: induction of "Self"-reactive, tumoricidal T cells using high-affinity, altered peptide ligand. *J. Exp. Med.*, **188**, 277–286.

Panelli, M. C., Bettinotti, M. P., Lally, K., Ohnmacht, G. A., Li, Y., Robbins, P. *et al.* (2000a) A tumor-infiltrating lymphocyte from a melanoma metastasis with decreased expression of melanoma differentiation antigens recognizes MAGE-12. *J. Immunol.*, **164**, 4382–4392.

Panelli, M. C., Riker, A., Kammula, U., Wang, E., Lee, K. H., Rosenberg, S. A. *et al.* (2000b) Expansion of tumor-T cell pairs from fine needle aspirates of melanoma metastases. *J. Immunol.*, **164**, 495–504.

Parkhurst, P., Fitzgerald, E., Southwood, S., Sette, A., Rosenberg, S. and Kawakami, Y. (1998) Identification of a shared HLA-A*0201 restricted T cell epitope from the melanoma antigen tyrosinase related protein 2 (TRP2). *Cancer Res.*, **58**, 4895–4901.

Parkhurst, M. R., Salgaller, M., Southwood, S., Robbins, P., Sette, A., Rosenberg, S. A. *et al.* (1996) Improved induction of melanoma reactive CTL with peptides from the melanoma antigen gp100 modified at HLA-A0201 binding residues. *J. Immunol.*, **157**, 2539–2548.

Pass, H., Schwarz, S., Wunderlich, J. and Rosenberg, S. (1998) Immunization of patients with melanoma peptide vaccines: immunologic assessment using the ELISPOT assay. *Cancer J. Sci. Am.*, **4**, 316–323.

Perez-Diez, A., Butterfield, L. H., Li, L., Chakraborty, N. G., Economou, J. S. and Mukherji, B. (1998) Generation of CD8+ and CD4+ T-cell response to dendritic cells genetically engineered to express the *MART-1/melan-A* gene. *Cancer Res.*, **58**, 5305–5309.

Pittet, M. J., Speiser, D. E., Lienard, D., Valmori, D., Guillaume, P., Dutoit, V. *et al.* (2001a) Expansion and functional maturation of human tumor antigen-specific CD8+ T cells after vaccination with antigenic peptide. *Clin. Cancer Res.*, **7**, 796s–803s.

Pittet, M. J., Zippelius, A., Speiser, D. E., Assenmacher, M., Guillaume, P., Valmori, D. *et al.* (2001b) *Ex vivo* IFN-gamma secretion by circulating CD8 T lymphocytes: implications of a novel approach for T cell monitoring in infectious and malignant diseases. *J. Immunol.*, **166**, 7634–7640.

Pittet, M., Valmori, D., Dunbar, P., Speiser, D., Lienard, D., Lejeune, F. *et al.* (1999) High frequencies of native melan-A/MART-1-specific CD8+ T cells in a large proportion of human histocompatibility leukocyte antigen (HLA)-A2 individuals. *J. Exp. Med.*, **190**, 705–715.

Riker, A., Cormier, J., Panelli, M., Kammula, U., Wang, E., Abati, A. *et al.* (1999) Immune selection after antigen-specific immunotherapy of melanoma. *Surgery*, **126**, 112–120.

Rivoltini, L., Kawakami, Y., Sakaguchi, K., Southwood, S., Sette, A., Robbins, P. F. *et al.* (1995) Induction of tumor-reactive CTL from peripheral blood and tumor-infiltrating lymphocytes of melanoma patients by *in vitro* stimulation with an immunodominant peptide of the human melanoma antigen MART-1. *J. Immunol.*, **154**, 2257–2265.

Robbins, P. F., El-Gamil, M., Kawakami, Y., Stevens, E., Yannelli, J. and Rosenberg, S. A. (1994) Recognition of tyrosinase by tumor infiltrating lymphocytes from a patient responding to immunotherapy. *Cancer Res.*, **54**, 3124–3126.

Robbins, P., El-Gamil, M., Li, Y., Fitzgerald, E., Kawakami, Y. and Rosenberg, S. (1997) The intronic region of an incompletely-spliced *gp100* gene transcript encodes an epitope recognized by melanoma reactive tumor infiltrating lymphocytes. *J. Immunol.*, **159**, 303–308.

Robbins, P. F., El-Gamil, M., Li, Y. F., Kawakami, Y., Loftus, D., Appella, E. *et al.* (1996) A mutated β-catenin gene encodes a melanoma-specific antigen recognized by tumor infiltrating lymphocytes. *J. Exp. Med.*, **183**, 1185–1192.

Robbins, P. F., El-Gamil, M., Li, Y. F., Topalian, S. L., Rivoltini, L., Sakaguchi, K. *et al.* (1995) Cloning of a new gene encoding an antigen recognized by melanoma-specific HLA-A24 restricted tumor-infiltrating lymphocytes. *J. Immunol.*, **154**, 5944–5950.

Romero, P., Cerottini, J. C. and Waanders, G. A. (1998) Novel methods to monitor antigen-specific cytotoxic T-cell responses in cancer immunotherapy. *Mol. Med. Today*, **4**, 305–312.

Romero, P., Dunbar, P. R., Valmori, D., Pittet, M., Ogg, G. S., Rimoldi, D. *et al.* (1998) *Ex vivo* staining of metastatic lymph nodes by class I major histocompatibility complex tetramers reveals high numbers of antigen-experienced tumor-specific cytolytic T lymphocytes. *J. Exp. Med.*, **188**, 1641–1650.

Romero, P., Gervois, N., Schneider, J., Escobar, P., Valmori, D., Pannetier, C. *et al.* (1997) Cytolytic T lymphocyte recognition of the immunodominant HLA-A*0201-restricted Melan-A/MART-1 antigenic peptide in melanoma. *J. Immunol.*, **159**, 2366–2374.

Rosenberg, S. A. and White, D. E. (1996) Vitiligo in patients with melanoma: normal tissue antigens can be target for cancer immunotherapy. *J. Immunother*, **19**, 81–84.

Rosenberg, S., Yang, J., Schwartzentruber, D., Hwu, P., Marincola, F., Topalian, S. *et al.* (1998) Immunologic and therapeutic evaluation of a synthetic peptide vaccine for the treatment of patients with metastatic melanoma. *Natl. Med.*, **4**, 321–327.

Rosenberg, S., Yang, J., Schwartzentruber, D., Hwu, P., Marincola, F., Topalian, S. *et al.* (1999) Impact of cytokine administration on the generation of antitumor reactivity in patients with metastatic melanoma receiving a peptide vaccine. *J. Immunol.*, **163**, 1690–1695.

Rosenberg, S. A., Yang, J. C., Topalian, S. L., Schwartzentruber, D. J., Weber, J. S., Parkinson, D. R. *et al.* (1994) Treatment of 283 consecutive patients with metastatic melanoma or renal cell cancer using high-dose bolus interleukin-2. *JAMA*, **271**, 907–913.

Rosenberg, S. A., Yannelli, J. R., Yang, J. C., Topalian, S. L., Schwartzentruber, D. J., Weber, J. S. *et al.* (1995) Treatment of patients with metastatic melanoma with autologous tumor-infiltrating lymphocytes and interleukin 2. *J. Natl. Cancer Inst.*, **86**, 1159–1166.

Rosenberg, S., Zhai, Y., Yang, J., Schwartzentruber, D., Hwu, P., Mrincola, F. *et al.* (1998) Immunizing patients with metastatic melanoma using recombinant adenoviruses encoding MART-1 or gp100 melanoma antigens. *J. Natl. Cancer Inst.*, **90**, 1894–1900.

Rubin, J. T., Elwood, L. J., Rosenberg, S. A. and Lotze, M. T. (1989) Immunohistochemical correlates of response to recombinant interleukin-2 based immunotherapy in humans. *Cancer Res.*, **49**, 7086–7092.

Rubinfeld, B., Robbins, P., El-Gamil, M., Albert, I., Prfiri, E. and Polakis, P. (1997) Stabilization of beta-catenin by genetic defects in melanoma cell lines. *Science*, **275**, 1790–1792.

Sahin, U., Tureci, O., Schmitt, H., Cochlovius, B., Johannes, T., Schmits, R. *et al.* (1995) Human neoplasms elicit multiple specific immune responses in the autologous host. *Proc. Natl. Acad. Sci. USA*, **92**, 11810–11813.

Salgaller, M., Marincola, F., Cormier, J. and Rosenberg, S. (1996) Immunization against epitopes in the human melanoma antigen gp100 following patient immunization with synthetic peptides. *Cancer Res.*, **56**, 4749–4757.

Salgaller, M., Marincola, F., Rivoltini, L., Kawakami, Y. and Rosenberg, S. (1995) Recognition of multiple epitopes in the human melanoma antigen gp100 by antigen specific peripheral blood lymphocytes stimulated with synthetic peptides. *Cancer Res.*, **55**, 4972–4979.

Salgaller, M. L., Weber, J. S., Koenig, S., Yanelli, J. R. and Rosenberg, S. A. (1994) Generation of specific anti-melanoma reactivity by stimulation of human tumor-infiltrating lymphocytes with MAGE-1 synthetic peptide. *Cancer Immunol. Immunoth.*, **39**, 105–116.

Scheibenbogen, C. (1997) Analysis of the T cell response to tumor and viral peptide antigens by an IFNgamma-ELISPOT assay. *Int. J. Cancer*, **71**, 932–936.

Sensi, M., Traversari, C., Radrizzani, M., Salvi, S., Maccalli, C., Mortarini, R. *et al.* (1995) Cytotoxic T lymphocyte clones from different patients display limited T-cell receptors variable gene usage in HLA-A2 restricted recognition of Melan/Mart-1 melanoma antigen. *Proc. Natl. Acad. Sci. USA*, **92**, 5674–5678.

Skinner, P. J., Daniels, M. A., Schmidt, C. S., Jameson, S. C. and Haase, A. T. (2000) Cutting edge: *in situ* tetramer staining of antigen-specific T cells in tissues. *J. Immunol.*, **165**, 613–617.

Skipper, J. C., Gulden, P. H., Hendrickson, R. C., Harthun, N., Caldwell, J. A., Shabanowitz, J. *et al.* (1999) Mass-spectrometric evaluation of HLA-A*0201-associated peptides identifies dominant naturally processed forms of CTL epitopes from MART-1 and gp100. *Int. J. Cancer*, **82**, 669–677.

Skipper, J. C. A., Hendrickson, R. C., Gulden, P. H., Brichard, V., Van Pel, A., Chen, Y. *et al.* (1996a) An HLA-A2-restricted tyrosinase antigen on melanoma cells results from posttranslational modification and suggests a novel pathway for processing of membrane proteins. *J. Exp. Med.*, **183**, 527–534.

Skipper, J. C. A., Kittlesen, R. C., Hendrickson, D. D., Deacon, N. L., Harthun, S. N., Wagner, D. F. *et al.* (1996b) Shared epitopes for HLA-A3 restricted melanoma reactive human CTL include a naturally processed epitope from Pmel17/gp100. *J. Immunol.*, **157**, 5027–5033.

Srivastava, P. (1996) Do human cancers express shared protective antigens? Or the necessity of remembrance of things past. *Semin. Immunol.*, **8**, 295–302.

Sugita, S., Sagawa, K., Mochizuki, M., Shichijo, S. and Itoh, K. (1996) Melanocyte lysis by cytotoxic T lymphocytes recognizing the MART-1 melanoma antigen in HLA-A2 patients with Vogt–Koyanagi–Harada disease. *Int. Immunol.*, **8**, 799–803.

Sun, Y., Song, M., Stevanovic, S., Jankowiak, C., Paschen, A., Rammensee, H. G. *et al.* (2000) Identification of a new HLA-A(*)0201-restricted T-cell epitope from the tyrosinase-related protein 2 (TRP2) melanoma antigen. *Int. J. Cancer*, **87**, 399–404.

Takahashi, K., Shichijo, S., Noguchi, M., Hirohata, M. and Itoh, K. (1995) Identification of MAGE-1 and MAGE-4 proteins in spermatogonia and primary spermatocytes of testis. *Cancer Res.*, **55**, 3478–3482.

Thurner, B., Haendle, I., Roder, C., Dieckmann, D., Keikavoussi, P., Jonuleit, H. *et al.* (1999) Vaccination with Mage-3A1 peptide pulsed mature, monocyte derived dendritic cells expands specific cytotoxic T cells and induces regression of some metastases in advanced stage IV melanoma. *J. Exp. Med.*, **190**, 1669–1678.

Topalian, S., Solomon, D., Avis, F. P., Chang, A. E., Freerksen, D. L., Linehan, W. M. *et al.* (1988) Immunotherapy of patients with advanced cancer using tumor infiltrating lymphocytes and recombinant interleukin-2: a pilot study. *J. Clin. Oncol.*, **6**, 839–853.

Tsai, V., Southwood, S., Sidney, J., Sakaguchi, K., Kawakami, Y., Appella, E. *et al.* (1997) Induction of subdominant CTL epitopes of the gp100 melanoma associated tumor antigen by primary *in vitro* immunization with peptide pulsed dendritic cells. *J. Immunol.*, **158**, 1796–1802.

Tureci, O., Sahin, U. and Pfreundschuh, M. (1997) Serological analysis of human tumor antigens: molecular definition and implications. *Mol. Med. Today*, **3**, 342–349.

Tureci, O., Sahin, U., Zwick, C., Koslowski, M., Seitz, G. and Pfreundschuh, M. (1998) Identification of a meiosis-specific protein as a member of the class of cancer/testis antigens. *Proc. Natl. Acad. Sci. USA*, **95**, 5211–5216.

Valmori, D., Dutoit, V., Lienard, D., Lejeune, F., Speiser, D., Rimoldi, D. *et al.* (2000a) Tetramer-guided analysis of TCR beta-chain usage reveals a large repertoire of melan-A-specific CD8+ T cells in melanoma patients. *J. Immunol.*, **165**, 533–538.

Valmori, D., Dutoit, V., Lienard, D., Rimoldi, D., Pittet, M. J., Champagne, P. *et al.* (2000b) Naturally occurring human lymphocyte antigen-A2 restricted CD8+ T-cell response to the cancer testis antigen NY-ESO-1 in melanoma patients. *Cancer Res.*, **60**, 4499–4506.

Valmori, D., Gervois, N., Rimoldi, D., Fonteneau, J. F., Bonelo, A., Lienard, D. *et al.* (1998) Diversity of the fine specificity displayed by HLA-A *0201-restricted CTL Specific for the immunodominant melan-A/MART-1 antigenic peptide. *J. Immunol.*, **161**, 6956–6962.

Valmori, D., Gileadi, U., Servis, C., Dunbar, P. R., Cerottini, J. C., Romero, P. *et al.* (1999) Modulation of proteasomal activity required for the generation of a cytotoxic T lymphocyte-defined peptide derived from the tumor antigen MAGE-3. *J. Exp. Med.*, **189**, 895–906.

Van der Bruggen, P., Traversari, C., Chomez, P., Lurquin, C., DePlaen, E., Van Den Eynde, B. *et al.* (1991) A gene encoding an antigen recognized by cytolytic T lymphocytes on a human melanoma. *Science*, **254**, 1643–1647.

Wang, R., Johnson, S., Southwood, S., Sette, A. and Rosenberg, S. (1998a) Recognition of an antigenic peptide derived from tyrosinase-related protein-2 by CTL in the context of HLA-A31 and -A33. *J. Immunol.*, **160**, 890–897.

Wang, R.-F., Johnston, S., Zeng, G., Topalian, S., Schwartzentruber, D. and Rosenberg, S. (1998b) A breast and melanoma-shared tumor antigen: T cell responses to antigenic peptides translated from different open reading frames. *J. Immunol.*, **161**, 3596–3606.

Wang, R., Wang, X., Johnston, S., Zeng, G., Robbins, P. and Rosenberg, S. (1998c) Development of a retrovirus-based complementary DNA expression system for the cloning of tumor antigens. *Cancer Res.*, **58**, 3519–3525.

Wang, R.-F., Appella, E., Kawakami, Y., Kang, X. and Rosenberg, S. A. (1996a) Identification of TRP2 as a human tumor antigen recognized by cytotoxic T lymphocytes. *J. Exp. Med.*, **184**, 2207–2216.

Wang, R.-F., Parkhurst, M., Kawakami, Y., Robbins, P. F. and Rosenberg, S. A. (1996b) Utilization of an alternative open reading frame of a normal gene in generating a novel human cancer antigen. *J. Exp. Med.*, **183**, 1131–1140.

Wang, R. F., Robbins, P. F., Kawakami, Y., Kang, X. Q. and Rosenberg, S. A. (1995) Identification of a gene encoding a melanoma tumor antigen recognized by HLA-A31-restricted tumor-infiltrating lymphocytes. *J. Exp. Med.*, **181**, 799–804.

Wiedenfeld, E. A., Fernandez-Vina, M., Berzofsky, J. A. and Carbone, D. P. (1994) Evidence for selection against human lung cancers bearing p53 missense mutations which occur within the HLA A-0201 peptide consensus motif. *Cancer Res.*, **54**, 1175–1177.

Wolfel, T., Hauer, M., Schneider, J., Serrano, M., Wolfel, C., Klehmann-Hieb, E. *et al.* (1995) A p16INK4a-insensitive CDK4 mutant targeted by cytolytic T lymphocytes in a human melanoma. *Science*, **269**, 1281–1284.

Wolfel, T., Van Pel, A., Brichard, V., Schneider, J., Seliger, B., Meyer Zum Buschenfelde, K.-H. *et al.* (1994) Two tyrosinase nonapeptides recognized on HLA-A2 melanomas by autologous cytolytic T lymphocytes. *Eur. J. Immunol.*, **24**, 759–764.

Yee, C., Savage, P. A., Lee, P. P., Davis, M. M. and Greenberg, P. D. (1999) Isolation of high avidity melanoma-reactive CTL from heterogeneous populations using peptide-MHC tetramers. *J. Immunol.*, **162**, 2227–2234.

Zorn, E. and Hercend, T. (1999a) A MAGE-6-encoded peptide is recognized by expanded lymphocytes infiltrating a spontaneously regressing human primary melanoma lesion. *Eur. J. Immunol.*, **29**, 602–607.

Zorn, E. and Hercend, T. (1999b) A natural cytotoxic T cell response in a spontaneously regressing human melanoma targets a neoantigen resulting from a somatic point mutation. *Eur. J. Immunol.*, **29**, 592–601.

Zuo, L., Weger, J., Yang, Q., Goldstein, A. M., Tucker, M. A., Walker, G. J. *et al.* (1996) Germline mutations in the p16INK4a binding domain of CDK4 in familial melanoma. *Nat. Genet.*, **12**, 97–99.

Chapter 5

Squamous cell and adeno cancer antigens recognized by cytotoxic T lymphocytes

Kyogo Itoh, Shigeki Shichijo, Akira Yamada, Masaaki Ito, Takashi Mine, Kazuko Katagiri and Mamoru Harada

Summary

Recent advances of molecular immunology allowed us to identify genes encoding human tumor-associated antigens (TAAs) and peptides that are recognized by $CD8^+$ cytotoxic T lymphocytes (CTLs) of patients with various types of cancers. In this chapter we review the current status of TAAs and their CTL epitopes expressed in epithelial cancers and introduce our recent data concerning seven new TAAs. The first TAAs to be described are cancer–testis antigens which were mainly cloned from melanoma cDNA libraries but were reported to be expressed in various types of epithelial cancers. The second TAAs are non-mutated self-antigens that are over-expressed in both cancer cells and proliferating normal cells. Most TAAs which were identified at out laboratory belong to this group. The third TAAs to be described are mutated antigens which are expressed only in cancer cells but not in normal tissue. The fourth TAAs are some viral antigens which are selectively expressed in tumor cells. In the latter of this review, we introduce six new TAAs that were identified from pancreatic cancer cells. We also introduce that multidrug resistance-associated protein 3 (MRP3) can be a new TAA capable of inducing tumor-reactive CTLs from HLA-$A24^+$ patients with epithelial cancers. The majority of human malignant tumors consists of epithelial cancer. We hope that this chapter could provide the update information of molecular basis for $CD8^+$ T cell-mediated recognition of epithelial cancer cells, and thereby promote the development of specific cancer immunotherapy.

Introduction

Recent advances of molecular immunology allowed us to identify genes encoding human TAAs and peptides that were recognized by the $CD8^+$ CTLs of patients with various types of human cancers, including melanomas and epithelial cancers. The majority of human malignant tumors consist of epithelial cancers (squamous cell carcinoma (SCC) or adenocarcinoma). This chapter at first reviews the current status of TAAs and their CTL-epitopes expressed in SCC and adenocarcinoma, and then describes six recently defined TAAs of pancreatic cancer and an MRP3 as new TAAs, both of which were recognized by HLA-A2- and -A24-restricted CTLs, respectively. The main objective of this chapter is to provide updated information on the molecular basis of $CD8^+$ T cell mediated recognition of epithelial cancer cells, and thereby to promote the development of new treatment modalities of specific cancer immunotherapy.

TAAs of SCC and adenocarcinoma

Most HLA-class I-restricted TAAs have been identified by the cDNA-expression cloning method originally developed by Boon and his colleagues [1]. Alternatively, peptides have been eluted from tumor cells, loaded on target cells bearing an appropriate HLA molecule followed by testing their ability to stimulate IFN-γ production by the CTLs [2]. With these approaches, approximately 50 TAAs and more than 100 CTL-directed peptide antigens derived from SCC and adenocarcinomas have been identified as far as searched at the literature levels (Table 5.1) [3–70]. It has been thought for a long time that TAAs would be either

Table 5.1 Tumor-associated antigens recognized by HLA-class I-restricted CTLs

Tumor-associated antigen	*HLA-restriction*	*Reference*
Shared cancer–testis antigens		
MAGE-1	A1	[3]
MAGE-1	A3, A28, B53, Cw2, Cw3	[4]
MAGE-1	A24	[5]
MAGE-1, MAGE-2, MAGE-3, MAGE-6	B37	[6]
MAGE-1	Cw1601	[7]
MAGE-2, MAGE-3, Her2/neu, CEA	A2	[8]
MAGE-2	A24	[9]
MAGE-3	A1, B35	[10]
MAGE-3	A2	[11]
MAGE-3	A24	[12]
MAGE-3	A24	[13]
MAGE-3	B44	[14]
MAGE-A10	A2	[15]
MAGE-12	Cw7	[16]
BAGE	Cw1601	[17]
GAGE-1	Cw6	[18]
NY-ESO-1	A2	[19]
NY-ESO-1	A31	[20]
PRAME	A24	[21]
PRAME	A2	[22]
Non-mutated self antigens over-expressed in epithelial cancers		
(Widely-expressed)		
SART1	A26	[23]
SART1	A24	[24]
SART2	A24	[25]
SART3	A24	[26]
SART3	A2	[27]
ART1	A24	[28]
ART4	A24	[29]
Cyclophilin B	A24	[30]
Cyclophilin B	A2	[31]
Lck	A24	[32]
Lck	A2	[33]
EIF4E-BP, ppMAPkkk, WHSC2, UBE2V, HNRPL	A2	[34]
MDR3	A24	[35]
Her2/neu	A24	[36]
Her2/neu	A2	[37]
Her2/neu	A2	[38]
Her2/neu	A2	[39]
Her2/neu	A2	[40]
Her2/neu	A2	[41]
Her2/neu, CEA	A3	[42]

(Continued)

Table 5.1 (Continued)

Tumor-associated antigen	*HLA-restriction*	*Reference*
P53	A24	[43]
P53	A2	[44]
hTERT	A24	[45]
hTERT	A2	[46]
M-CSF	B35	[47]
Survivin	A2	[48]
(Selectively-expressed)		
CEA	A24	[49]
CEA	A24	[50]
α-fetoprotein	A2	[51]
MUC-1	A2	[52]
MUC-2	A2	[53]
PSA	A2	[54]
PSM	A2	[55]
PSCA	A2	[56]
PAP	A24	[57]
Recoverin	A24	[58]
G250	A2	[59]
RAGE	B8	[60]
Intestinal carboxyl esterase	B7	[61]
Mutated antigens		
Caspase-8	B35	[62]
α-actinin-4	A2	[63]
Malic enzyme	A2	[64]
hsp 70-2	A2	[65]
Viral antigens		
Human papilloma virus type 16(E6/E7)	A2	[66]
EBV(LMP-1)	A2	[67]
EBV(LMP-2)	A2	[68]
EBV(EBNA-3)	B27	[69]
EBV(EBNA-3)	B7	[70]

tumor-specific antigens or mutated-self-antigens. However, the majority of TAAs defined by the methods mentioned above have been non-mutated self-antigens, and only a few of TAAs are mutated self-antigens or viral antigens.

The first TAAs to be described were shared cancer–testis antigens [3–19]. These antigens were originally cloned from melanoma cDNAs, and reported to be expressed in various types of cancer cells including in SCC and adenocarcinomas. Several families of these cancer–testis antigens are also expressed in various types of cancer cells and in the testis but not in either the other normal cells or normal tissues [1, 3–18]. The second TAAs to be described were non-mutated self-antigens that are over-expressed in both cancer cells and proliferating normal cells. These non-mutated self-antigens (SART1 to SART3, ART1, ART4 and the others listed in Table 5.1) were mostly cloned from cDNA libraries of epithelial cancers (esophageal SCC, lung and pancreatic adenocarcinomas and bladder transitional adenocarcinoma) (23–48). Among these TAAs, SART1, SART3, HNRPL and EIF4EBP1 are known as RNA- or DNA-binding proteins that are involved in cellular proliferation in the nucleus. Cyclophilin B, lck, HER2/neu, and ppMAPkkk might also be involved in cellular proliferation. These growth-related proteins would be vigorously synthesized, utilized and then processed in cancer cells. Subsequently, the processed peptides might be loaded onto HLA-class I molecules over the level of immunological tolerance or ignorance, and these molecules in turn be possibly recognized by HLA-class I-restricted and tumor-reactive CTLs. MUC1, CEA, PSA, PAP and several others listed in Table 5.1 have a characteristic of antigens selectively expressed in epithelial tissues or cells [49–61].

In contrast to these non-mutated self-antigens, caspase 8 and several others listed in Table 5.1 have been described as mutated forms of TAAs, and thus represent antigens expressed only on cancer cells and not on normal cells [62–65]. The use of these antigens in the development of generally applicable cancer vaccines could be limited, since each of them is unique to the individual patient or individual cells in one patient. Several viral antigenic epitopes are known to be expressed on tumor cells and therefore to be recognized by the CTLs. TAAs in this category include antigens from Epstein–Barr virus, and E6 and E7 proteins from human papillomavirus type 16 [66–70].

HLA-class I-A alleles capable of presenting these TAA-directed peptides include HLA-A1, -A2, -A3, -A11, -A24, -A26 and -A31 alleles (Table 5.1). These HLA-class I-A alleles are commonly found in various ethnic populations. For example, HLA-A2, -A24 and -A26 alleles are expressed in 40%, 60% and 22% of Japanese, 50%, 20% and 20% of Caucasians, and 22%, 12% and 14% of Africans, respectively. In addition, several HLA-class I-B and C alleles capable of presenting TAA-directed peptides have been reported (Table 5.1). With regard to origins of cancers, these TAAs are highly expressed in the majority of SCC and adenocarcinomas from various organs, including lung, esophagus, stomach, pancreas, colon, breast, prostate, ovary and uterine. Therefore, these TAAs and peptides could be applicable for almost all epithelial cancer patients in the world.

Tumor-associated antigens of pancreatic cancer

Pancreatic cancer continues to be a major unsolved health problem in the world. The prognosis of pancreatic cancer is extremely poor with a median survival of 3–4 months and the 5-year survival being 1–4%. This poor prognosis is primarily due to lack of effective therapies, and thus development of new treatment modalities is needed. One of these treatments could involve specific immunotherapy, for which elucidation of the molecular basis of T cell-mediated recognition of cancer cells is required. We have recently reported 6 different genes and 19 immunogenic epitopes from pancreatic adenocarcinoma cells, and T-cell receptor β usage of HLA-A2-restricted CTL clones reacting to some of these epitopes [34]. Sixteen of 19 epitopes were found to possess the ability to induce HLA-A2-restricted CTL activity in the peripheral blood lymphocytes of patients with pancreatic and also colon adenocarcinomas. For the cloning of these genes, the HLA-A2-restricted and tumor-specific CTL line (OK-CTL) that was established by a long-term incubation of tumor-infiltrating lymphocytes (TILs) from a patient (HLA-A0207/3101) with colon adenocarcinoma was used as indicator cells. The details of characteristics of the OK-CTL line were reported elsewhere [34].

After repeated experiments, six cDNA clones, *no. 1* to *no. 6*, were identified (Figure 5.1). The nucleotide (nt) sequence of cDNA clones *no. 1* or *no. 2* was almost identical to that of the *ubiquitin-conjugated enzyme variant Kua* (*UBE2* gene) (*no. 1*) or the *heterogeneous nuclear ribonucleoprotein L* gene (*HNRPL* gene) (*no. 2*), respectively. *UBE2* gene encodes one of the heterogeneous nuclear ribonucleoprotein complexes providing a substrate for the processing events that pre-mRNAs undergo before becoming functional and translatable mRNAs in the

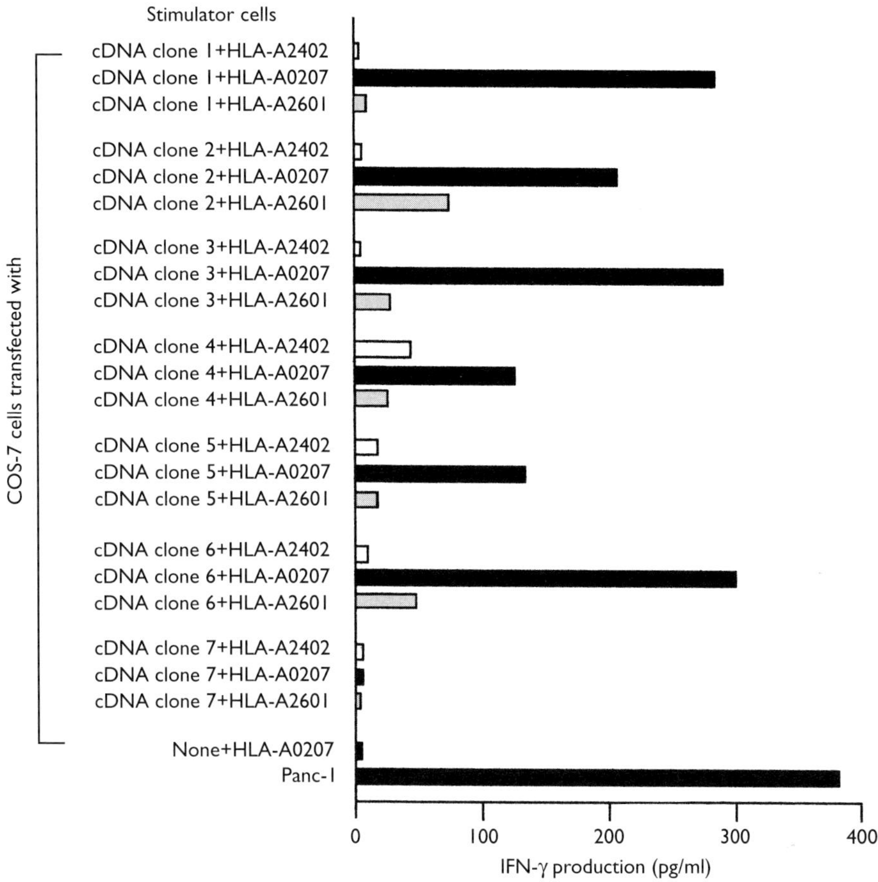

Figure 5.1 Six genes coding for tumor epitopes. One-hundred ng of each of cDNA clones (*no. 1 to no. 6*) derived from Panc-1 tumor cells and 100 ng of *HLA-A0207, -A2402, or -A2601* cDNA were co-transfected into COS-7 cells, incubated for 48 h, and then tested for their ability to stimulate IFN-γ release by the OK-CTLs. The background of IFN-γ release by the CTLs in response to COS-7 cells (under 100 pg/ml) was subtracted from the value in the figure. cDNA *clone* 7 represents the irrelevant clones that were not recognized by the OK-CTLs.

cytoplasm [71–73]. The nt sequence of cDNA clone *no. 3* was identical to that of the *Wolf–Hirschhorn syndrome candidate 2*(*WHSC2*) gene. The *WHSC2* seems to play a role in the phenotype of WHS, a multiple malformation syndrome characterized by mental and developmental defects resulting from a partial deletion of the short arm of chromosome 4 [74]. The nt sequence of cDNA *no. 4* was identical to that of *eIF-4E-binding protein 1* gene (*EIF4EBP1*). This protein is known as a translation initiation factor that initiates insulin-dependent phosphorylation of 4E-BP1, making it available to form an active cap-binding complex [75]. The nt sequence of cDNA clone *no. 5* or *no. 6* was almost identical to that of the *partial putative mitogen-activated protein kinase kinase kinase* (*ppMAPkkk*) gene with unreported function or identical to that of the *2′,5′-oligoadenylate synthetase 3* gene (*2–5 OAS3*), respectively.

The 2–5 OAS3 is known as an IFN-induced protein, that plays an important role in immuno-protection from microbacterial infection [76, 77]. These six genes, except with *ppMAPkkk*, were ubiquitously expressed in both cancer and normal cells, and their expression levels in cancer cells, including Panc-1, SW620, and CA9-22 tumors, were significantly higher than those in the normal cells, including PHA-blastoid T cells and EBV-B cells (data not shown). mRNA expression of *ppMAPkkk* was scarcely detectable under the employed conditions as reported previously [34].

Three-hundred $CD8^+$ CTL clones were established from the parental OK-CTL line by a limiting dilution culture. Eighty CTL clones among them showed HLA-A2-restricted and tumor-specific CTL activity. All these 80 CTL clones expressing $CD3^+CD4^-CD8^+$ and TCR $\alpha\beta^+$ phenotypes were tested for their reactivity to the six gene products. Among them, 2, 3, 1, 3, 2 and 4 CTL clones were reactive to *UBE2V, HNRPL, WHSC2, EIF4EBP1, ppMAPkkk* and *2–5 OAS3* gene products, respectively. Each of 27, 17, 21, 5, 19 or 39 different synthesized peptides with HLA-A2 molecules-binding motifs derived from the six gene products, respectively, was loaded onto the T2 cells followed by testing for their ability to induce IFN-γ release by the OK-CTLp and its CTL clones. Five peptides of UBE2V at positions 43–51, 64–73, 85–93, 201–209 and 208–216 were recognized by the OK-CTLp, while the two peptides at positions 43–51 and 64–73, but not any of the other 25 peptides, were strongly and dimly recognized by the CTL clone 2-2-H3, respectively (Figure 5.2, upper left column). Four peptides of HNRPL at positions 140–148, 404–412, 443–451 and 501–510 were recognized by the OK-CTLp, while one peptide at positions 140–148, but not any of the other 16 peptides, was recognized by the CTL clone 1-2-D12 (Figure 5.2, middle left column). Similarly, the peptides recognized by the OK-CTLp were as follows; four peptides of WHSC2 at positions 103–111, 141–149, 157–165 and 267–275, two peptides of EIF4EBP1 51–59 and 52–60, three peptides of ppMAPkkk 290–298, 294–302 and 432–440, and one 2–5 OAS3 peptide 666–674. Further, CTL clones, 4-2-A11, 4-2-B3, 0.5-1-H2, and 1-2-D1 recognized a WHSC2 peptide 103–111, an EIF4EBP1 51–59, a ppMAPkkk 432–440, and a 2–5 OAS3 666–674, respectively (Figure 5.2).

To confirm a different TCR usage in each CTL clone with different specificity, TCR Vβ usage of these peptide-specific CTL clones was determined by amplification of the TCR Vβ chain by the RT-PCR method using the specific primer sets of TCR Vβ1-20 and TCR Cβ1-2 [78, 79]. Two each of the CTL clones reacting to a UBE2V, an HNRPL and a 2–5 OAS3 peptide used the TCR Vβ 8.1, Vβ 3.2 and Vβ 14, respectively (Table 5.2). Each CTL clone recognizing a WHSC2, an EIF4EBP1, and a ppMAPkkk peptide used TCR Vβ 13.1, Vβ 8.1 and Vβ 18, respectively. These amplified products were further provided for direct sequencing of the TCR β chain to address if these CTL clones reacting to different tumor epitopes possess similar or different complementarity-determining regions 3 (CDR3), an element responsible for binding to antigenic epitopes on the groove of HLA-class I molecules [80]. Two each of CTL clones recognizing a UBE2V, an HNRPL or a 2–5 OAS3 peptide used the different CDR3, respectively (Table 5.2). Similarly, each CTL clone reacting to a WHSC2, an EIF4EBP1 or a ppMAPkkk peptide used the different CDR3, respectively.

Because of its wide reactivity to tumor cells with different HLA-A2 subtypes and with different histologies, the OK-CTLp used for the study likely consisted of a mixture of CTL clones recognizing the shared tumor epitopes capable of binding to the HLA-A2 subtypes, and were expressed on various cancers originating from different organs. Indeed, the six genes and 19 immunogenic epitopes were identified with this CTL line. Further, the

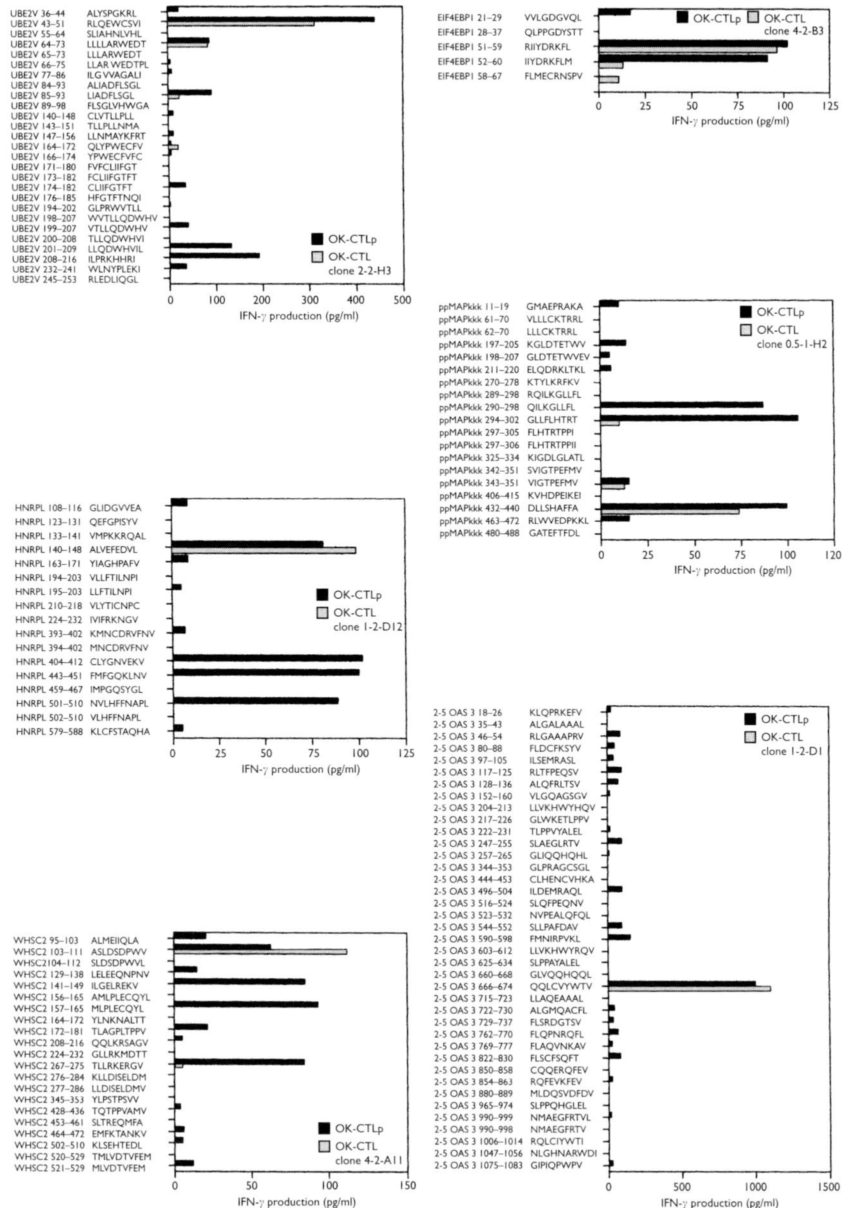

Figure 5.2 Determination of CTL epitopes. Each of the 27 UBE2V-derived peptides, 17 HNRPL-derived peptides, 21 WHSC2-derived peptides, 5 EIF4EBP1-derived peptides, 19 ppMAPkkk-derived peptides and 39 2–5 OAS3- derived peptides (9–11 mer) were loaded onto T2 cells at a concentration of 10 μM for 2 h. The OK-CTLp or its CTL clones were then added at an E/T ratio of 10 or 2, respectively, and incubated for 18 h followed by collection of the supernatant for measurement of IFN-γ. Values indicate the mean of triplicate assays. The background of IFN-γ release by the CTLs (under 100 pg/ml) in response to the T2 cells alone was subtracted from the values in the figure.

Table 5.2 TCRβ usage of the OK-CTL clones

CTL clone	*Epitopes*[a]	*V*β	*D*β	*J*β	*C*β	*V*β	*D*β/*J*β[b]	*C*β
2-2-H3	UBE2V 43–51	8.1	2.1	2.3	2	IYFNNNVPIDDSGMPEDRFSAKMPNASFSTLKIQPSEPRDSAVYFCAS	<u>SLGLAGGEQF</u>FGPGTRLTVL	EDLKNVFPPE
2-1-H12	UBE2V 43–51	8.1	2.1	2.3	2	IYFNNNVPIDDSGMPEDRFSAKMPNASFSTLKIQPSEPRDSAVYFCAS	<u>SLGLAGGEQF</u>FGPGTRLTVL	EDLKNVFPPE
1-2-D7	HNRPL 140–148	3.2	1.1	2.7	2	VSREKKERFSLILESASTNQTSMYLCAS	<u>SLDRSYEQY</u>FGPGTRLTVT	EDLKNVFPPE
1-2-D12	HNRPL 140–148	3.2	1.1	2.7	2	VSREKKERFSLILESASTNQTSMYLCAS	<u>SLDRSYEQY</u>FGPGTRLTVT	EDLKNVFPPE
4-2-A11	WHSC2 103–111	13.1	2.1	2.7	2	QGEVPNGYNVSRSTTEDFPLRLLSAAPSQTSVYFCAS	<u>SYGGGSSY</u>EQYFGPGTRLTVT	EDLKNVFPPE
4-2-B3	EIF4EBP1 51–59	8.1	1.1	1.1	1	IYFNNNVPIDDSGMPEDRFSAKMPNASFSTLKIQPSEPRDSAVYFCAS	<u>SRVSG</u>EAFFGQGTRLTVV	EDLKNVFPPE
0.5-1-H2	ppMAPkkk 432–440	18	1.1	1.1	1	DESGMPKERFSAEFPKEGPSILRIQQVVRGDSAAYFCAS	<u>SPTELDT</u>EAFFGQGTRLTVV	EDLKNVFPPE
1-2-D1	2–5 OAS3 666–674	14	2.1	2.3	2	VSRKEKRNFPLILESPSPNQTSLYFCAS	<u>GGSTDTQY</u>FGPGTRLTVL	EDLKNVFPPE
2-2-B4	2–5 OAS3 666–674	14	2.1	2.3	2	VSRKEKRNFPLILESPSPNQTSLYFCAS	<u>GGSTDTQY</u>FGPGTRLTVL	EDLKNVFPPE

Notes

a Each immunogenic epitope reactive to each CTL clone in shown. The data of reactivity are presented in Figure 5.2.

b The underline shows the CDR3 of the TCRβ of each CTL clone.

results provide clear evidence that different CDR3 of the CTL clones are responsible for the recognition of different epitopes from the six gene products. There have been conflicting reports regarding the usage of TCR of CTLs reacting to melanoma cells in the early 1990s. Several studies have resulted in the observation of clonal usage [81, 82], whereas others, including our study [79], have observed polyclonal usage of CTLs [83, 84]. However, recent studies utilizing HLA-class I-restricted and peptide-specific CTL clones have found evidence of multiple specificities in the repertoire of tumor-reactive CTLs [85, 86]. Therefore, CTLs at the tumor sites would consist of a mixture of CTLs with many different TCR usages reacting to many different epitopes on HLA-class I-A alleles of tumor cells.

Nineteen peptides recognized by the OK-CTLp were then tested for their ability to induce HLA-A2-restricted and tumor-specific CTLs from the autologous PBMCs (OK) and two HLA-A0201$^+$ patients (one with pancreatic cancer and one with colon cancer). These PBMCs produced significant levels of IFN-γ in response to HLA-A2$^+$ SW620, CA9-22 and Panc-1, but not to the HLA-A2$^-$ tumor cells, when stimulated three times *in vitro* with the following 13 peptides; UBE2V 43–51, 85–93 and 208–216, HNRPL 140–148, 443–451 and 501–510, WHSC2 103–111, 141–149 and 267–275, EIF4EBP1 51–59 and 52–60, ppMAPkkk 294–302 and 432–440. These CTL activities were inhibited by anti-HLA-class I, anti-CD8, or anti-HLA-A2 mAb, but not by any of the other mAbs tested. Similar results were obtained for all three patients, and representative results from the autologous PBMCs are shown in Table 5.3. An HNRPL 404–412 or a WHSC2 157–165 peptide induced the CTLs reactive to only SW620 or Panc-1 tumor cells, respectively. A ppMAPkkk 290–298 induced the CTLs reactive CA9-22 and Panc-1, but not SW620 tumor cells (Table 5.3). In contrast, the UBE2V 64–73 and 201–209 and 2–5 OAS3 666–674 peptides induced no CTL activity. The levels of binding affinity for these 19 peptides, though different from each other, did not correlate well with the ability to induce CTLs (Table 5.3).

Among the six identified gene products, HNRPL and EIF4EBP1 are known as RNA and DNA-binding proteins, respectively, both of which are involved in cellular proliferation [71–73, 75]. A *ppMAPkkk* gene might also be involved in cellular proliferation if involved in the regulation of the *MAPk* gene [87]. A mutated *MAPk* gene encodes tumor epitopes recognized by the CTLs in a murine model [88]. We have reported other growth-related proteins (cyclophilin B, SART1, SART3), all non-mutated forms that also include immunogenic epitopes recognized by the HLA-A24-restricted CTLs [23, 26, 30]. These growth-related proteins would be vigorously synthesized, utilized and then processed in cancer cells. Subsequently, the processed peptides might be loaded onto HLA-A2 molecules from tumor cells over the level of immunological ignorance, with these molecules in turn possibly being recognized by the T cells.

Collectively, this study reported six genes and 16 immunogenic epitopes capable of inducing HLA-A2-restricted and tumor-specific CTLs in PBMCs from pancreatic and/or colon cancer patients. These results suggest that pancreatic and colon cancers share the same tumor epitopes recognized by the host CTLs. The incidence and number of cancer deaths from colon cancer are 5–6 times and 2–3 times higher than those of pancreatic cancers [89, 90]. Although resection for cure is possible in 70–75% of all colon cancer patients, 50% still die from their disease regardless of the many different treatments, and thus, the development of new treatment modalities is needed. The HLA-A2 allele is found in 23% of African Blacks, 53% of Chinese, 40% of Japanese, 49% of Northern Caucasians and 38% of Southern Caucasians [91]. The information presented in this chapter should provide a better understanding of the molecular basis of T cell-mediated recognition of pancreatic

Table 5.3 Induction of HLA-A2-restricted CTL activity by the peptides in PBMCs

Peptide	*MFI*	*CD4(%)*[b]	*CD8(%)*[b]	*IFN-γ production (pg/ml) in response to*[a]										
				QG56 (HLA-A26/26)	*RERF-LC-MS (HLA-A11/11)*	*COLO320 (HLA-A24/24)*	*SW620 (HLA-A0201/24)*	*CA9-22 (HLA-A0207/24)*	*Panc-1 (HLA-A0201/11)*	*Panc-1*[c] *+anti-class I*	*Panc-1 +anti-class II*	*Panc-1 +anti-CD4*	*Panc-1 +anti-CD8*	*Panc-1 +anti-HLA-A2*
UBE2V 43–51	571	9.9	84.3	0	26	50	235	81	492	59	468	442	128	178
UBE2V 64–73	607	12.0	83.9	0	0	5	53	0	0					
UBE2V 85–93	910	21.6	75.8	0	0	44	188	58	289					
UBE2V 201–209	1008	15.6	81.2	0	0	0	60	0	0					
UBE2V 208–216	637	16.6	81.0	0	0	38	500	96	638	242	596	602	310	302
HNRPL 140–148	819	7.7	85.8	0	8	40	344	863	527	162	540	542	318	383
HNRPL 404–412	783	15.2	80.7	0	0	0	344	0	54					
HNRPL 443–451	499	14.2	79.0	0	26	0	142	165	186					
HNRPL 501–510	832	18.1	78.1	0	27	0	194	98	339					
WHSC2 103–111	504	10.8	75.4	0	0	0	108	130	163	15	159	148	50	92
WHSC2 141–149	1089	10.7	83.1	0	0	0	893	62	>1000	106	732	766	63	444
WHSC2 157–165	780	9.1	87.9	0	0	40	46	0	197					
WHSC2 267–275	656	19.7	77.1	0	15	0	151	95	115					
EIF4EBP1 51–59	591	13.2	86.8	0	0	0	112	184	265	42	212	179	23	130
EIF4EBP1 52–60	789	13.0	85.6	0	32	0	199	219	502	129	402	395	140	232
ppMAPkkk 290–298	887	12.7	79.5	0	0	0	0	147	113					
ppMAPkkk 294–302	660	25.4	64.2	0	0	0	>1000	>1000	691	162	614	618	251	340
ppMAPkkk 432–440	657	44.5	53.0	29	0	0	>1000	70	>1000					
2–5 OAS3 666–674	775	92.3	3.0	0	30	66	55	48	105	65	74	63	52	82
No peptide	491	18.0	72.3	0	0	36	0	0	17					

Notes

a The PBMCs of a patient with colon adenocarcinoma, from which the OK-CTL clones were established, were stimulated *in vitro* with a peptide (10 μM) three times every seven days followed by a test for their ability to produce IFN-γ at day 21 of culture in response to various target cells at an E/T ratio of 5.

b Percentage of $CD3^+$ $CD4^+$ $CD8^-$ or $CD3^+$ $CD4^-$ $CD8^+$ T cells of the peptide-stimulated PBMCs at the time of assay.

c For inhibition assay, the OK-CTL-mediated IFN-γ production by recognition of Panc-1 tumor cells at an E/T ratio of 5 was tested in the presence of 20 μg/ml of mAbs shown in the table.

cancer cells and also of colon cancer cells. Further, these peptides could be applicable in use for peptide-based specific immunotherapy of these cancers.

Multidrug resistance-associated protein 3 (MRP3) is a new TAA

We have recently determined that MRP3 is a new TAA recognized by HLA-A2402-restricted CTLs established from T cells infiltrating into lung adenocarcinoma [35]. Although the details of the results are reported elsewhere [35], this chapter briefly presents the results and discusses the potential usefulness of these antigens and peptides as cancer vaccines. Four dominant MRP3-derived antigenic peptides recognized by the CTLs have been identified, each possessing *in vitro* immunogenicity (Figure 5.3). Further, these four peptides (MRP3-503, MRP3-692, MRP3-765, and MRP3-1293) can induce peptide-specific CTLs after stimulation by these peptides in PBMCs of HLA-A24^{+} cancer patients, with the CTLs expressing cytotoxicity against HLA-A2402^{+} MRP3^{+} tumor cells but not against either HLA-A2402^{-} or MRP3^{-} target cells. Widespread MRP3 expression in various tumor cell lines and tumor tissues at the mRNA level was confirmed (Figure 5.4). Furthermore, reactivity of the MRP3-peptide-induced CTLs against tumor cells correlated with MRP3 expression in the tumor cells (Figure 5.5). These results suggest that MRP3 and its peptides shown above are potential candidates for cancer vaccines in regard to HLA-A24^{+} patients with various tumors, particularly for those tumors that show anti-cancer drug resistance.

The MRP family consists of at least seven ATP-binding cassette (ABC) transporters, several of which have been demonstrated to transport amphipathic anions and to confer *in vitro* resistance to chemotherapeutic agents [92–95]. Two prominent members of the ABC superfamily of transmembrane proteins, MDR1 P-glycoprotein (ABCB1) and MRP1 (ABCC1), can mediate the cellular extrusion of xenobiotics and anti-cancer agents from normal and tumor cells [92–94, 96, 97]. The roles of other members (MRP2-6, ABCC2-6) of the MRP family in MDR have been reported [98–100]: MRP2 has been shown to confer low-level resistance to the anticancer drug cisplatin, etoposide, vincristine and methotrexate [101–103], MRP3 to etoposide, vincristine and methotrexate [98, 104], MRP4 to acyclic nucleotide phosphonates, such as 9-(2-phosphonylmethoxyethyl) guanine, and anti-HIV drug 9-(2-phosphonylmethoxyethyl) adenine [105], and MRP5 to thiopurine drugs, 6-mercaptopurine and thioguanine and 9-(2-phosphonylmethoxyethyl) adenine [106].

The expression of several MRP genes at mRNA levels can be up-regulated after selection by anticancer drugs [92–94]. Up-regulation of MRP3 expression has been observed in several cell lines after selection with doxorubicin, regardless of the apparent lack of correlation of the mRNA levels with resistance to either doxorubicin or cisplatin [98]. We have demonstrated that *MRP3* was expressed in most cell lines derived from lung cancers, ovarian cancers and renal cancers at the mRNA levels. In contrast, the *MRP3* message was very low in non-tumorous cell lines (COS-7, VA13, 293T) or EBV-transformed B cells. MRP1 and MRP5 are ubiquitously expressed in normal tissues, whereas MRP3 expression in normal tissues is restricted to the liver, duodenum, colon and adrenal gland at relatively high levels and to the lung, kidney, bladder, spleen, stomach, pancreas and tonsil at low levels [99]. The MRP3 shall be a unique target molecule for cancer vaccines since the expression of MRP3 is associated with MDR, the most important problem in chemotherapy. These results suggest that immunotherapy with MRP3-derived peptide vaccine is advantageous for tumors with acquired MDR. Patients with renal cancer may be particularly suitable subjects

Peptide name	*Length*	*Sequence*	*A24-binding score*	IFN-γ (pg/ml) 0 20 40 60 80 100
HIV	9	RYPLTFGWCF		
MRP3-555	9	VYVDPNNVL	432	
MRP3-174	9	FYIHFALVL	300	
MRP3-765	9	VYSDADIFL	240	*
MRP3-1128	9	FYAATSRQL	200	
MRP3-503	9	LYAWEPSFL	200	*
MRP3-529	9	AYLHTTTTF	150	
MRP3-692	9	AWPQQAWI	90	*
MRP3-419	9	RFMDLAPFL	86	
MRP3-896	9	TYVVQKQFM	45	
MRP3-574	9	LFNILRLPL	36	
MRP3-457	9	AFMVLLIPL	36	
MRP3-475	9	AFQVKQMKL	33	
MRP3-902	9	QFMRQLSAL	30	
MRP3-1187	9	PYIISNRWL	30	
MRP3-1200	9	EFVGNCWL	30	
MRP3-177	9	HFALVLSAL	28	
MRP3-1110	9	LFTVVILPL	28	
MRP3-316	9	CFKLIQDLL	28	
MRP3-1297	10	RYRPGLDLVL	576	
MRP3-1231	10	SYSLQVTFAL	280	
MRP3-1517	10	FYGMARDAGL	200	
MRP3-1293	10	NYSVRYRPGL	200	*
MRP3-372	10	HYIFVTGVKF	165	
MRP3-1163	10	AYNRSRDFEI	82	
MRP3-977	10	LYVGQSAAAI	75	
MRP3-356	10	MFLCSMMQSL	36	
MRP3-206	10	PYPETSAGFL	36	
MRP3-310	10	SFLISACFKL	33	
MRP3-349	10	GFLVAGLMFL	30	
MRP3-1406	10	TFVSSQPAGL	30	
MRP3-1375	10	LFSGTLRMNL	24	

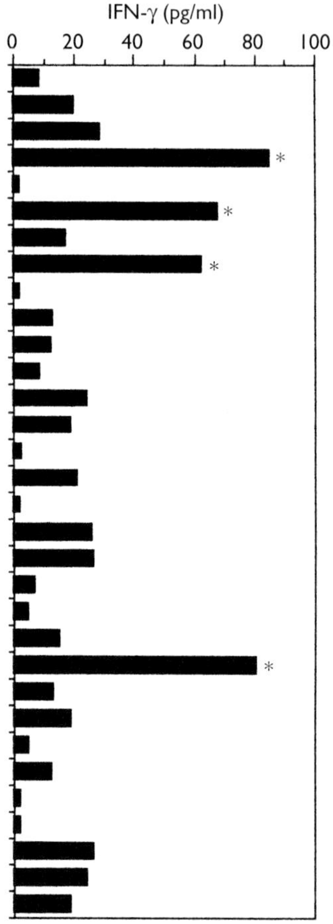

Figure 5.3 Identification of MRP3-derived antigenic peptides recognized by the GK-CTLs. Each of the 31 different MRP3-derived peptides was loaded onto C1R-A2402 cells at a concentration of 10 μM. The GK-CTLs were cultured with the peptide-loaded C1R-A2402 for 18 h, and the culture supernatant was harvested to measure IFN-γ production using ELISA. Values represent the means of triplicate assays. The background of IFN-γ production by the GK-CTLs in response to peptide-unloaded C1R-A2402 cells was subtracted from the values. The two-tailed Student's *t* test was used for the statistical analysis between the IFN-γ production by the GK-CTLs in response to peptide-loaded C1R-A2402 cells and that in response to unloaded C1R-A2402 cells. * indicates $P < 0.05$. The A24-binding score shows estimated score of half time of dissociation of each peptide for HLA-A24 molecules.

for the MRP3-peptide vaccine, since renal cancer is generally resistant to chemotherapy and radiation therapy. Indeed, our results regarding CTL induction by MRP3-peptides in the PBMC cultures of patients with renal cancer supported this suggestion. Namely the MRP3-peptides induced HLA-A24-restricted and tumor-reactive CTLs in the PBMC from three out of four patients with renal cancer tested. Furthermore, immunotherapy with MRP3-peptides in combination with chemotherapy might be possible, if the immunosuppression induced by the chemotherapeutic agents is not severe in the patient. The effectiveness of the

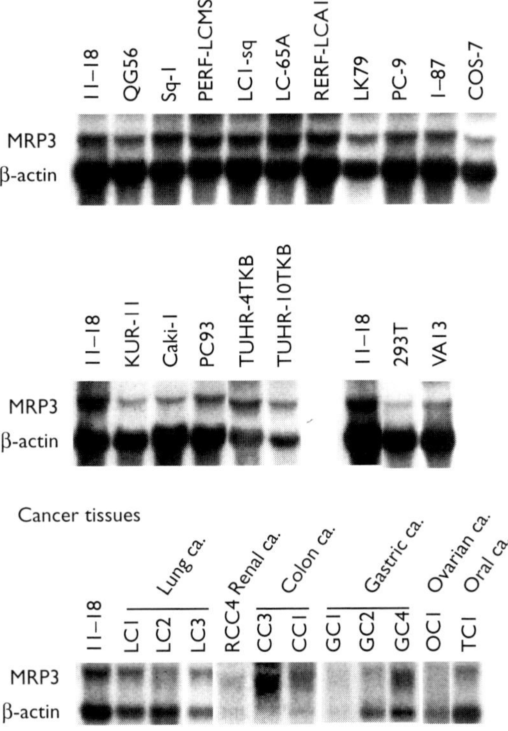

Figure 5.4 Northern blot analysis of *MRP3* expression in various tumor cell lines and tissues. Total RNA was separated on formaldehyde-agarose gel and transferred to nylon membranes. The membranes were further hybridized with ^{32}P-labeled fragment of *clone 5* and control β-*actin* cDNA. Representative results are shown in the figure.

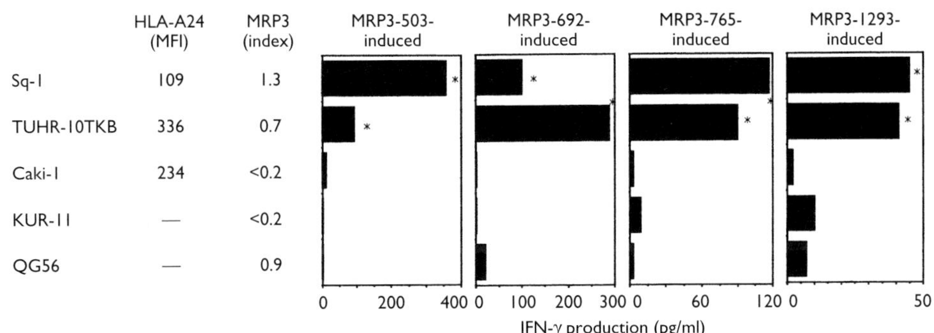

Figure 5.5 Reactivity of the MRP3-peptide-induced CTLs against MRP3^{+} and MRP3^{-} tumor cells. Representative reactivities of PBMC culture of lung cancer patient LC2 (MRP3-501, MRP3-692, MRP-765, and MRP-1293-induced CTLs) against HLA-A24^{+} MRP3^{+} (Sq-1, TUHR-10TKB), HLA-A24^{+} MRP3^{-} (Caki-1), HLA-A24^{-} MRP3^{+} (QG56), and HLA-A24^{-} MRP3^{-} (KUR-11) tumor cells are shown. Values represent the means of triplicate assays. Two-tailed Student's *t* test was used for the statistical analysis between IFN-γ production by the CTLs in response to indicated cells and that in response to QG56 cells. * indicates $P < 0.05$. MFI, mean fluorescence intensity.

combination of humanized mAb-mediated immunotherapy in accompaniment with chemotherapy for treatment of breast cancer and B-cell lymphoma has already been reported [107, 108].

Conclusion

As shown in the previous section, a large number of genes encoding tumor antigens and peptides recognized by CTLs were identified from cDNAs of SCC and adenocarcinoma cells in the past decade. These scientific advances in the field of cancer immunology allowed the specific-immunotherapy of cancer. Indeed, the preceding peptide-based immunotherapy for melanoma patients induced the increased CTL responses in the post-vaccination PBMCs in the majority of melanoma patients [109–114]. However, those immunotherapies rarely resulted in tumor-regression in these patients. Consequently, how to get tumor-regression by peptide-based specific cancer immunotherapy is one of the most important issues to be addressed. A main goal of vaccine protocol to virus or the other infectious diseases is the development of "preventive vaccine," whereas that to cancer is the development of "therapeutic vaccine." Consequently, translational studies of clinical studies shall be needed to develop the protocol of therapeutic vaccine.

From this point of view, many translational studies are in progress in the world. Some of our peptides with the ability to induce CTLs from PBMCs of cancer patients are also under clinical trials as peptide vaccines at our Kurume University Hospital by Dr Yamana and other clinicians. SART3 peptides have been used as cancer vaccines for HLA-A24$^+$colon cancer patients, and significant levels of increased CTL responded to both HLA-A24$^+$colon cancer cells and the vaccinated peptide were observed in the post-vaccination PBMCs in the majority of patients (Miyagi *et al.*, manuscript under submission). Significant levels of increased CTL responded to both HLA-A24$^+$ lung cancer cells and the vaccinated peptide $CypB_{91-99}$ peptide (Gouhara *et al.*, manuscript under submission). The currently ongoing

Table 5.4 A panel of peptides in use for CTL precursor-oriented peptide vaccine at Kurume University

HLA-A24 binding peptide		*Reference*	*HLA-A2 binding peptide*		*Reference*
$SART1_{690}$	EYRGFTQDF	[24]	$SART3_{302}$	LLQAEAPRL	[27]
$SART2_{93}$	DYSARWNEI	[25]	$SART3_{309}$	RLAEYQAYI	[27]
$SART2_{161}$	AYDFLYNYL	[25]	Cyp B_{129}	KLKHYGPGWV	[31]
$SART2_{889}$	SYTRLFLIL	[25]	Cyp B_{172}	VLEGMEVV	[31]
$SART3_{109}$	VYDYNCHVDL	[26]	lck_{246}	KLVERLGAA	[33]
$SART3_{315}$	AYIDFEMKI	[26]	lck_{422}	DVWSFGILL	[33]
$CypB_{84}$	KFHRVIKDF	[30]	$EIF4E\text{-}BP_{51}$	RIIYDRKFL	[34]
Cyp B_{91}	DFMIQGGDF	[30]	$ppMAPkkk_{294}$	GLLFLHTRT	[34]
lck_{208}	HYTNASDGL	[32]	$ppMAPkkk_{432}$	DLLSHAFFA	[34]
lck_{486}	TFDYLRSVL	[32]	$WHSC2_{103}$	ASLDSDPWV	[34]
lck_{488}	DYLRSVLEDF	[32]	$WHSC2_{141}$	ILGELREKV	[34]
$ART1_{170}$	EYCLKFTKL	[28]	$UBE2V_{43}$	RLQEWCSVI	[34]
$ART4_{13}$	AFLRHAAL	[29]	$UBE2V_{85}$	LIADFLSGL	[34]
$ART4_{75}$	DYPSLSATDI	[29]	$UBE2V_{208}$	ILPRKHHRI	[34]
			$HNRPL_{141}$	ALVEFEDVL	[34]
			$HNRPL_{501}$	NVLHFFNAPL	[34]

trial is the-peptide-specific CTLprecursor-oriented peptide vaccines for HLA-A24 and -A2 cancer patients with 14 and 16 different peptides, respectively. A list of peptides is given in Table 5.4. An important clinical endpoint of these translational studies will be determining the correlation between immune response and overall survival to clarify whether augmented peptide-induced immunity can provide a clinical benefit in order to develop an appropriate protocol of therapeutic cancer vaccine.

References

[1] Boon, T., Coulie, P. G. and Van den Eynde, B. (1997) Tumor antigens recognized by T cells. *Immunology Today*, **18**, 267–268.

[2] Cox, A. L., Skipper, J., Chen, Y., Henderson, R. A., Darrow, T. L., Shabanowitz, J. *et al.* (1994) Identification of a peptide recognized by five melanoma-specific human cytotoxic T cell lines. *Science*, **264**, 716–719.

[3] Traversari, C., van der Bruggen, P., Luescher, I. F., Lurquin, C., Chomez, P., Van Pel, A. *et al.* (1992) A nonapeptide encoded by human gene *MAGE-1* is recognized on HLA-A1 by cytolytic T lymphocytes directed against tumor antigen MZ2-E. *Journal of Experimental Medicine*, **176**, 1453–1457.

[4] Chaux, P., Luiten, R., Demotte, N., Vantomme, V., Stroobant, V., Traversari, C. *et al.* (1999) Identification of five MAGE-A1 epitopes recognized by cytolytic T lymphocytes obtained by *in vitro* stimulation with dendritic cells transduced with MAGE-A1. *Journal of Immunology*, **163**, 2928–2936.

[5] Fujie, T., Tahara, K., Tanaka, F., Mori, M., Takesako, K. and Akiyoshi, T. (1999) A MAGE-1-encoded HLA-A24-binding synthetic peptide induces specific anti-tumor cytotoxic T lymphocytes. *International Journal of Cancer*, **80**, 169–172.

[6] Tanzarella, S., Russo, V., Lionello, I., Dalerba, P., Rigatti, D., Bordignon, C. *et al.* (1999) Identification of a promiscuous T-cell epitope encoded by multiple members of the MAGE family. *Cancer Research*, **59**, 2668–2674.

[7] van der Bruggen, P., Szikora, J. P., Boel, P., Wildmann, C., Somville, M., Sensi, M. *et al.* (1994) Autologous cytolytic T lymphocytes recognize a MAGE-1 nonapeptide on melanomas expressing HLA-Cw*1601. *European Journal of Immunology*, **24**, 2134–2140.

[8] Kawashima, I., Hudson, S.J., Tsai, V., Southwood, S., Takesako, K., Appella, E. *et al.* (1998) The multi-epitope approach for immunotherapy for cancer: identification of several CTL epitopes from various tumor-associated antigens expressed on solid epithelial tumors. *Human Immunology*, **59**, 1–14.

[9] Tahara, K., Takesako, K., Sette, A., Celis, E., Kitano, S. and Akiyoshi, T. (1999) Identification of a MAGE-2-encoded human leukocyte antigen-A24-binding synthetic peptide that induces specific antitumor cytotoxic T lymphocytes. *Clinical Cancer Research*, **5**, 2236–2241.

[10] Schultz, E. S., Zhang, Y., Knowles, R., Tine, J., Traversari, C., Boon, T. *et al.* (2001) A MAGE-3 peptide recognized on HLA-B35 and HLA-A1 by cytolytic T lymphocytes. *Tissue Antigens*, **57**, 103–109.

[11] van der Bruggen, P., Bastin, J., Gajewski, T., Coulie, P. G., Boel, P., De Smet, C. *et al.* A peptide encoded by human gene *MAGE-3* and presented by HLA-A2 induces cytolytic T lymphocytes that recognize tumor cells expressing MAGE-3. *European Journal of Immunology*, **24**, 3038–3043.

[12] Tanaka, F., Fujie, T., Tahara, K., Mori, M., Takesako, K., Sette, A. *et al.* (1997) Induction of antitumor cytotoxic T lymphocytes with a MAGE-3-encoded synthetic peptide presented by human leukocytes antigen-A24. *Cancer Research*, **57**, 4465–4468.

[13] Oiso, M., Eura, M., Katsura, F., Takiguchi, M., Sobao, Y., Masuyama, K. *et al.* (1999) A newly identified MAGE-3-derived epitope recognized by HLA-A24-restricted cytotoxic T lymphocytes. *International Journal of Cancer*, **81**, 387–394.

[14] Herman, J., van der Bruggen, P., Luescher, I. F., Mandruzzato, S., Romero, P., Thonnard, J. *et al.* (1996) A peptide encoded by the human *MAGE3* gene and presented by HLA-B44 induces cytolytic T lymphocytes that recognize tumor cells expressing MAGE3. *Immunogenetics*, **43**, 377–383.

[15] Huang, L. Q., Brasseur, F., Serrano, A., De Plaen, E., van der Bruggen, P., Boon, T. *et al.* (1999) Cytolytic T lymphocytes recognize an antigen encoded by MAGE-A10 on a human melanoma. *Journal of Immunology*, **162**, 6849–6854.

[16] Panelli, M. C., Bettinotti, M. P., Lally, K., Ohnmacht, G. A., Li, Y. and Robbins, P. *et al.* (2000) A tumor-infiltrating lymphocyte from a melanoma metastasis with decreased expression of melanoma differentiation antigens recognizes MAGE-12. *Journal of Immunology*, **164**, 4382–4392.

[17] Boel, P., Wildmann, C., Sensi, M. L., Brasseur, R., Renauld, J. C., Coulie, P. *et al.* (1995) *BAGE*: a new gene encoding an antigen recognized on human melanomas by cytolytic T lymphocytes. *Immunity*, **2**, 167–175.

[18] Van den Eynde, B., Peeters, O., De Backer, O., Gaugler, B., Lucas, S., Boon, T. (1995) A new family of genes coding for an antigen recognized by autologous cytolytic T lymphocytes on a human melanoma. *Journal of Experimental Medicine*, **182**, 689–698.

[19] Jager, E., Chen, Y. T., Drijfhout, J. W., Karbach, J., Ringhoffer, M., Jager, D. *et al.* (1998) Simultaneous humoral and cellular immune response against cancer–testis antigen NY-ESO-1: definition of human histocompatibility leukocyte antigen (HLA)-A2-binding peptide epitopes. *Journal of Experimental Medicine*, **187**, 265–270.

[20] Wang, R. F., Johnston, S. L., Zeng, G., Topalian, S. L., Schwartzentruber, D. J. and Rosenberg, S. A. (1998) A breast and melanoma-shared tumor antigen: T cell responses to antigenic peptides translated from different open reading frames. *Journal of Immunology*, **161**, 3598–3606.

[21] Ikeda, H., Lethe, B., Lehmann, F., van Baren, N., Baurain, J. F., de Smet, C. *et al.* (1997) Characterization of an antigen that is recognized on a melanoma showing partial HLA loss by CTL expressing an NK inhibitory receptor. *Immunity*, **6**, 199–208.

[22] Kessler, J. H., Beekman, N. J., Bres-Vloemans, S. A., Verdijk, P., van Veelen, P. A., Kloosterman-Joosten, A. M. *et al.* (2001) Efficient identification of novel HLA-A(*)0201-presented cytotoxic T lymphocyte epitopes in the widely expressed tumor antigen PRAME by proteasome-mediated digestion analysis. *Journal of Experimental Medicine*, **193**, 73–88.

[23] Shichijo, S., Nakao, M., Imai, Y., Takasu, H., Kawamoto, M., Niiya, F. *et al.* (1998) A gene encoding antigenic peptides of human squamous cell carcinoma recognized by cytotoxic T lymphocytes. *Journal of Experimental Medicine*, **187**, 277–288.

[24] Kikuchi, M., Nakao, M., Inoue, Y., Matsunaga, K., Shichijo, S., Yamana, H. *et al.* (1999) Identification of a SART-1-derived peptide capable of inducing HLA-A24-restricted and tumor-specific cytotoxic T lymphocytes. *International Journal of Cancer*, **81**, 459–466.

[25] Nakao, M., Shichijo, S., Imaizumi, T., Inoue, Y., Matsunaga, K., Yamada, A. *et al.* (2000) Identification of a gene coding for a new squamous cell carcinoma antigen recognized by the CTL. *Journal of Immunology*, **164**, 2565–2574.

[26] Yang, D., Nakao, M., Shichijo, S., Sasatomi, T., Takasu, H., Matsumoto, H. *et al.* (1999) Identification of a gene coding for a protein possessing shared tumor epitopes capable of inducing HLA-A24-restricted cytotoxic T lymphocytes in cancer patients. *Cancer Research*, **59**, 4056–4063.

[27] Ito, M., Shichijo, S., Miyagi, Y., Kobayashi, T., Tsuda, N., Yamada, A. *et al.* (2000) Identification of SART3-derived peptides capable of inducing HLA-A2-restricted and tumor-specific CTLs in cancer patients with different HLA-A2 subtypes. *International Journal of Cancer*, **88**, 633–639.

[28] Nishizaka, S., Gomi, S., Harada, K., Oizumi, K., Itoh, K. and Shichijo, S. (2000) A new tumor-rejection antigen recognized by cytotoxic T lymphocytes infiltrating into a lung adenocarcinoma. *Cancer Research*, **60**, 4830–4837.

[29] Kawano, K., Gomi, S., Tanaka, K., Tsuda, N., Kamura, T., Itoh, K. *et al.* (2000) Identification of a new endoplasmic reticulum-resident protein recognized by HLA-A24-restricted tumor-infiltrating lymphocytes of lung cancer. *Cancer Research*, **60**, 3550–3558.

[30] Gomi, S., Nakao, M., Niiya, F., Imamura, Y., Kawano, K., Nishizaka, S. *et al.* (1999) A *cyclophilin B* gene encodes antigenic epitopes recognized by HLA-A24-restricted and tumor-specific CTLs. *Journal of Immunology*, **163**, 4994–5004.

[31] Tamura, M., Nishizaka, S., Maeda, Y., Ito, M., Harashima, N., Harada, M. *et al.* (2001) Identification of cyclophilin B-derived peptides capable of inducing histocompatibility leukocyte antigen-A2-restricted and tumor-specific cytotoxic T lymphocytes. *Japanese Journal of Cancer Research*, **92**, 762–767.

[32] Harashima, N., Tanaka, K., Sasatomi, T., Shimizu, K., Miyagi, Y., Yamada, A. *et al.* (2001) Recognition of the Lck tyrosine kinase as a tumor antigen by cytotoxic T lymphocytes of cancer patients with distant metastases. *European Journal of Immunology*, **31**, 323–332.

[33] Imai, N., Harashima, N., Ito, M., Miyagi, Y., Harada, M., Yamada, A. *et al.* (2001) Identification of Lck-derived peptides capable of inducing HLA-A2-restricted and tumor-specific CTLs in cancer patients with distant metastases. *International Journal of Cancer*, **94**, 237–242.

[34] Ito, M., Shichijo, S., Tsuda, N., Ochi, M., Harashima, N., Saito, N. *et al.* (2001) Molecular basis of T cell-mediated recognition of pancreatic cancer cells. *Cancer Research*, **61**, 2038–2046.

[35] Yamada, A., Kawano, K., Koga, M., Matsumoto, T. and Itoh, K. (2001) Multidrug resistance-associated protein 3 (MRP3) is a tumor rejection antigen recognized by HLA-A2402-restricted cytotoxic T lymphocytes, *Cancer Research*, **61**, 6459–6466.

[36] Okugawa, T., Ikuta, Y., Takahashi, Y., Obata, H., Tanida, K., Watanabe, M. *et al.* (2000) A novel human HER2-derived peptide homologous to the mouse K(d)-restricted tumor rejection antigen can induce HLA-A24-restricted cytotoxic T lymphocytes in ovarian cancer patients and healthy individuals. *European Journal of Immunology*, **30**, 3338–3346.

[37] Disis, M. L., Smith, J. W., Murphy, A. E., Chen, W. and Cheever, M. A. (1994) *In vitro* generation of human cytolytic T-cells specific for peptides derived from the HER-2/neu protooncogene protein. *Cancer Research*, **54**, 1071–1076.

[38] Fisk, B., Blevins, T. L., Wharton, J. T. and Ioannides, C. G. (1995) Identification of an immunodominant peptide of HER-2/neu protooncogene recognized by ovarian tumor-specific cytotoxic T lymphocyte lines. *Journal of Experimental Medicine*, **181**, 2109–2117.

[39] Peiper, M., Goedegebuure, P. S., Linehan, D. C., Ganguly, E., Douville, C. C. and Eberlein, T. J. (1997) The HER2/neu-derived peptide p654–662 is a tumor-associated antigen in human pancreatic cancer recognized by cytotoxic T lymphocytes. *European Journal of Immunology*, **27**, 1115–1123.

[40] Rongcun, Y., Salazar-Onfray, F., Charo, J., Malmberg, K. J., Evrin, K., Maes, H. *et al.* (1999) Identification of new HER2/neu-derived peptide epitopes that can elicit specific CTL against autologous and allogeneic carcinomas and melanomas. *Journal of Immunology*, **163**, 1037–1044.

[41] Kono, K., Rongcun, Y., Charo, J., Ichihara, F., Celis, E., Sette, A. *et al.* (1998) Identification of HER2/neu-derived peptide epitopes recognized by gastric cancer-specific cytotoxic T lymphocytes. *International Journal of Cancer*, **78**, 202–208.

[42] Kawashima, I., Tsai, V., Southwood, S., Takesako, K., Sette, A. and Celis, E. (1999) Identification of HLA-A3-restricted cytotoxic T lymphocyte epitopes from carcinoembryonic antigen and HER-2/neu by primary *in vitro* immunization with peptide-pulsed dendritic cells. *Cancer Research*, **59**, 431–435.

[43] Umano, Y., Tsunoda, T., Tanaka, H., Matsuda, K., Yamaue, H. and Tanimura, H. (2001) Generation of cytotoxic T cell responses to an HLA-A24 restricted epitope peptide derived from wild-type p53. *British Journal of Cancer*, **84**, 1052–1057.

[44] Nijman, H. W., Houbiers, J. G., van der Burg, S. H., Vierboom, M. P., Kenemans, P., Kast, W. M. *et al.* (1993) Characterization of cytotoxic T lymphocyte epitopes of a self-protein, p53, and a non-self-protein, influenza matrix: relationship between major histocompatibility

complex peptide binding affinity and immune responsiveness to peptides. *Journal of Immunotherapy*, **14**, 121–126.

[45] Arai, J., Yasukawa, M., Ohminami, H., Kakimoto, M., Hasegawa, A. and Fujita, S. (2001) Identification of human telomerase reverse transcriptase-derived peptides that induce HLA-A24-restricted antileukemia cytotoxic T lymphocytes. *Blood*, **97**, 2903–2907.

[46] Vonderheide, R. H., Hahn, W. C., Schultze, J. L. and Nadler, L. M. (1999) The telomerase catalytic subunit is a widely expressed tumor-associated antigen recognized by cytotoxic T lymphocytes. *Immunity*, **10**, 673–679.

[47] Probst-Kepper, M., Stroobant, V., Kridel, R., Gaugler, B., Landry, C., Brasseur, F. *et al.* (2001) An alternative open reading frame of the human macrophage colony-stimulating factor gene is independently translated and codes for an antigenic peptide of 14 amino acids recognized by tumor-infiltrating CD8 T lymphocytes. *Journal of Experimental Medicine*, **193**, 1189–1198.

[48] Schmitz, M., Diestelkoetter, P., Weigle, B., Schmachtenberg, F., Stevanovic, S., Ockert, D. *et al.* (2000) Generation of survivin-specific CD8+ T effector cells by dendritic cells pulsed with protein or selected peptides. *Cancer Research*, **60**, 4845–4849.

[49] Kim, C., Matsumura, M., Saijo, K. and Ohno, T. (1998) *In vitro* induction of HLA-A2402-restricted and carcinoembryonic-antigen-specific cytotoxic T lymphocytes on fixed autologous peripheral blood cells. *Cancer Immunological Immunotherapy*, **47**, 90–96.

[50] Nukaya, I., Yasumoto, M., Iwasaki, T., Ideno, M., Sette, A., Celis, E. *et al.* (1999) Identification of HLA-A24 epitope peptides of carcinoembryonic antigen which induce tumor-reactive cytotoxic T lymphocyte. *International Journal of Cancer*, **80**, 92–97.

[51] Butterfield, L. H., Koh, A., Meng, W., Vollmer, C. M., Ribas, A., Dissette, V. *et al.* (1999) Generation of human T-cell responses to an HLA-A2.1-restricted peptide epitope derived from alpha-fetoprotein. *Cancer Research*, **59**, 3134–3142.

[52] Heukamp, L. C., van der Burg, S. H., Drijfhout, J. W., Melief, C. J., Taylor-Papadimitriou, J. and Offringa, R. (2001) Identification of three non-VNTR MUC1-derived HLA-A*0201-restricted T-cell epitopes that induce protective anti-tumor immunity in HLA-A2/K(b)-transgenic mice. *International Journal of Cancer*, **91**, 385–392.

[53] Bohm, C. M., Hanski, M. L., Stefanovic, S., Rammensee, H. G., Stein, H., Taylor-Papadimitriou, J. *et al.* (1998) Identification of HLA-A2-restricted epitopes of the tumor-associated antigen MUC2 recognized by human cytotoxic T cells. *International Journal of Cancer*, **75**, 688–693.

[54] Correale, P., Walmsley, K., Nieroda, C., Zaremba, S., Zhu, M., Schlom, J. *et al.* (1997) *In vitro* generation of human cytotoxic T lymphocytes specific for peptides derived from prostate-specific antigen. *Journal of the National Cancer Institute*, **89**, 293–300.

[55] Lodge, P. A., Jones, L. A., Bader, R. A., Murphy, G. P. and Salgaller, M. L. (2000) Dendritic cell-based immunotherapy of prostate cancer: immune monitoring of a phase II clinical trial. *Cancer Research*, **60**, 829–833.

[56] Dannull, J., Diener, P. A., Prikler, L., Furstenberger, G., Cerny, T., Schmid, U. *et al.* (2000) Prostate stem cell antigen is a promising candidate for immunotherapy of advanced prostate cancer. *Cancer Research*, **60**, 5522–5528.

[57] Inoue, Y., Katou, K., Takei, M., Tobisu, K., Takaue, Y., Kakizoe, T. *et al.* (2001) Induction of the tumor specific cytotoxic T lymphocyte from prostate cancer patients using prostatic acid phosphate(PAP)derived HLA-A2402 binding peptide. *Journal of Urology*, **166**, 1508–1513.

[58] Maeda, A., Ohguro, H., Nabeta, Y., Hirohashi, Y., Sahara, H., Maeda, T. *et al.* (2001) Identification of human antitumor cytotoxic T lymphocytes epitopes of recoverin, a cancer-associated retinopathy antigen, possibly related with a better prognosis in a paraneoplastic syndrome. *European Journal of Immunology*, **31**, 563–572.

[59] Vissers, J. L., De Vries, I. J., Schreurs, M. W., Engelen, L. P., Oosterwijk, E.,Figdor, C. G. *et al.* (1999) Adema GJ. The renal cell carcinoma-associated antigen G250 encodes a human leukocyte antigen (HLA)-A2.1-restricted epitope recognized by cytotoxic T lymphocytes. *Cancer Research*, **59**, 5554–5559.

[60] Flad, T., Spengler, B., Kalbacher, H., Brossart, P., Baier, D., Kaufmann, R. *et al.* (1998) Direct identification of major histocompatibility complex class I-bound tumor-associated peptide antigens of a renal carcinoma cell line by a novel mass spectrometric method. *Cancer Research*, **58**, 5803–5811.

[61] Ronsin, C., Chung-Scott, V., Poullion, I., Aknouche, N., Gaudin, C. and Triebel, F. (1999) A non-AUG-defined alternative open reading frame of the intestinal carboxyl esterase mRNA generates an epitope recognized by renal cell carcinoma-reactive tumor-infiltrating lymphocytes *in situ*. *Journal of Immunology*, **163**, 483–490.

[62] Mandruzzato, S., Stroobant, V., Demotte, N. and van der Bruggen, P. (2000) A human CTL recognizes a caspase-8-derived peptide on autologous HLA-B*3503 molecules and two unrelated peptides on allogeneic HLA-B*3501 molecules. *Journal of Immunology*, **164**, 4130–4134.

[63] Echchakir, H., Mami-Chouaib, F., Vergnon, I., Baurain, J. F., Karanikas, V., Chouaib, S. *et al.* (2001) A point mutation in the *alpha-actinin-4* gene generates an antigenic peptide recognized by autologous cytolytic T lymphocytes on a human lung carcinoma. *Cancer Research*, **61**, 4078–4083.

[64] Karanikas, V., Colau, D., Baurain, J. F., Chiari, R., Thonnard, J., Gutierrez-Roelens, I. *et al.* (2001) High frequency of cytolytic T lymphocytes directed against a tumor-specific mutated antigen detectable with HLA tetramers in the blood of a lung carcinoma patient with long survival. *Cancer Research*, **61**, 3718–3724.

[65] Gaudin, C., Kremer, F., Angevin, E., Scott, V. and Triebel, F. (1999) A hsp70-2 mutation recognized by CTL on a human renal cell carcinoma. *Journal of Immunology*, **162**, 1730–1738.

[66] Murakami, M., Gurski, K. J., Marincola, F. M., Ackland, J. and Steller, M. A. (1999) Induction of specific CD8+ T-lymphocyte responses using a human papillomavirus-16 E6/E7 fusion protein and autologous dendritic cells. *Cancer Research*, **59**, 1184–1187.

[67] Khanna, R., Burrows, S. R., Nicholls, J. and Poulsen, L. M. (1998) Identification of cytotoxic T cell epitopes within Epstein–Barr virus (EBV) oncogene latent membrane protein 1 (LMP1): evidence for HLA A2 supertype-restricted immune recognition of EBV-infected cells by LMP1-specific cytotoxic T lymphocytes. *European Journal of Immunology*, **28**, 451–458.

[68] Lee, S. P., Thomas, W. A., Murray, R. J., Khanim, F., Kaur, S., Young, L. S. *et al.* (1993) HLA A2.1-restricted cytotoxic T cells recognizing a range of Epstein–Barr virus isolates through a defined epitope in latent membrane protein LMP2. *Journal of Virology*, **67**, 7428–7435.

[69] Brooks, J. M., Murray, R. J., Thomas, W. A., Kurilla, M. G. and Rickinson, A. B. (1993) Different HLA-B27 subtypes present the same immunodominant Epstein–Barr virus peptide. *Journal of Experimental Medicine*, **178**, 879–887.

[70] Hill, A., Worth, A., Elliott, T., Rowland-Jones, S., Brooks, J., Rickinson, A. *et al.* (1995) Characterization of two Epstein–Barr virus epitopes restricted by HLA-B7. *European Journal of Immunology*, **25**, 18–24.

[71] Dreyfuss, G., Choi, Y. D. and Adam, S. A. (1989) The ribonucleoprotein structures along the pathway of mRNA formation. *Endocrine Research*, **15**, 441–474.

[72] Pinol-Roma, S., Swanson, M. S., Gall, J. G. and Dreyfuss, G. (1989) A novel heterogeneous nuclear RNP protein with a unique distribution on nascent transcripts. *Journal of Cell Biology*, **109**, 2575–2587.

[73] Shih, S. C. and Claffey, K. P. (1999) Regulation of human vascular endothelial growth factor mRNA stability in hypoxia by heterogeneous nuclear ribonucleoprotein L. *Journal of Biological Chemistry*, **274**, 1359–1365.

[74] Wright, T. J., Costa, J. L., Naranjo, C., Francis-West, P. and Altherr, M. R. (1999) Comparative analysis of a novel gene from the Wolf-Hirschhorn/Pitt-Rogers-Danks syndrome critical region. *Genomics*, **59**, 203–212.

[75] Pause, A., Belsham, G. J., Gingras, A. C., Donze, O., Lin, T. A., Lawrence, J. C. Jr *et al.* (1994) Insulin-dependent stimulation of protein synthesis by phosphorylation of a regulator of 5′-cap function. *Nature*, **371**, 762–767.

[76] Hovnanian, A., Rebouillat, D., Mattei, M. G., Levy, E. R., Marie, I., Monaco, A. P. *et al.* (1998) Hovanessian AG. The human 2′,5′-oligoadenylate synthetase locus is composed of three distinct genes clustered on chromosome 12q24.2 encoding the 100-, 69-, and 40-kDa forms. *Genomics*, **52**, 267–277.

[77] Rebouillat, D., Hovnanian, A., Marie, I. and Hovanessian, A. G. (1999) The 100-kDa 2′,5′-oligoadenylate synthetase catalyzing preferentially the synthesis of dimeric pppA2′p5′A molecules is composed of three homologous domains. *Journal of Biological Chemistry*, **274**, 1557–1565.

[78] Choi, Y. W., Kotzin, B., Herron, L., Callahan, J., Marrack, P. and Kappler, J. (1989) Interaction of *Staphylococcus aureus* toxin "superantigens" with human T cells. *Proceedings of the National Academy of Sciences of the USA*, **86**, 8941–8945.

[79] Seito, D., Morita, T., Masuoka, K., Maeda, T., Saya, H. and Itoh, K. (1994) Polyclonal uses of T-cell receptor (TCR) alpha and beta genes for cytotoxic T lymphocytes in human metastatic melanoma: possible involvement of TCR alpha in tumor-cell recognition. *International Journal of Cancer*, **58**, 497–502.

[80] Chothia, C., Boswell, D. R. and Lesk, A. M. (1988) The outline structure of the T-cell alpha beta receptor. *EMBO Journal*, **7**, 3745–3755.

[81] Nitta, T., Oksenberg, J. R., Rao, N. A. and Steinman, L. (1990) Predominant expression of T cell receptor V alpha 7 in tumor-infiltrating lymphocytes of uveal melanoma. *Science*, **249**, 672–674.

[82] Sensi, M., Salvi, S., Castelli, C., Maccalli, C., Mazzocchi, A., Mortarini, R. *et al.* (1993) T cell receptor (TCR) structure of autologous melanoma-reactive cytotoxic T lymphocyte (CTL) clones: tumor-infiltrating lymphocytes overexpress *in vivo* the TCR beta chain sequence used by an HLA-A2-restricted and melanocyte-lineage-specific CTL clone. *Journal of Experimental Medicine*, **178**, 1231–1246.

[83] Ferradini, L., Roman-Roman, S., Azocar, J., Avril, M. F., Viel, S., Triebel, F. *et al.* (1992) Analysis of T-cell receptor alpha/beta variability in lymphocytes infiltrating a melanoma metastasis. *Cancer Research*, **52**, 4649–4654.

[84] Sensi, M., Castelli, C., Anichini, A., Grossberger, D., Mazzocchi, A., Mortarini, R. *et al.* (1991) Two autologous melanoma-specific and MHC-restricted human T cell clones with identical intra-tumour reactivity do not share the same TCR V alpha and V beta gene families. *Melanoma Research*, **1**, 261–271.

[85] Romero, P., Pannetier, C., Herman, J., Jongeneel, C. V., Cerottini, J. C. and Coulie, P. G. (1995) Multiple specificities in the repertoire of a melanoma patient's cytolytic T lymphocytes directed against tumor antigen MAGE-1.A1. *Journal of Experimental Medicine*, **182**, 1019–1028.

[86] Sensi, M., Traversari, C., Radrizzani, M., Salvi, S., Maccalli, C., Mortarini, R. *et al.* (1995) Cytotoxic T-lymphocyte clones from different patients display limited T-cell-receptor variable-region gene usage in HLA-A2-restricted recognition of the melanoma antigen Melan-A/MART-1. *Proceedings of the National Academy of Sciences of the USA*, **92**, 5674–5678.

[87] Thomas, G. (1992) MAP kinase by any other name smells just as sweet. *Cell*, **68**, 3–6.

[88] Ikeda, H., Ohta, N., Furukawa, K., Miyazaki, H., Wang, L., Kuribayashi, K. *et al.* (1997) Mutated mitogen-activated protein kinase: a tumor rejection antigen of mouse sarcoma. *Proceedings of the National Academy of Sciences of the USA*, **94**, 6375–6379.

[89] Cohen, A. M. (1997) Cancer of the colon. In *Cancer Principles and Practice of Oncology*, edited by V. T. De Vita, Jr, S. Hellman and S. A. Rosenberg, pp. 1144–1197. Philadelphia: Lippincott-Raven Publishers.

[90] Joseph, G. and Tuckson, W. (1997) Colon Cancer. In *Oncology Evidence-based Management*, edited by B. Djulbegovic and D. M. Sullivan, pp. 205–209. Philadelphia: Curchill Livingstone.

[91] Browning, M. and Krausa, P. (1996) Genetic diversity of HLA-A2: evolutionary and functional significance. *Immunology Today*, **17**, 165–170.

[92] Loe, D. W., Deeley, R. G. and Cole, S. P. (1996) Biology of the multidrug resistance-associated protein, MRP. *European Journal of Cancer*, **32A**, 945–957.

[93] Borst, P., Evers, R., Kool, M. and Wijnholds, J. (2000) A family of drug transporters: the multidrug resistance-associated proteins. *Journal of the National Cancer Institute*, **92**, 1295–1302.

[94] Higgins, C. F. (1992) ABC transporters: from microorganisms to man. *Annual Review of Cell Biology*, **8**, 67–113.

[95] Kondo, A., Sidney, J., Southwood, S., del Guercio, M.F., Appella, E., Sakamoto, H. *et al.* (1995) Prominent roles of secondary anchor residues in peptide binding to HLA-A24 human class I molecules. *Journal of Immunology*, **155**, 4307–4312.

[96] Ishikawa, T., Li, Z. S., Lu, Y. P. and Rea, P. A. (1997) The GS-X pump in plant, yeast, and animal cells: structure, function, and gene expression. *Bioscience Reports*, **17**, 189–207.

[97] Young, L. C., Campling, B. G., Voskoglou-Nomikos, T., Cole, S. P., Deeley, R. G. and Gerlach, J. H. (1999) Expression of multidrug resistance protein-related genes in lung cancer: correlation with drug response. *Clinical Cancer Research*, **5**, 673–680.

[98] Kool, M., de Haas, M., Scheffer, G. L., Scheper, R. J., van Eijk, M. J., Juijn, J. A. *et al.* (1997) Analysis of expression of cMOAT (MRP2), MRP3, MRP4 and MRP5, homologues of the multidrug resistance-associated protein gene (MRP 1), in human cancer cell lines. *Cancer Research*, **57**, 3537–3547.

[99] Allikmets, R., Gerrard, B., Hutchinson, A. and Dean, M. (1996) Characterization of the human ABC superfamily: isolation and mapping of 21 new genes using the expressed sequence tags database. *Human Molecular Genetics*, **5**, 1649–1655.

[100] Cui, Y., Konig, J., Buchholz, J. K., Spring, H., Leier, I. and Keppler, D. (1999) Drug resistance and ATP-dependent conjugate transport mediated by the apical multidrug resistance protein, MRP2, permanently expressed in human and canine cells. *Molecular Pharmacology*, **55**, 929–937.

[101] Evers, R., Kool, M., van Deemter, L., Janssen, H., Calafat, J., Oomen, L. C. *et al.* (1998) Drug export activity of the human canalicular multispecific organic anion transporter in polarized kidney MDCK cells expressing cMOAT (MRP2) cDNA. *Journal of Clinical Investigation*, **101**, 1310–1319.

[102] Koike, K., Kawabe, T., Tanaka, T., Toh, S., Uchiumi, T., Wada, M. *et al.* (1997) A canalicular multispecific organic anion transporter (cMOAT) antisense cDNA enhances drug sensitivity in human hepatic cancer cells. *Cancer Research*, **57**, 5475–5479.

[103] Kool, M., van der Linden, M., de Haas, M., Scheffer, G. L., de Vree, J. M., Smith, A. J. *et al.* (1999) MRP3, an organic anion transporter able to transport anti-cancer drugs. *Proceedings of the National Academy of Sciences of the USA*, **96**, 6914–6919.

[104] Zeng, H., Bain, L. J., Belinsky, M. G. and Kruh, G. D. (1999) Expression of multidrug resistance protein-3 (multispecific organic anion transporter-D) in human embryonic kidney 293 cells confers resistance to anticancer agents. *Cancer Research*, **59**, 5964–5967.

[105] Schuetz, J. D., Connelly, M. C., Sun, D., Paibir, S. G., Flynn, P. M., Srinivas, R. V. *et al.* (1999) MRP4: a previously unidentified factor in resistance to nucleoside-based antiviral drugs. *Nature Medicine*, **5**, 1048–1051.

[106] Wijnholds, J., Mol, C. A., van Deemter, L., de Haas, M., Scheffer, G. L., Baas, F. *et al.* (2000) Multidrug-resistance protein 5 is a multispecific organic anion transporter able to transport nucleotide analogs. *Proceedings of the National Academy of Sciences of the USA*, **97**, 7476–7481.

[107] Pegram, M. D., Lipton, A., Hayes, D. F., Weber, B. L., Baselga, J. M., Tripathy, D. *et al.* (1998) Phase II study of receptor-enhanced chemosensitivity using recombinant humanized anti-p185HER2/neu monoclonal antibody plus cisplatin in patients with HER2/neu-overexpressing metastatic breast cancer refractory to chemotherapy treatment. *Journal of Clinical Oncology*, **16**, 2659–2671.

[108] Pegram, M., Hsu, S., Lewis, G., Pietras, R., Beryt, M., Sliwkowski, M. *et al.* (1999) Inhibitory effects of combinations of HER-2/neu antibody and chemotherapeutic agents used for treatment of human breast cancers. *Oncogene*, **18**, 2241–2251.

[109] Jager, D., Jager, E. and Knuth, A. (2001) Vaccination for malignant melanoma: recent developments. *Oncology*, **60**, 1–7.

[110] Rosenberg, S. A., Yang, J. C., Schwartzentruber, D. J., Hwu, P., Marincola, F. M., Topalian, S. L. *et al.* (1998) Immunologic and therapeutic evaluation of a synthetic peptide vaccine for the treatment of patients with metastatic melanoma. *Nature Medicine*, **4**, 321–327.

[111] Nestle, F. O., Alijagic, S., Gilliet, M., Sun, Y., Grabbe, S., Dummer, R. *et al.* (1998) Vaccination of melanoma patients with peptide- or tumor lysate-pulsed dendritic cells. *Nature Medicine*, **4**, 328–332.

[112] Marchand, M., van Baren, N., Weynants, P., Brichard, V., Dreno, B., Tessier, M. H. *et al.* (1999) Tumor regressions observed in patients with metastatic melanoma treated with an antigenic peptide encoded by gene *MAGE-3* and presented by HLA-A1. *International Journal of Cancer*, **80**, 219–230.

[113] Wang, F., Bade, E., Kuniyoshi, C., Spears, L., Jeffery, G., Marty, V. *et al.* (1999) Phase I trial of a MART-1 peptide vaccine with incomplete Freund's adjuvant for resected high-risk melanoma. *Clinical Cancer Research*, **5**, 2756–2765.

[114] Jager, E., Nagata, Y., Gnjatic, S., Wada, H., Stockert, E., Karbach, J. *et al.* (2000) Monitoring CD8 T cell responses to NY-ESO-1: correlation of humoral and cellular immune responses. *Proceedings of the National Academy of Sciences of the USA*, **97**, 4760–4765.

Chapter 6

Altered peptide ligands of tumor T-cell epitopes: implications for more effective vaccine therapy in human neoplasia

Licia Rivoltini, Matteo Carrabba, Lorenzo Pilla and Giorgio Parmiani

Summary

T lymphocytes can specifically recognize tumor cells, but this interaction is known to be rather inefficient *in vivo*. Among other mechanims, the inadequacy of tumor destruction by T cells could stem from suboptimal activation by antigenic determinants expressed by melanoma cells. T-cell receptor (TCR) is in fact a very flexible structure capable of interacting with a broad spectrum of ligands (i.e. HLA/peptide complexes) through a versatile signaling complex. T lymphocytes have thus no single ligand specificity, but recognize a large array of HLA/peptide complexes, termed "Altered Peptide Ligands" (APL). These APL can mediate a number of different outcomes in interacting T cells, ranging from inducing selective immunological functions (*partial agonist*) to completely turning off their functional capacity (*antagonist*). In addition to suboptimal ligands, TCR can also be triggered by superagonists, which are analogs that enhance T-cell stimulation by inducing immunological functions not detected with the cognate ligand. In this chapter we will discuss the rational for the usage of APL, that is, modified peptides from tumor antigens, to boost a potent anti-tumor T-cell reactivity cross-reacting with the native epitope expressed by cancer cell. Clinical results obtained with vaccination based on the usage of APL from tumor antigens will be additionally described. Finally, we will discuss the potential role of APL in inducing heterogeneous immune responses, including suppressive cytokine profiles, T-cell anergy and antigen-driven T-cell apoptosis, in anti-tumor lymphocytes.

Introduction

Due to the limited efficacy and the high toxicity of conventional treatments in certain human cancers, scientists have focused their efforts over the last two decades on the study of the interactions between cancer cells and immune system, in order to identify new and more "physiological" tools for controlling cancer growth *in vivo*. It is becoming clear that tumors, especially melanoma, do express antigens that can be recognized by T lymphocytes. These antigens have been recently identified as belonging to different protein families, including tumor-specific antigens, differentiation antigens and unique, mutated proteins (Renkvist *et al.*, 2001). With variations depending on their nature and immunogenicity, these antigens have been shown to be targets of specific anti-tumor T cells that can be found in tumor lesions, invaded or tumor-free lymph nodes and peripheral blood of cancer patients. Several data now support the evidence that melanoma patients have significantly higher frequency of such lymphocytes as compared to healthy donors (Marincola *et al.*, 1996b), thus suggesting that the immune system has been able to recognize neoplastic cells and to respond with specific T-cell expansion. What is still unclear is, however, the reason(s) why such immune reactions are, or become during the course of the disease, ineffective in mediating a significant impairment of *in vivo* tumor development. The limited results in terms of clinical efficacy obtained with the first attempts to vaccinate cancer patients with molecularly defined tumor antigens (see following section) further underline the possibility that interfering with cancer cell growth *in vivo* may not be as easy as it could be predicted from *in vitro* studies.

One of the new strategies recently tested in the search of more effective cancer immunotherapies is represented by the use of "modified peptides" for improving either HLA binding or TCR interaction (Parkhurst *et al.*, 1996; Valmori *et al.*, 1998a; Rivoltini *et al.*, 1999). Since most of the tumor antigens are self-proteins, low affinity T cells recognizing such antigens should be "turned off" by peripheral tolerance. The aim of any vaccine treatment is thus to break tolerance state and trigger powerful and effective immune responses able to control tumor growth *in vivo*.

This chapter will report about the rational for the usage of APL, that is, modified peptides from tumor antigens, that can boost a potent anti-tumor T-cell reactivity cross-reacting with the native epitope expressed by cancer cells. While once APL referred to "analogues of immunogenic peptides in which the TCR contact sites have been manipulated" (Evavold *et al.*, 1993), this term is currently used for indicating any modified peptide that can lead to modulation of T-cell activity. *In vitro* results available in different tumor antigen systems and preliminary clinical data will be discussed. On the basis of studies performed mostly by our group on the melanoma antigen MART-1, we will also discuss the possibility that APL may be involved in T-cell anergy induced *in situ* by tumor cells and the potential implications of such findings in the development of new and more effective strategies of immune intervention in cancer treatment.

The TCR: a flexible and tunable structure for antigen recognition by T cells

T cells are engaged in target recognition by the interaction of their TCR with a specific ligand composed by peptide-complexed HLA molecules. Classes I and II HLA are molecules responsible for binding peptides from endogenously or exogenously synthetized proteins and displaying them on the cell surface, for T-cell detection. Specific T cells can then respond to this interaction by producing various cytokines, killing antigen-bearing targets and/or activating other immune cells.

T-cell recognition was once considered as an "all or nothing" type event, but recent studies have shown that indeed T cells can respond to TCR triggering with several types of activation. TCR is in fact a very flexible structure capable of interacting with a broad spectrum of ligands (i.e. HLA/peptide complexes) through a versatile signaling complex (Hennecke and Wiley, 2001). Thus, the degree of T-cell responses ranging from full activation to partial activation, anergy or even apoptosis, is tuned by the fine structure of HLA/peptide complexes presented to the specific lymphocyte on target-cell surface. Antagonist, partial agonist and full T-cell agonist peptides define HLA-presented protein fragments capable of hierarchical recruitment of T-cell functions ranging from anergy induction to optimal T-cell response (Figure 6.1).

TCR flexibility has been found to play a central role in T-cell development (Kersh and Allen, 1996). The fate of developing thymocytes depends in fact on their level of avidity for HLA/peptide complexes expressed at thymic level. Developing T cells will undergo cell death if expressing TCR with no (positive selection) or high (negative selection) affinity for HLA/peptide ligands. The intrathymic negative selection process ensures the removal of

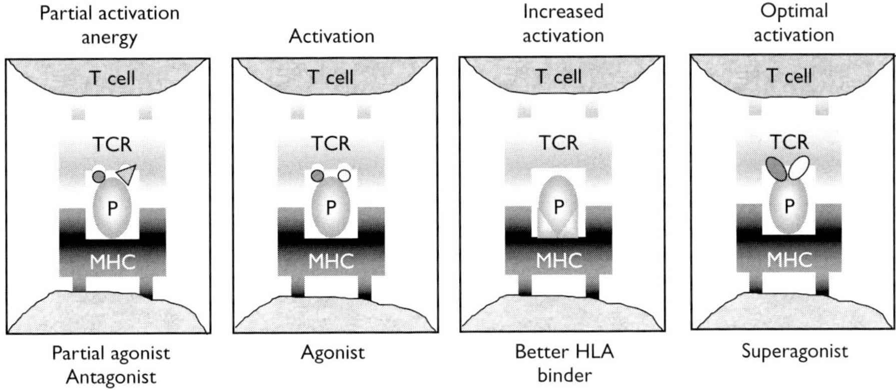

Figure 6.1 Altered peptide ligands and their differential activity on T cells.

high-avidity self-specific T cells from the T-cell repertoire, allowing thus immune tolerance to normal determinants. However, tolerance state is not complete but must be kept under control by different peripheral mechanisms (Marincola *et al.*, 2000). Powerful antigenic challenges, such as the ones occurring in the presence of strong inflammatory reactions, can then break peripheral tolerance and trigger low-avidity self-specific T cells, which may cause autoimmunity. Indeed, tumor immunotherapy is perhaps the one clinical setting in which the induction of an autoaggressive cellular response appears to be desirable (Parmiani, 1993). In fact, given the self-nature of most tumor antigens recognized by T cells, the possibility of breaking tolerance and activating immune responses against self-proteins overexpressed by tumor cells through vaccine-based strategies is being extensively evaluated.

Interaction between T cells and their targets: structure of the HLA/peptide complex

T cells interact with precise complexes expressed on tumor cell-surface, represented by short peptides (products of specific protein cellular processing) displayed on the binding groove of surface-expressed HLA molecules. CD8+ T cells recognize peptides of 8–10 amino acids in length, digested in proteasomes and presented *via* the endoplasmic reticulum on cell-surface class I HLA molecules. CD4 T cells interact with longer (13–20 aa) peptides mostly deriving from extracellular proteins digested at endosome level and presented on cell-surface in the context of HLA class II molecules.

HLA/peptide complex thus represents the minimal structure required to initiate, regulate and sustain a specific immune response and the peptide allocated inside the binding groove of a given HLA allele is the molecular determinant specifically recognized by the TCR of CD8 and CD4 T cells (Ruppert *et al.*, 1993; Sette *et al.*, 2001). The immunogenicity of a given peptide, that is, the capacity to stimulate a T-cell-mediated response *in vitro* and *in vivo*, is therefore directly affected by both its ability to bind the presenting HLA allele and by its capacity to stabilize the HLA/peptide complex with the TCR expressed by the T cells.

Amino acid sequencing of naturally processed peptides eluted from a given HLA allele revealed that each particular allelic variant efficiently binds only a subset of peptides sharing conserved amino acid residues in specific fixed positions (Ruppert *et al.*, 1993).

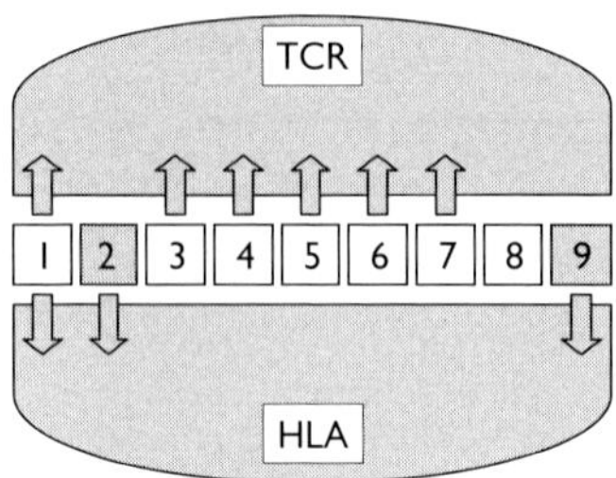

Figure 6.2 Interaction between T cells and their targets: structure of the HLA/peptide complex.

The position and the chemical nature of the amino acid directly affecting the ability of a given peptide to bind the corresponding HLA allele are defined as "peptide binding motif." The crystallographic structure of MHC molecules together with the allele polymorphism of HLA molecules provides the direct structural explanation for the existence of the peptide-binding motif (Ruppert *et al.*, 1993). In fact, the peptide binding groove of each MHC allele does contain a variable number of pockets differing in their chemical nature and ability to allocate distinctive side chains of the antigenic peptides. A network of multiple hydrogen bonds fastens the peptide backbone to the main groove of the MHC molecules, while the side chains of the anchor residues establish specific chemical links with the HLA positioned inside the binding pockets. In this field, HLA-A2.1–peptide binding requirements have been widely investigated and may represent a model for peptide/HLA interactions (Ruppert *et al.*, 1993). As shown in Figure 6.2, features required for binding to this allele of a putative nonamer include defined amino acid at anchor positions (P2 and P9) and only peptides with permissive residues (e.g. L, M at P2 or L, V, I at P9) at these positions have optimized HLA-A2.1 binding. Other residues, however, can modulate the efficacy of this interaction; for example, Y, F or W in P1 or P3 usually increase binding affinity.

The interaction between T cells and targets, which is regulated by the affinity of binding between the ligand (HLA/peptide) and its specific TCR, represents the other face of peptide requirements for immunogenicity. For the MHC class I presented peptides, the residues directly involved in the TCR triggering lay in the center of the peptide sequence, in amino acid positions included between P3 and P7. However, examples of amino acid substitutions at other positions (e.g. P1) have been shown to affect affinity of TCR binding, possibly through indirect effects of the tridimensional epitope structure (Rivoltini *et al.*, 1999). All the TCR studied so far have been found to bind to peptide–MHC complexes in a similar way, positioned across the HLA/peptide complex at an angle between 45° and 80° (Hennecke and Wiley, 2001). The similarity in binding mode is apparently achieved without using conserved contacts, which suggests its crucial importance in initiating signaling events in the responding T cells.

The functional outcome of T-cell interaction with HLA/peptide complex seems to be strictly related to half-life or dwell-time of TCR/HLA/peptide contact more than to macroscopic structural differences. Studies have shown that peptide substitutions inducing minor conformational refitting of the TCR/HLA/peptide complexes lead to very different functional outcomes, and no major differences in crystallographic structures of TCR/HLA/peptide interfaces of weak and strong agonist peptides can always be detected (Hennecke and Wiley, 2001; van der Merwe, 2001).

The recent identification of the so called "immunological synapse," the specialized junction between T cells and targets whose stability is a determinative event for T-cell activation, provides further confirmation to this hypothesis (Grakoui *et al.*, 1999; Lanzavecchia and Sallusto, 2001).

APL for finely tuning anti-tumor T-cell reactivity

T lymphocytes have thus no single ligand specificity, but recognize a large array of HLA/peptide complexes, termed "APL" (Loftus *et al.*, 1996, 1998; Sloan-Lancaster *et al.*, 1996). These APL can mediate a number of different outcomes in interacting T cells, ranging from inducing selective immunological functions (*partial agonist*) to completely turning off their functional capacity (*antagonist*). In addition to suboptimal ligands, TCR can also be triggered by superagonists, which are analogs that enhance T-cell stimulation by inducing immunological functions not detected with the cognate ligand (Figure 6.2) (Rivoltini *et al.*, 1999).

During the last few years different approaches have successfully led to the definition of tumor-associated antigens and to the identification of immunogenic peptides recognized by tumor-specific T cells (Rosenberg, 2001). The tumor-associated, T-cell defined epitopes known up to now include peptides mostly derived from normal proteins either selectively expressed or over-expressed by tumor cells. However, the role of tumor-derived antigenic peptides in inducing specific T-cell responses depends on various factors that altogether control what is termed "immunogenicity." It is a general belief that the failure of the immune system to control cancer growth *in vivo* stems, among other reasons, from the poor immunogenicity of natural epitopes expressed by tumor cells. With the only exception of the immunodominant MART-1_{27-35} peptide, which has been shown to readily induce specific T-cell reactivities both *in vitro* and *in vivo* (Kawakami *et al.*, 1994a,b), most tumor-derived epitopes require in fact repeated *in vitro* stimulations for boosting T-cell responses and show limited immunogenicity when used for patient vaccination (Salgaller *et al.*, 1995, 1996; Parkhurst *et al.*, 1996; Rivoltini *et al.*, 1996). This phenomenon has been associated to several mechanisms, including a limited HLA-binding ability of tumor-derived peptide or a low TCR affinity of anti-tumor T cells for the HLA/peptide complex (Marincola *et al.*, 2000).

In order to overcome the reduced immunogenicity, and thus the limited clinical efficacy, of natural tumor-derived peptides, the strategy consisting on slight modifications of peptide sequence at amino acid residues crucial for the interaction either with the HLA molecule or with the specific TCR has been recently evaluated both *in vitro* and *in vivo* (Table 6.1).

Peptide optimized for HLA binding or bioavailability

Most of the natural tumor-derived epitopes (including those derived from the melanoma antigen Melan A/MART-1 and other differentiation antigens) do not bind with high affinity to HLA-class I molecules, mainly due to the lack of optimal amino acidic residues at anchor positions (Parkhurst *et al.*, 1996; Valmori *et al.*, 1998a,b). Despite the fact that self-protein-derived peptides are believed to be specifically selected for intermediate binding affinity during immune system development (Sloan-Lancaster and Allen, 1996), attempts to improve HLA/peptide interaction by introducing single amino acid substitutions at anchor positions have thus been performed with the intent of increasing peptide immunogenicity (Ruppert *et al.*, 1993; Parkhurst *et al.*, 1996; Valmori *et al.*, 1998).

Table 6.1 Strategies for improving peptide immunogenicity by identifying altered peptide ligands

Approach	*Modifications*	*Examples*
Increase binding to HLA	P2, P9 (P1)[a]	MART-$1_{26\text{–}35}$2L gp100–209/2M
Improve fitting with TCR	P3, P5, P7 (P6, P1)[b]	CEA.691/5H MAGE-A3.112/5I, 7W MAGE-A2.157/5F, 5I CEA-CAP1/6D MART-$1_{27\text{–}35}$1L
Enhance bioavailability		
Reduce degradation by seric peptidases	C-terminal aminidation C-terminal PEGylation N-terminal acetylation	MART-1/MAGE
Improve peptide biostability	e.g. substitution of C residues	NY-ESO-1

Notes
a Primary anchor positions are in P2 and P9. P1 is a secondary position.
b "Odd-numbered position" rule refers to modifications at P3, 5 or 7. Other positions (such as P6 or even P1) can generate superagonist analogs when modified.

The MART-$1_{26\text{–}35}$2L analog (where the more hydrophobic Leu in pos 2 replaces the native Ala) (Valmori *et al.*, 1999) has been shown to display much improved HLA-A0201 binding and consequently more efficient generation of melanoma-specific T cells from peripheral blood of melanoma patients. 2L modified analog, as a consequence of the M introduction at P2, associates a better binding with a "zig-zag" HLA-A2.1–peptide complex configuration. This conformation probably reproduces native epitope-binding mode and contributes to cross-reactivity between MART-$1_{27\text{–}35}$ family peptides (Sliz *et al.*, 2001). Functionally, T cells generated with this modified epitope display improved induction of MART-1 specific T cells from patient PBMC in terms of lysis and IFNγ release. The 2L modification of MART-1 epitope has additionally provided advantages for the *in vitro* generation of immunological tools such as HLA-tetramers. These reagents, which allow cytofluorimetric detection of antigen-specific T cells by mimicking the HLA/peptide complex, can in fact be synthetized only in the presence of peptides with high and stable binding affinity for the HLA-molecule (Pittet *et al.*, 1999).

One of the HLA-A2-binding epitopes of the differentiation melanoma antigen gp100(209–217) has been also modified in P2 (modified at pos 2 with a Met substituting a Thr) (gp100–209/2M) for ameliorating binding affinity (Parkhurst *et al.*, 1996). As compared to the native peptide, gp100–209/2M mediates more efficient *in vitro* generation of melanoma-specific T cells, requiring shorter culture times for eliciting from peripheral blood of melanoma patients, T cells releasing IFNγ in response to gp100-expressing targets.

Using computer-prediction programs, modified epitopes with increased HLA-A2.1 binding have been identified also for the epithelial antigen EP-CAM. Introduction of a V at P9 in three EPCAM-derived HLA-A2.1-restricted peptides (YQLDPKFIT, ILYENNVIT and ILYENNVITI) increased HLA-binding affinity and ameliorated the generation of T cells with lytic activity against HLA-A2.1^+/EPCAM$^+$ tumor cell lines (Trojan *et al.*, 2001).

Another potential approach, aimed to improve antigenicity of low-affinity peptides and possibly identify cryptic epitopes of tumor antigens, consists in the introduction of an aromatic amino acid, such as Tyr, in the secondary anchor position 1 (P1Y), which has been proved to enhance HLA-A2 binding and stability in several tumor peptides, including those derived from her2-neu (Tourdot *et al.*, 2000).

Enhanced immunogenicity could additionally be achieved by improving bioavailability and stability *in vivo*. For example, MART-1 and MAGE1-derived peptides were modified by introducing terminal modifications (C-terminal aminidation or PEGylation and N-terminal acetylation) designed to inhibit proteolytic degradation by seric peptidases and thus enhance *in vivo* peptide stability, without interference with CTL recognition (Brinckerhoff *et al.*, 1999). Similar increases of peptide stability and thus immunogenicity have also been reported to occur through substitution of cystein residues (known to easily dimerize with consequent reduction of bioavailability) in peptides derived from the tumor antigen NY-ESO-1 (Chen *et al.*, 2000).

Peptide optimized at TCR contact residues

A more efficient interaction of the HLA/peptide complex with specific T cells can furthermore be achieved by modifying individual amino-acidic residues not involved in the peptide-MHC bond and thus expected to directly or indirectly interact with the TCR. This strategy, though more complex to be applied due to the difficulty of predicting crucial residues for TCR/peptide MHC complexes, allows the identification of heteroclitic peptides also called "enhancer agonist" or "superagonist" analogs, that is, epitopes capable of fully triggering specific T cells to optimally recognize native peptides expressed by tumor cells. This approach, which is not due to better HLA binding, has the theoretical advantage of inducing qualitatively improved immune responses that can include production of cytokines, such as IL-2, crucial for *in vivo* lymphocyte survival and expansion (Tangri *et al.*, 2001). In experimental models, heteroclitic analogs have been shown to break/overcome tolerance by reversing a state of T-cell anergy or recruiting new T-cell specificities (Zugel *et al.*, 1998; Wang *et al.*, 1999). In addition, heteroclitic variants may allow to finely modulate cytokine production profiles in response to certain antigen, and thus be useful in several disease, such as cancer, where generation of specific subsets of Th cell response is required (Slanky *et al.*, 2000).

More recently, some discrete structural features associated with heteroclicity have been identified and used for rational design of heteroclitic analogs of known HLA-A2.1 restricted tumor epitopes (Tangri *et al.*, 2001). It has been proven that the occurrence of conservative or semiconservative substitutions at odd-numbered positions (P3, P5 and P7) in the middle of peptide is a feature associated with heteroclitic effects for HLA-A2.1 epitopes. Since heteroclicity is ultimately dependent on the orientation of specific peptide side chains in the structure of the trimeric complex (TCR/peptide–MHC), insertion of substitutions at odd-numbered positions represents an empirical guide for identification of superagonist analogs that enhance T-cell activation. These observations are in agreement with the report that minor structural changes at TCR/peptide MHC interface are responsible for major difference in TCR triggering. Such rational approach to heteroclitic analog design have been useful for identification of six HLA-A2.1 tumor epitope heteroclitic analogs, including M3 and H5 variants for CEA.691 peptide, I5 and W7 for MAGE-A3.112, F5 and I5 for MAGE-A2.157 (Tangri *et al.*, 2001).

Although this strategy appears to be very efficient in detecting superagonist variants of given HLA-A2.1 epitopes, the role of "central odd positions" should not be considered as exclusive since other substitutions, such as the ones in P1 or P6 may result in heteroclitic activity as well (Zaremba *et al.*, 1997; Rivoltini *et al.*, 1999).

Indeed, the heteroclitic analog of CAP1 peptide (CAP1-6D) has been obtained from the CEA-derived HLA-A2 binding peptide 605–613 by introducing an Asp in P6 that results in as much as a 1000-fold increase in the levels of cytokine released as compared to the native peptide (Zaremba *et al.*, 1997). This peptide shows clear superagonist activity, mediating significantly improved potency in the generation of CTL recognizing CEA+ targets as compared to the native peptide.

We have described a superagonist variant of the immunodominant peptide MART-1_{27-35}, which contains a Leu in pos 1, replacing an Ala (27–35/1L). This modification, though not directly involving TCR/peptide contact residue, induces qualitatively and quantitatively improved anti-MART-1 T-cell response, without any enhancement in HLA-binding affinity or stability, and it is thus believed to improve immunogenicity through a more efficient interaction at TCR level (Rivoltini *et al.*, 1999). Specific T cells generated by *in vitro* sensitization with the 27–35/1L superagonists were also found to release high amounts of IL-2 in response to MART-1+ melanoma cells, a feature that could not be observed in T cells raised with the native peptide.

Potential limitations to the use of tumor antigen-derived APL

The strategy of utilizing optimized peptide analogs for up-regulating and potentiating anti-tumor T-cell responses is thus considered a new tool for the identification of more immunogenic and therapeutically effective cancer vaccines. However, the TCR is a very sensitive structure that can be finely tuned by the HLA/peptide complex, with multiple and heterogeneous functional effects, ranging from optimal activation to T-cell anergy and antigen-driven T-cell apoptosis (Hennecke and Wiley, 2001).

A common assumption about peptide binding to class I MHC is that each residue binds independently. In addition, it is frequently assumed that anchor substitutions do not affect TCR contact residues. However, crystal structures of Her2-neu peptide APL showed that the central residues change position depending on the identity of the anchor residue(s). Thus, it is clear that subtle changes in the identity of anchor residues may have significant effects on the positions of the TCR contact residues, and on T-cell function (Sharma *et al.*, 2001). Results obtained analyzing clonal T-cell responses induced by the modified gp100–209/2M (analog with increased HLA binding) showed in fact high variability both in terms of avidity toward the peptide and recognition of tumor cells (Dudley *et al.*, 1999). This evidence suggests that modifications at HLA contact residues can indeed alter TCR contacts and lead to the potential expansion of T-cell populations not necessarily able to homogeneously cross-react with the native epitope expressed by tumor cells.

Single amino acid substitution can not only induce improved T-cell responses, but also trigger a vast array of immunological functions, which may influence negatively the final outcome of a certain immune response. Specific substitutions aimed to improved HLA binding of MART-1 immunodominant peptides (e.g. E to F at pos 26 of the MART-1_{26-35}) have been shown to enhance production of type 2 cytokines (such as IL-10 and IL-13), whose effects on the *in vivo* generation and activation of anti-tumor T cells may result as detrimental (Nielsen *et al.*, 2000). T-cell production of IL-10 by naturally occurring altered peptide ligands have been indeed reported as likely utilized by malaria parasite to escape immune recognition and to mediate T-cell immunosuppression in patients affected by *Plasmodium falciparum* (Plebanski *et al.*, 1999). These and other findings suggest some caution in the usage of modified epitopes, in order to avoid the *in vivo* induction of unwanted responses, which can unfavorably influence the ability of the immune system to control tumor growth.

In vivo efficacy of APL-based vaccine in neoplastic disease

Several experimental studies have been performed in mouse models for testing the actual benefits in terms of immunogenicity and anti-tumor efficacy of optimized peptides. From data raised in different tumor and viral systems, it has been proved that vaccination with higher affinity ligands (to improve stability of the MHC/peptide/TCR complexes) promotes *in vivo* expansion of specific T cells without affecting their antigen specificity or sensitivity (Slanky *et al.*, 2000). This phenomenon, mainly tested with "self" antigens, could be related to the ability of optimized peptides to break peripheral tolerance and activate low-affinity T cells to efficiently destroy target cells bearing the parental epitope. Animal models have also shown that optimizing peptide interaction with the TCR, by using "superagonist" variants, leads to the additional advantage of triggering qualitatively improved T-cell functions (e.g. IL-2 release) which can further help overcoming the anergic or tolerant state often affecting tumor-specific T lymphocytes *in vivo* (Overwijk *et al.*, 1998).

The possibility of increasing *in vivo* immunogenicity and thus anti-tumor efficacy of peptide-based vaccination by using optimized tumor epitopes has been also evaluated in clinical trials (Table 6.2).

An experimental protocol has been tested in melanoma patients consisting in the vaccination with an analog of the gp100/HLA-A2.1 peptide, modified in position 2 (gp100–209/2M) for better HLA binding (Rosenberg *et al.*, 1998). Treatment with this optimized epitope mediated much improved clinical activity as compared to the native peptide, in terms of tumor regressions of metastatic lesions. The addition of high dose IL-2 therapy was however needed to achieve clinical efficacy. This raises, among others, the hypothesis

Table 6.2 Clinical results of vaccination trials with APL of tumor peptides

Tumor	*N. patients*	*Vaccine*	*Clinical responses*	*Immunological responses*[a]	*Assay*[b]
Melanoma[c]	8	gp100/209 + IFA	1 PR	2/8	IFNγ ELISA[d]
	11	gp100/209-2M + IFA	—	10/11	
	31	gp100/209-2M + IFA + L-2	1 CR, 12 PR	3/19	
Colon ca. and NSCLC[f]	12	DC + CAP1-6D + KLH	2 CR, 1 PR, 1 MR	7/12	Lysis[e]
				5/12	HLA tetramer

Notes

a Number of patients showing immunological responses to the vaccine.

b Immunological assay used for *in vitro* immune monitoring.

c From Rosenberg *et al.*, *Natl. Med.* **4**, 321–7 (1998).

d Immune responses were detected as IFNγ release (by ELISA) after repeated *in vitro* peptide stimulations.

e Immune responses were measured by ^{51}Cr release assay on fresh PBMC (detecting lytic activity on CEA positive cells), and by cytofluorimetric staining with fluorocrome-labeled HLA/CAP1-6D tetramers.

f From Fong L. *et al.*, *Proc. Natl. Acad. Sci.*, **98**, 8809–14 (2001).

that peptide analogs with improved HLA-binding affinity may not provide sufficient T-cell triggering for production of this crucial cytokine at tumor site. Additionally, at difference with animal studies, no clear evidence of powerful peptide-specific T-cell responses was indeed detectable in the peripheral blood of vaccinated patients (Rosenberg *et al.*, 1998).

Data concerning a second clinical trial based on vaccination with optimized tumor peptides have been recently published, showing a good clinical efficacy of immunotherapy with autologous DC pulsed with a superagonist variant of the HLA-A2 binding peptide CEA-CAP1 (Fong *et al.*, 2001). This modified epitope, containing a D in position 6 (CAP1-6D), has been shown to exert "superagonist" activities, that is, more powerful induction of CEA-specific T cells through improved interactions with the TCR, as compared to the natural CEA epitope. Potentiated T cells raised *in vitro* with CAP1 retain strong and efficient ability to lysed tumor cells expressing native CEA (Zaremba *et al.*, 1997; Fong *et al.*, 2001), supporting the possibility that CAP1-6D is able to overcome peripheral immune tolerance toward this self molecule. The effect of vaccination with CAP1-6D has been recently tested in a clinical trial performed in patients with CEA-expressing metastatic carcinomas (colon carcinoma and NSCLC) (Fong *et al.*, 2001). The vaccination protocol used consisted of peptide-pulsed autologous DC expanded *in vivo* by treatment with the hematopoietic growth factor Flt3. The vaccine was given in conjunction with local KHL, a powerful CD4 non-specific activator often used as adjuvant in cancer immunization therapies. More than 50% of patients developed significant increase of CEA-specific CD8+ T cells in PBMC, as detected by *in vitro* lysis of CEA-expressing tumor cells or by staining with HLA-A2 tetramers bearing CAP1 peptide (as native and modified form). Tetramer-stained CD8+ lymphocytes in PBMC possessed an "effector cytotoxic" phenotype (CD27+, CD45RA+, CCR7−), suggesting the *in vivo* expansion of lytic T cells by the vaccine. Such immune responses seemed to correlate with clinical regressions (detected in 5 out of 12 patients treated), including two complete and long-lasting tumor regressions.

Role of APL in tumor-induced T-cell anergy and immune escape

Self-peptides play a more direct role in immune function when serving as target epitopes of autoimmune or autoreactive T-cell responses. Included in this latter category are the descriptions of CTL reactivity to unmodified self-epitopes present on tumors in melanoma patients (Parmiani, 1993). Although autoreactive $CD8^+$ cells may be recovered from tumor-infiltrating lymphocytes (TIL) or peripheral blood of patients, direct evidence about their functional anti-tumor activity at tumor site has yet to be obtained. Among $HLA\text{-}A2.1^+$ patients in particular, CTL reactivity to the peptide ($MART\text{-}1_{27\text{–}35}$) derived from melanocyte/melanoma differentiation protein MART-1 is frequently and readily observed in *ex vivo* cultured TIL (Kawakami *et al.*, 1994a). CTL reactivity to this epitope can be additionally enhanced by *in vivo* administration of peptide (Kawakami *et al.*, 1996; Marincola *et al.*, 1996a,b; Skipper *et al.*, 1999). However, this CTL activity does not appear to lead to tumor regression. The co-existence of tumor with apparently robust CTL reactivity to the $MART\text{-}1_{27\text{–}35}$ epitope, at least as isolated and characterized *ex vivo*, has thus to be viewed as a sort of paradox (Rivoltini *et al.*, 1998).

We found that CTL nominally specific for the $MART\text{-}1_{27\text{–}35}$ epitope might be expanded *in vivo* by the encountering with $MART\text{-}1_{27\text{–}35}$-like sequences derived from environmental antigens (Loftus *et al.*, 1996). The general motif embodied in the $MART\text{-}1_{27\text{–}35}$ sequence,

AAGIGILTV, indeed occurs frequently within a variety of foreign and endogenous proteins. Several naturally occurring peptides can in fact sensitize HLA-A2.1+ cells for lysis by melanoma patient-derived CTL. This observation prompted us to speculate that endogenous MART-1_{27-35}-like peptides might also play a role in shaping this CTL response by contributing to tolerance, acting as APL (Loftus *et al.*, 1996, 1998) and perhaps negatively modulating CTL function. Although other factors may affect the anti-tumor CTL response to MART-1_{27-35}, our results suggest that MART-1 analogs derived from self-proteins act as partial agonist or antagonist of anti-MART-1 T cells, by mediating suboptimal activation and partially triggering anti-tumor functions in such T cells. This phenomenon may not only contribute to the maintenance of MART-1_{27-35}-reactive CTL but can also have a net negative influence on anti-MART-1_{27-35} T-cell functional efficacy, possibly accounting for the paradoxical nature of this response in melanoma patients.

Additionally, more recent data (Carrabba and Rivoltini, manuscript in preparation) suggest that peptides with partial agonist or antagonist functions can also be found in HPLC peptide fractions eluted from HLA-A2+ melanoma cells, that induce *in vitro* T-cell anergy in tumor-specific lymphocytes. This phenomenon can be overcome by generating anti-MART-1_{27-35} T cells using the superagonist peptide analog 1L (described in the previous section). These latter data suggest that the usage of modified tumor peptide analogs could thus represent a promising approach for overcoming tumor-induced immunosuppression and possibly designing more successful vaccinations in cancer patients.

Conclusions

Significant and durable tumor regressions can be achieved in metastatic cancer patients refractory to conventional therapies, by boosting anti-tumor T-cell responses *in vivo* through specific tumor vaccines. However, clinical efficacy of the immunological approaches tested thus far must be improved by identifying new strategies based on more effective immunization protocols. The usage of altered peptide ligands derived from tumor epitopes, that can quantitatively and qualitatively improve anti-tumor T-cell responses, may represent a valuable clinical approach. The search for more powerful immunization tools must however be paralleled by *ex vivo* studies leading to the identification of immunological mechanisms used by tumor cells to escape immune recognition. In this view, vaccination with peptide analogs with superagonist features may result more effective, thanks to their additional ability of generating immune responses potentially able to resist immunosuppression and anergy induction mediated by tumor cells.

References

Brinckerhoff, L. H., Kalashnikov, V. V., Thompson, L. W., Yamshchikov, G. V., Pierce, R. A., Galavotti, H. S., Engelhard, V. H. and Slingluff, C. L., Jr. (1999) Terminal modifications inhibit proteolytic degradation of an immunogenic MART-1(27–35) peptide: implications for peptide vaccines. *Int. J. Cancer*, **83**, 326–34.

Chen, J. L., Dunbar, P. R., Gileadi, U., Jager, E., Gnjatic, S., Nagata, Y., Stockert, E., Panicali, D. L., Chen, Y. T., Knuth, A., Old, L. J. and Cerundolo, V. (2000) Identification of NY-ESO-1 peptide analogues capable of improved stimulation of tumor-reactive CTL. *J. Immunol.*, **165**, 948–55.

Dudley, M. E., Nishimura, M. I., Holt, A. K. and Rosenberg, S. A. (1999) Antitumor immunization with a minimal peptide epitope (G9-209-2M) leads to a functionally heterogeneous CTL response. *J. Immunother.*, **22**, 288–98.

Evavold, B. D., Sloan-Lancaster, J. and Allen, P. M. (1993) Tickling the TCR: selective T-cell functions stimulated by altered peptide ligands. *Immunol. Today*, **14**, 602–9.

Fong, L., Hou, Y., Rivas, A., Benike, C., Yuen, A., Fisher, G. A., Davis, M. M. and Engleman, E. G. (2001) Altered peptide ligand vaccination with Flt3 ligand expanded dendritic cells for tumor immunotherapy. *Proc. Natl. Acad. Sci. USA*, **98**, 8809–14.

Grakoui, A., Bromley, S. K., Sumen, C., Davis, M. M., Shaw, A. S., Allen, P. M. and Dustin, M. L. (1999) The immunological synapse: a molecular machine controlling T cell activation. *Science*, **285**, 221–7.

Hennecke, J. and Wiley, D. C. (2001) T-cell receptor–MHC interactions up close. *Cell*, **104**, 1–4.

Kawakami, Y., Eliyahu, S., Delgado, C. H., Robbins, P. F., Rivoltini, L., Topalian, S. L., Miki, T. and Rosenberg, S. A. (1994a) Cloning of the gene coding for a shared human melanoma antigen recognized by autologous T cells infiltrating into tumor. *Proc. Natl. Acad. Sci. USA*, **91**, 3515–9.

Kawakami, Y., Eliyahu, S., Sakaguchi, K., Robbins, P. F., Rivoltini, L., Yannelli, J. R., Appella, E. and Rosenberg, S. A. (1994b) Identification of the immunodominant peptides of the MART-1 human melanoma antigen recognized by the majority of HLA-A2-restricted tumor infiltrating lymphocytes. *J. Exp. Med.*, **180**, 347–52.

Kawakami, Y., Robbins, P. F. and Rosenberg, S. A. (1996) Human melanoma antigens recognized by T lymphocytes. *Keio. J. Med.*, **45**, 100–8.

Kersh, G. J. and Allen, P. M. (1996) Essential flexibility in the T-cell recognition of antigen. *Nature*, **380**, 495–8.

Lanzavecchia, A. and Sallusto, F. (2001) Antigen decoding by T lymphocytes: from synapses to fate determination. *Nat. Immunol.*, **2**, 487–92.

Loftus, D. J., Castelli, C., Clay, T. M., Squarcina, P., Marincola, F. M., Nishimura, M. I., Parmiani, G., Appella, E. and Rivoltini, L. (1996) Identification of epitope mimics recognized by CTL reactive to the melanoma/melanocyte-derived peptide MART-1(27–35). *J. Exp. Med.*, **184**, 647–57.

Loftus, D. J., Squarcina, P., Nielsen, M. B., Geisler, C., Castelli, C., Odum, N., Appella, E., Parmiani, G. and Rivoltini, L. (1998) Peptides derived from self-proteins as partial agonists and antagonists of human CD8+ T-cell clones reactive to melanoma/melanocyte epitope MART1(27–35). *Cancer Res.*, **58**, 2433–9.

Marincola, F. M., Hijazi, Y. M., Fetsch, P., Salgaller, M. L., Rivoltini, L., Cormier, J., Simonis, T. B., Duray, P. H., Herlyn, M., Kawakami, Y. and Rosenberg, S. A. (1996a) Analysis of expression of the melanoma-associated antigens MART-1 and gp100 in metastatic melanoma cell lines and in in situ lesions. *J. Immunother. Emphasis Tumor. Immunol.*, **19**, 192–205.

Marincola, F. M., Rivoltini, L., Salgaller, M. L., Player, M. and Rosenberg, S. A. (1996b) Differential anti-MART-1/MelanA CTL activity in peripheral blood of HLA-A2 melanoma patients in comparison to healthy donors: evidence of *in vivo* priming by tumor cells. *J. Immunother. Emphasis Tumor Immunol.*, **19**, 266–77.

Marincola, F. M., Jaffee, E. M., Hicklin, D. J. and Ferrone, S. (2000) Escape of human solid tumors from T-cell recognition: molecular mechanisms and functional significance. *Adv. Immunol.*, **74**, 181–273.

van der Merwe, P. A. (2001) The TCR triggering puzzle. *Immunity*, **14**, 665–8.

Nielsen, M. B., Kirkin, A. F., Loftus, D., Nissen, M. H., Rivoltini, L., Zeuthen, J., Geisler, C. and Odum, N. (2000) Amino acid substitutions in the melanoma antigen recognized by T cell 1 peptide modulate cytokine responses in melanoma-specific T cells. *J. Immunother.*, **23**, 405–11.

Overwijk, W. W., Tsung, A., Irvine, K. R., Parkhurst, M. R., Goletz, T. J., Tsung, K., Carroll, M. W., Liu, C., Moss, B., Rosenberg, S. A. and Restifo, N. P. (1998) gp100/pmel 17 is a murine tumor rejection antigen: induction of "self"-reactive, tumoricidial T cells using high-affinity, altered peptide ligand. *J. Exp. Med.*, **188**, 277–86.

Parkhurst, M. R., Salgaller, M. L., Southwood, S., Robbins, P. F., Sette, A., Rosenberg, S. A. and Kawakami, Y. (1996) Improved induction of melanoma-reactive CTL with peptides from the melanoma antigen gp100 modified at HLA-A*0201-binding residues. *J. Immunol.*, **157**, 2539–48.

Parmiani, G. (1993) Tumor immunity as autoimmunity: tumor antigens include normal self proteins which stimulate anergic peripheral T cells. *Immunol. Today*, **14**, 536–8.

Pittet, M. J., Valmori, D., Dunbar, P. R., Speiser, D. E., Lienard, D., Lejeune, F., Fleischhauer, K., Cerundolo, V., Cerottini, J. C. and Romero, P. (1999) High frequencies of naive Melan-A/MART-1-specific CD8(+) T cells in a large proportion of human histocompatibility leukocyte antigen (HLA)-A2 individuals. *J. Exp. Med.*, **190**, 705–15.

Plebanski, M., Flanagan, K. L., Lee, E. A., Reece, W. H., Hart, K., Gelder, C., Gillespie, G., Pinder, M. and Hill, A. V. (1999) Interleukin 10-mediated immunosuppression by a variant CD4 T cell epitope of Plasmodium falciparum. *Immunity*, **10**, 651–60.

Renkvist, N., Castelli, C., Robbins, P. F. and Parmiani, G. (2001) A listing of human tumor antigens recognized by T cells. *Cancer Immunol. Immunother.*, **50**, 3–15.

Rivoltini, L., Loftus, D. J., Barracchini, K., Arienti, F., Mazzocchi, A., Biddison, W. E., Salgaller, M. L., Appella, E., Parmiani, G. and Marincola, F. M. (1996) Binding and presentation of peptides derived from melanoma antigens MART-1 and glycoprotein-100 by HLA-A2 subtypes. Implications for peptide-based immunotherapy. *J. Immunol.*, **156**, 3882–91.

Rivoltini, L., Loftus, D. J., Squarcina, P., Castelli, C., Rini, F., Arienti, F., Belli, F., Marincola, F. M., Geisler, C., Borsatti, A., Appella, E. and Parmiani, G. (1998) Recognition of melanoma-derived antigens by CTL: possible mechanisms involved in down-regulating anti-tumor T-cell reactivity. *Crit. Rev. Immunol.*, **18**, 55–63.

Rivoltini, L., Squarcina, P., Loftus, D. J., Castelli, C., Tarsini, P., Mazzocchi, A., Rini, F., Viggiano, V., Belli, F. and Parmiani, G. (1999) A superagonist variant of peptide MART1/Melan A27–35 elicits anti-melanoma CD8+ T cells with enhanced functional characteristics: implication for more effective immunotherapy. *Cancer Res.*, **59**, 301–6.

Rosenberg, S. A. (2001) Progress in human tumour immunology and immunotherapy. *Nature*, **411**, 380–4.

Rosenberg, S. A., Yang, J. C., Schwartzentruber, D. J., Hwu, P., Marincola, F. M., Topalian, S. L., Restifo, N. P., Dudley, M. E., Schwarz, S. L., Spiess, P. J., Wunderlich, J. R., Parkhurst, M. R., Kawakami, Y., Seipp, C. A., Einhorn, J. H. and White, D. E. (1998) Immunologic and therapeutic evaluation of a synthetic peptide vaccine for the treatment of patients with metastatic melanoma. *Nat. Med.*, **4**, 321–7.

Ruppert, J., Sidney, J., Celis, E., Kubo, R. T., Grey, H. M. and Sette, A. (1993) Prominent role of secondary anchor residues in peptide binding to HLA-A2.1 molecules. *Cell*, **74**, 929–37.

Salgaller, M. L., Afshar, A., Marincola, F. M., Rivoltini, L., Kawakami, Y. and Rosenberg, S. A. (1995) Recognition of multiple epitopes in the human melanoma antigen gp100 by peripheral blood lymphocytes stimulated *in vitro* with synthetic peptides. *Cancer Res.*, **55**, 4972–9.

Salgaller, M. L., Marincola, F. M., Cormier, J. N. and Rosenberg, S. A. (1996) Immunization against epitopes in the human melanoma antigen gp100 following patient immunization with synthetic peptides. *Cancer Res.*, **56**, 4749–57.

Sette, A., Chesnut, R. and Fikes, J. (2001) HLA expression in cancer: implications for T cell-based immunotherapy. *Immunogenetics*, **53**, 255–63.

Sharma, A. K., Kuhns, J. J., Yan, S., Friedline, R. H., Long, B., Tisch, R. and Collins, E. J. (2001) Class I major histocompatibility complex anchor substitutions alter the conformation of T cell receptor contacts. *J. Biol. Chem.*, **276**, 21443–9.

Skipper, J. C., Gulden, P. H., Hendrickson, R. C., Harthun, N., Caldwell, J. A., Shabanowitz, J., Engelhard, V. H., Hunt, D. F. and Slingluff, C. L. (1999) Mass-spectrometric evaluation of HLA-A*0201-associated peptides identifies dominant naturally processed forms of CTL epitopes from MART-1 and gp100. *Int. J. Cancer*, **82**, 669–77.

Slansky, E. J., Rattis, M. F., Boyd, F. L., Fahmy, T., Jaffee, M. E., Schneck, P. J., Margulies, H. D. and Pardoll, M. D. (2000) Enhanced antigen-specific antitumor immunity with altered peptide ligands that stabilize the MHC–peptide–TCR complex. *Immunity*, **13**, 529–38.

Sliz, P., Michielin, O., Cerottini, J. C., Luescher, I., Romero, P., Karplus, M. and Wiley, D. C. (2001) Crystal structures of two closely related but antigenically distinct HLA-A2/melanocyte–melanoma tumor-antigen peptide complexes. *J. Immunol.*, **167**, 3276–84.

Sloan-Lancaster, J. and Allen, P. M. (1996) Altered peptide ligand-induced partial T cell activation: molecular mechanisms and role in T cell biology. *Annu. Rev. Immunol.*, **14**, 1–27.

Tangri, S., Ishioka, G. Y., Huang, X., Sidney, J., Southwood, S., Fikes, J. and Sette, A. (2001) Structural features of peptide analogs of human histocompatibility leukocyte antigen class I epitopes that are more potent and immunogenic than wild-type peptide. *J. Exp. Med.*, **194**, 833–46.

Tourdot, S., Scardino, A., Saloustrou, E., Gross, D. A., Pascolo, S., Cordopatis, P., Lemonnier, F. A. and Kosmatopoulos, K. (2000) A general strategy to enhance immunogenicity of low-affinity HLA-A2.1-associated peptides: implication in the identification of cryptic tumor epitopes. *Eur. J. Immunol.*, **30**, 3411–21.

Trojan, A., Witzens, M., Schultze, J. L., Vonderheide, R. H., Harig, S., Krackhardt, A. M., Stahel, R. A. and Gribben, J. G. (2001) Generation of cytotoxic T lymphocytes against native and altered peptides of human leukocyte antigen-A*0201 restricted epitopes from the human epithelial cell adhesion molecule. *Cancer Res.*, **61**, 4761–5.

Valmori, D., Fonteneau, J. F., Lizana, C. M., Gervois, N., Lienard, D., Rimoldi, D., Jongeneel, V., Jotereau, F., Cerottini, J. C. and Romero, P. (1998a) Enhanced generation of specific tumor-reactive CTL *in vitro* by selected Melan-A/MART-1 immunodominant peptide analogues. *J. Immunol.*, **160**, 1750–8.

Valmori, D., Gervois, N., Rimoldi, D., Fonteneau, J. F., Bonelo, A., Lienard, D., Rivoltini, L., Jotereau, F., Cerottini, J. C. and Romero, P. (1998b) Diversity of the fine specificity displayed by HLA-A*0201-restricted CTL specific for the immunodominant Melan-A/MART-1 antigenic peptide. *J. Immunol.*, **161**, 6956–62.

Valmori, D., Pittet, M. J., Rimoldi, D., Lienard, D., Dunbar, R., Cerundolo, V., Lejeune, F., Cerottini, J. C. and Romero, P. (1999) An antigen-targeted approach to adoptive transfer therapy of cancer. *Cancer Res.*, **59**, 2167–73.

Wang, R., Wang-Zhu, Y., Gabaglia, C. R., Kimachi, K. and Grey, H. M. (1999) The stimulation of low-affinity, nontolerized clones by heteroclitic antigen analogues causes the breaking of tolerance established to an immunodominant T cell epitope. *J. Exp. Med.*, **190**, 983–94.

Zaremba, S., Barzaga, E., Zhu, M., Soares, N., Tsang, K. Y. and Schlom, J. (1997) Identification of an enhancer agonist cytotoxic T lymphocyte peptide from human carcinoembryonic antigen. *Cancer Res.*, **57**, 4570–7.

Zugel, U., Wang, R., Shih, G., Sette, A., Alexander, J. and Grey, H. M. (1998) Termination of peripheral tolerance to a T cell epitope by heteroclitic antigen analogues. *J. Immunol.*, **161**, 1705–9.

Chapter 7

Ex vivo and *in situ* detection of tumor-specific T-cell immunity with MHC tetramers

Ton N. M. Schumacher and John B. A. G. Haanen

Summary

MHC tetramer analysis has revealed pronounced spontaneous and therapy-induced expansions of tumor-specific T cells in cancer patients. This allows the coupling of T-cell responses to clinical course, and has, in particular for hematological malignancies, provided significant evidence for a role of these T-cell expansions in tumor control. A limitation of the currently available data is its focus on $CD8^+$ immunity, ignoring the possible contribution of $CD4^+$ T cells. However, with the increasing availability of multimeric MHC class II reagents this is likely to change in the foreseeable future. A second limitation is formed by the fact that T-cell responses have primarily been monitored in peripheral blood. The development of technology to monitor T-cell responses in small amounts of tissue material by CLSM or by other means will be essential for our understanding between T-cell immunity and clinical course for solid tumors. On a similar note, in future studies it will be important to establish not only the magnitude of (vaccine-induced) tumor-specific T-cell responses, but also the functional activity and migration of these cells, and T-cell phenotypes that correlate with this behavior. Finally, the use of MHC tetramers to selectively isolate naturally occurring or *in vitro*-generated TCRs may give us the tools to induce sufficiently strong tumor-specific T-cell immunity in cases in which the natural T-cell repertoire has failed.

Detection of antigen-specific T-cell immunity with multimeric MHC technology

Since the discovery of MHC-restricted recognition of peptide antigens by T cells, one of the main aims of immunologists has been to visualize disease- and vaccine-induced T-cell immunity. Traditional approaches for the quantification of antigen-specific T-cell immunity monitored functional responses of T cells upon antigen encounter. This involved either antigen-driven proliferation (as measured by ^{3}H-thymidine uptake or limiting dilution assays), or T-cell effector functions such as antigen-induced cytokine secretion or cytolysis. However, by definition these methods could only monitor T cells that were active in a particular functional assay and could therefore only provide a relative measure of the magnitude of a T-cell response. Consequently, aspects such as the phenotype, temporal and spatial distribution of antigen-specific T cells, but also the absolute magnitude of T-cell immunity could not readily be addressed. Early attempts to directly visualize antigen-specific T cells by flow cytometry focused on the detection of antigen-specific T-cell responses that were characterized by a very homogenous Vα-Vβ usage, and that could therefore be detected by a combination of Vα-Vβ-specific antibodies. These experiments provided a first glimpse of the development and magnitude of antigen-specific T-cell responses (McHeyzer-Williams and Davis, 1995). However, the value of this approach was limited by the fact that in the vast majority of T-cell responses, Vα-Vβ usage is highly diverse. The real breakthrough came when John Altman in the lab of Mark Davis prepared multivalent complexes of HLA-A2.1 molecules complexed with HIV-derived epitopes and showed that these polyvalent MHC–peptide reagents could bind with sufficient avidity to HIV-specific $CD8^+$ T cells to allow their detection by flow cytometry (Altman *et al.*, 1996). Since then MHC tetramers and other types of MHC multimers (Altman *et al.*, 1996; Walter *et al.*, 1998; Malherbe *et al.*, 2000; Schneck, 2000) (here all referred to as MHC tetramers for the sake of simplicity) have been used extensively to visualize antigen-specific T-cell immunity in humans and in animal model systems. The main advantage of MHC tetramers is that their use allows detection of antigen-specific T-cell responses without any requirement for *in vitro* proliferation, or production of cytokines. The importance of this property has become very clear over the past years. Already the very first experiments using MHC tetramer technology in the study of murine and human infectious diseases demonstrated that prior analyses of antigen-specific T-cell responses by limiting dilution analysis had provided a very large underestimate of the magnitude of such responses (Callan *et al.*, 1998; Murali-Krishna *et al.*, 1998). In the current chapter we will describe the use of MHC tetramers in the study of tumor-specific T-cell immunity.

Production of MHC class I tetramers and detection of antigen-specific $CD8^+$ T cells

At present the NIH provides MHC class I tetramers at no cost to researchers upon request (see http://www.niaid.nih.gov/reposit/tetramer/overview.html). In addition, several companies in Europe and in the United States have started the production of MHC class I tetramers/ MHC dimers on a commercial scale. However, most MHC class I tetramers that are currently used in research are still produced by researchers themselves. The technology that is used to produce MHC tetramers in most labs is very similar to the strategy first worked out in the Davis lab ((Altman *et al.*, 1996), see Figure 7.1). In this strategy, the MHC

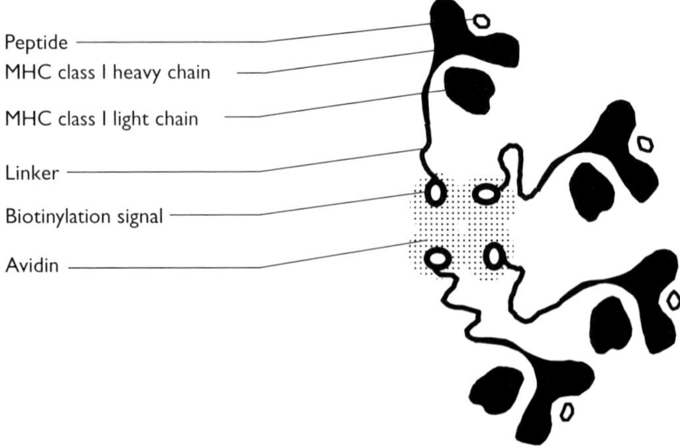

Figure 7.1 Structure of MHC class I tetramers.

class I heavy chain and β_2m (MHC class I light chain) are expressed separately in *Escherichia coli*. Both chains lack the signal peptide normally required for ER import and the MHC class I heavy chain is further modified by the removal of the transmembrane/cytoplasmic domain and COOH terminal addition of a recognition site for the enzyme biotin ligase (the biotag). Purified MHC class I heavy chain and light chain are folded with antigenic peptide to form monomeric MHC class I complexes. Conditions for folding can vary considerably between different MHC class I molecules, and although conditions have been established for a significant number of MHC class I alleles, some alleles remain problematic. Folded MHC-peptide monomers are purified and the complex is biotinylated by the enzyme BirA. Following removal of the biotin and biotin ligase, fluorochrome-labeled streptavidin or avidin is added. Since both streptavidin and avidin have four biotin binding sites, this results in the formation of tetravalent complexes of MHC molecules. Since BirA only couples biotin residues to lysine sidechains that form part of a defined peptide sequence present in the biotag, biotinylation of MHC complexes in this manner is strictly site-specific. As a consequence, all MHC molecules associate with streptavidin through the biotin residue present at the COOH terminus of the class I heavy chain, allowing the part of the MHC–peptide complex seen by the T-cell receptors (TCR) to remain freely available for binding. A detailed protocol for the preparation of MHC class I tetramers by this method is available at the NIAID tetramer site (http://www.emory.edu/WHSC/TETRAMER/protocol.html). Plasmids that encode β_2m, different MHC class I heavy chain alleles and biotin ligase are all freely available from several labs. In addition to this standard protocol, several other formats have been put forward. However, at present no systematic comparison has been made between different types of MHC multimers with respect to background issues, or sensitivity of detecting T cells with low affinity TCRs (see next paragraph).

The detection of antigen-specific T cells with MHC class I tetramers is at least in theory extremely straightforward. Cell suspensions are incubated with MHC tetramers in combination with lineage markers and following a few brief washes, samples are analyzed by flow cytometry. However, because the binding of MHC tetramers to cells displays a number of unusual characteristics as compared to antibody binding, additional care is required. Furthermore, in most situations the frequency of tumor-specific T cells will be low, often below 1% of $CD8^+$ T cells. Therefore, any strategy that can help distinguish between low frequencies of antigen-specific T cells and background binding is of importance.

Conditions of binding

Binding of MHC class I-restricted T cells to MHC–peptide complexes on target cells is mediated not only by the TCR, but also by the CD8 co-receptor. Likewise, binding of MHC class I tetramers to $CD8^+$ T cells also in part involves the interaction between MHC tetramers and CD8 molecules, and depending on the situation, this TCR-nonselective component can be favorable or unfavorable. Similar to its role in T-cell–target-cell encounter, interaction of CD8 molecules with MHC tetramers lowers the threshold for TCR binding. CD8 co-receptor binding leads to significant binding of H-$2K^b$ tetramers to $CD8^+$ T cells independent of their TCR specificity and this background binding can be reduced by inclusion of certain anti-CD8 antibodies (see http://www.niaid.nih.gov/reposit/tetramer/overview.html). Likewise, a significant CD8-binding component has been observed for human MHC tetramers at high tetramer concentrations, and this can be reduced through the use of mutant tetramers in which CD8 binding is decreased as a consequence of a mutation in the MHC class I heavy chain α3 domain (Bodinier *et al.*, 2000). At the other side of the spectrum, CD8 co-receptor binding appears essential for the binding of MHC tetramers to certain murine antigen-specific T cells (Daniels and Jameson, 2000) and a human T-cell clone specific for the melanoma antigen gp100 (Denkberg *et al.*, 2001). Likewise, we have observed that binding of A2.1-Tyr_{368} tetramers to polyclonal populations of tyrosinase-specific T cells in melanoma patients is CD8-dependent (van Oijen *et al.*, Ms submitted). In all these cases inclusion of (certain) anti-CD8 antibodies is sufficient to abrogate MHC tetramer binding. This negative effect on MHC tetramer binding can be avoided either through the use of anti-CD8 antibodies that do not block (Daniels and Jameson, 2000), or by defining $CD8^+$ T cells by exclusion of other cell lineages, using a panel of lineage marker antibodies (van Oijen *et al.*, Ms submitted). It may be speculated that the low tetramer binding observed for certain T-cell populations that recognize self-antigens such as tyrosinase (Yee *et al.*, 1999; van Oijen *et al.*, Ms submitted) is due to the effects of self-tolerance. A number of studies in transgenic and knockout mouse strains have directly demonstrated that in many cases T-cell tolerance towards self-antigens results in the deletion of high avidity T cells while low avidity T cells are spared (de Visser *et al.*, 2000; Hernandez *et al.*, 2000; Nugent *et al.*, 2000). These low avidity self-specific T cells show reduced levels of MHC tetramer binding, and antigen-specific effector function is only observed at high antigen concentrations. TCR expression level on these cells is indistinguishable from TCR levels of antigen-specific T cells from non-tolerant animals, implying a reduced affinity of the encoded TCRs for the self peptide–MHC complex as the mechanism involved. It is currently unclear to what extent or under which conditions such low avidity T cells can mediate significant anti-tumor effects. However, approaches to circumvent the deleterious effects of self-tolerance on the tumor/self antigen-specific T-cell repertoire are likely to be beneficial and will be discussed in one of the following sections.

As first pointed out by Cerundolo, Sewell and coworkers, the binding of MHC class I tetramers to antigen-specific T cells is a temperature-sensitive process (Whelan *et al.*, 1999). MHC tetramers bind with greater selectivity at higher temperature (i.e. 37 °C versus 4 °C), as significant binding of MHC tetramers complexed with partial agonist ligands is observed at reduced temperatures. This temperature dependency of MHC class I tetramer-binding fits well with a previous thermodynamic analysis of TCR–MHC interactions (Willcox *et al.*, 1999) and can probably be extrapolated to MHC class II tetramers (see further). It is possible that the observed temperature dependency of the specificity of MHC tetramer binding is also related to the propensity of TCRs to localize in lipid rafts upon antigen-specific T-cell activation. It has been observed that MHC tetramer binding is affected by disruption of lipid raft integrity (Drake and Braciale, 2001). Furthermore, confocal laser scanning microscopy of MHC tetramer binding to T cells reveals a pronounced punctate staining pattern, and this TCR clustering occurs more rapidly at higher temperatures (Haanen *et al.*, 2000). Further evidence for an effect of TCR–lipid raft interaction on MHC multimer binding comes from the observation that the binding of MHC dimers to antigen-specific T cells is influenced by the T-cell activation state (Cameron *et al.*, 2001). Upon activation, T cells bind MHC dimers at lower concentrations and this process can be mimicked by membrane cholesterol. Collectively, these experiments indicate that binding of MHC multimers to T cells is not a simple multivalent interaction between MHC molecules and TCRs, but is a dynamic process that is sensitive to alterations in membrane organization that affect TCR localization and can be significantly influenced by temperature and the CD8 co-receptor.

Data acquisition and analysis

In most clinically relevant settings the frequency of antigen-specific T cells is very low. It is therefore of importance to optimize MHC tetramer flow cytometry in order to reduce background binding as much as possible. In addition, in MHC tetramer analysis of human T-cell responses it is useful to define criteria that can help distinguish a true antigen-specific T-cell response from background binding.

To help increase signal to noise ratio and to distinguish true staining of antigen-specific T cells from background staining a set of simple rules can be helpful. (1) Count as many cells as possible. To reliably visualize $CD8^+$ T-cell frequencies in the range 0.1–0.5% it is necessary to count $\sim 5 \times 10^5$ peripheral blood lymphocytes. (2) Set both a well-defined life gate (i.e. select for propidium-iodide (PI) negative cells) and lymphocyte gate. Careful data analysis helps to significantly reduce background staining, and in some cases can prevent false positives. In particular early apoptotic cells that are PI-low, but have a slightly reduced forward scatter appear to contribute significantly to the background. (3) Background staining varies between tetramers and can be over 0.2% of $CD8^+$ T cells for certain tetramers and below 0.02% for others. Consequently, it is not good practice to rely on staining with control tetramers to set the background. Rather, levels of background staining should be determined by staining lymphocytes from a panel of healthy individuals, preferably matched for the relevant MHC allele. (4) The presence of a population of tetramer-positive cells is insufficient to document the existence of an antigen-specific T-cell response in the absence of any supporting data. There are still a number questions surrounding MHC tetramer technology that preclude its use as definitive proof for the existence of a T-cell response. Additional evidence may come from (lack of) staining of lymphocytes of healthy individuals, expansion of the tetramer-reactive population upon antigen encounter (which can readily be visualized by CFSE labeling (Novak *et al.*, 1999)), or functional responses, such as cytokine secretion or CD69 up-regulation.

Detection of naturally occurring tumor-specific $CD8^+$ T-cell immunity by tetramer flow cytometry

Magnitude of tumor-specific T-cell responses

Shortly after MHC tetramers had started to revolutionize our understanding of the size and kinetics of pathogen-specific immunity, this technology was employed for the detection of tumor-specific $CD8^+$ T cells. To date, the majority of studies has been on melanoma-specific T-cell immunity. In the past decade a large set of antigens that are recognized by melanoma-specific T cells has been identified (see Chapter 4 by Kawakami). The shared melanoma antigens identified in these studies can be divided into two groups. A first group is formed by the family of melanocyte differentiation antigens (MDA), and includes MART-1/MelanA, gp100, tyrosinase and the tyrosinase-related proteins 1 and 2 (TRP-1, 2). These antigens are lineage-specific proteins that are expressed in normal melanocytes and in most melanomas. A second, large group of antigens expressed in melanomas is derived from so-called Cancer/Testis genes. This group of antigens is expressed in testis and placenta and in a variety of different tumors. This group includes the MAGE family of antigens identified by Boon and colleagues, and antigens such as NY-ESO-1 that were identified by SEREX technology (see Chapter 10 by Pfreundschuh).

The first striking thing that emerged from MHC tetramer analysis of spontaneous tumor-specific T-cell immunity in melanoma patients is that – at least for certain antigens – tumor-specific T cells are not few and far between. Others and we have documented the presence of high numbers of T cells with specificity for MDA (Romero *et al.*, 1998; Lee *et al.*, 1999b, van Oijen *et al.*, Ms submitted). In more than 40% of stage IV melanoma patients, MART-1-specific T cells are detectable directly *ex vivo* in peripheral blood, sometimes in very high numbers. As a dramatic example, in one patient greater than 10% of peripheral blood $CD8^+$ T cells was found to be specific for a single HLA-A2.1-restricted epitope of MART-1 (van Oijen *et al.*, Ms submitted). In these patients T cells specific for tyrosinase or gp100 can be detected as well, although in fewer cases and at a lower frequency.

CT antigens have always been considered immunologically non-self, because in healthy individuals expression is limited to immunologically privileged sites (testicular germ cells and placenta). Remarkably, however, the immunogenicity of these antigens appears quite limited. NY-ESO-1, a CT antigen that was originally identified in esophageal carcinoma, and that is expressed in 20–30% of melanomas is the only antigen in this group that induces readily detectable T-cell responses in a *significant* fraction of patients. Several groups have reported naturally occurring T-cell immunity against this antigen (Jager *et al.*, 2000; Valmori *et al.*, 2000), although the frequency amongst $HLA\text{-}A2^+$ melanoma patients seems to be lower than for MART-1. Strikingly, with only few exceptions (MAGE-A10) (Valmori *et al.*, 2001) T-cell reactivity against melanoma-associated antigen gene (MAGE) family members is weak (Traversari *et al.*, 1992; van der Bruggen *et al.*, 1994; Gaugler *et al.*, 1994; Zorn and Hercend, 1999).

The available data therefore suggest a hierarchy in spontaneous tumor-specific T-cell immunity. MART-specific T-cell immunity is clearly dominant and MDA-specific immunity appears somewhat more prevalent than immunity against CT antigens. This hierarchy is not only observed in MHC tetramer analysis of peripheral blood, but also in cultures of tumor-infiltrating lymphocytes (Kawakami *et al.*, 2000). The factors that determine this hierarchy are currently unclear. Factors such as intracellular processing, MHC affinity of peptide antigens and the frequency of antigen-specific T cells within the naïve T-cell repertoire have previously been shown to determine immunodominance in a mouse model system (Chen *et al.*, 2000). These factors are likely to affect also the immunodominance of melanoma antigens. In addition, because all these antigens are (at least genetically) self-antigens, a varying degree of self-tolerance to different antigens may also affect the magnitude of responses. Despite the fact that some of the CT antigens are overexpressed in melanomas (Gibbs *et al.*, 2000), T-cell reactivity has remained a rare finding. It has been argued that whereas expression of MDA decreases as the tumor dedifferentiates – MDA expression is a remnant of the melanocytic origin of the tumor – expression of CT antigens remains the same or increases (Coulie *et al.*, 1997). Interestingly, in a patient with decreased MDA expression, tumor-infiltrating lymphocytes were collected with specificity for MAGE-12, which was interpreted as an indication of a "cryptic" or subdominant nature of this antigen compared to MDA (Panelli *et al.*, 2000). Finally, perhaps the most likely explanation for the skewing of spontaneous T-cell immunity towards the MART-1 antigen in melanoma patients is the observation that low numbers of MART-1-specific T cells have also been detected in a significant fraction of healthy individuals. It has been argued that these expansions may occur as a consequence of cross-reactivity of MART-1-specific T cells with pathogen-derived epitopes, such as from HSV-1. However, it was subsequently shown that MART-1-specific T cells in healthy individuals that can be identified by MHC tetramer staining have a naïve phenotype (Pittet *et al.*, 1999) (see following section), strongly suggesting that these cells are not the product of a cross-reactive T-cell response. It should be pointed out that dominance of certain antigen specificities in spontaneous melanoma-specific T-cell immunity does not indicate that these antigens will be superior tumor rejection antigens. The role of MART as a tumor rejection antigen is highly dubious (see below) and it remains possible that if strong T-cell responses would be induced against CT-derived epitopes that such T cells could mediate anti-tumor effects. The positive results recently obtained in a clinical study with MAGE-3 peptide vaccination support this notion, although also in this trial, MAGE-specific immunity remained difficult to detect (Marchand *et al.*, 1999).

The massive T-cell expansions found in melanoma patients correlate with the abundant presence of MDA antigens upon tumor growth or spreading. Strong expansions are only observed in the peripheral blood or tumor-infiltrated lymph nodes from stage III (regional lymph node involvement) and particularly stage IV (distant metastases) patients. In patients with tumors that are confined to the primary site (i.e. stage I and stage II melanoma), tetramer-reactive T cells have been found, but at significantly lower frequencies (Palermo *et al.*, 2001).

The reasons for the relative weak T-cell response towards locally confined melanoma are likely to be manifold. In early stage tumors only small numbers of tumor cells will drain to the lymph nodes where T-cell activation takes place. Consequently, the antigen load as perceived by the immune system will be limited. In addition, the absence of a profound inflammatory environment in tumor sites will not promote maturation of antigen-presenting cells (APCs) and subsequent cross-presentation of melanoma antigens. Nevertheless, it is clear that in at least a fraction of patients, the immune system is not fully ignorant of primary melanomas as evidenced by a variable degree of T-lymphocyte infiltration of these lesions. Furthermore, the degree of T-cell infiltration is an independent positive prognostic indicator in primary melanoma, providing indirect evidence for immune control (Clark *et al.*, 1989; Clemente *et al.*, 1996). Collectively, these data suggest that, the induction of (increased) MDA-specific T-cell immunity is most likely to be beneficial in stage I and stage II patient groups, both because of this indirect evidence for T-cell control and because spontaneous melanoma-specific immunity is weak and may easily be improved by vaccination.

To date, little has been published on naturally occurring T-cell immunity in other tumors. We speculate that the scarcity of published data on tetramer analysis of spontaneous T-cell immunity in other tumor types is at least in part due to the fact that tumor-specific T-cell expansions are of a much smaller magnitude in tumor types other than melanoma. In a limited set of experiments we screened peripheral blood samples of HLA-A2.1-positive metastatic renal cell carcinoma (RCC) patients with an array of MHC tetramers containing $CD8^+$ T-cell epitopes of G250, MAGE-3, WT-1 (Wilms tumor derived antigen) and hTERT (human telomerase reverse transcriptase), antigens that are all expressed by a proportion of RCC. In 2/17 patients low numbers of hTERT-specific $CD8^+$ T cells and in 1/17 patients a low number of WT-1-specific T cells was detected. However, the relevance of these MHC-tetramer-identified T-cell expansions remains to be further established.

Phenotype and function of melanoma-reactive T cells

As discussed in the preceding section, the most striking observation made in MHC tetramer analysis of melanoma-specific T cells is the magnitude of the MDA-specific T-cell responses in some patients. Despite these significant tumor-specific T-cell responses, spontaneous regression of melanoma remains a very rare event. Therefore, the question arises how in the majority of cases, these tumors can grow in the presence of such strong melanoma-specific T-cell responses. Explanations for this have been found both at the level of the tumor and at the level of the responding T-cell populations.

Correlation of MDA-specific T cells and melanoma progression

In a retrospective analysis of blood samples from 37 HLA-A2.1$^+$ stage IV melanoma patients drawn before treatment, naturally occurring CD8$^+$ T-cell immunity towards MDA (MART-1, tyrosinase and gp100) was determined. In over 40% of patients antigen-specific CD8$^+$ T cells for one or more of the MDA were detectable. All patients were subsequently treated in a phase II clinical trial with combined chemo-immunotherapy modality consisting of temozolomide and low dose IL-2, IFN-α and GM-CSF administered sequentially. After two courses, patients were evaluated for response to treatment. 50% of patients had progressive disease (PD). Surprisingly, in the vast majority (14/19) of patients with progressive disease MDA-specific T cells were detectable in peripheral blood by MHC tetramer analysis, sometimes in extremely high numbers. Statistical analysis of these data revealed a highly significant correlation between the presence of MDA-specific T cells and rapid disease progression despite immunotherapy-based treatment. Although MDA- and in particular MART-1-specific T-cell immunity is also found in healthy HLA-A2.1$^+$ individuals, the frequencies found in patients with advanced melanoma are much higher, indicating that these T-cell expansions are disease related and not a reflection of the normal distribution of MDA-specific T cells in HLA-A2.1$^+$ individuals (Romero *et al.*, 1998; Pittet *et al.*, 1999, van Oijen *et al.*, Ms submitted).

Phenotype and function of tumor-specific T cells

Examination of the phenotype of MDA-specific T cells from melanoma patients also provides evidence for a disease-related immune response. MART-1-specific T cells in healthy individuals display a naïve phenotype with high CD45RA, CCR7 and CD27 expression and low CD45RO expression. In contrast, MDA-specific T cells in about 50% of melanoma patients have an effector/memory phenotype based on high expression of CD45RO, and reduced expression of CD45RA, CD27 and CCR7 (D'Souza *et al.*, 1998; Romero *et al.*, 1998; Lee *et al.*, 1999b; Dunbar *et al.*, 2000; van Oijen *et al.*, Ms submitted), indicating that these cells are antigen-experienced. Furthermore, the extent of this phenotypic shift correlates with the magnitude of the MDA-specific T-cell response and with the ability of these cells to mediate *ex vivo* effector function (Dunbar *et al.*, 2000; van Oijen *et al.*, Ms submitted). Lee *et al.* (1999b) described one patient with high numbers of circulating (oligoclonal) tyrosinase-specific T cells (1 in 40 CD8$^+$ T cells was specific for tyrosinase) with a different cell surface phenotype. These cells expressed high levels of CD45RA, but no CD45RO, CD27 and CD28. This pattern of cell surface marker expression is thought to represent the phenotype of T cells that have fully differentiated towards effector cells (Hamann *et al.*, 1997). Despite this phenotype, these tyrosinase-specific T cells lacked cytolytic activity against peptide-pulsed or tyrosinase-expressing target cells, and did not produce IFN-γ or TNF-α upon activation. It seems however that the existence of this type of anergized melanoma-specific T cells is the exception rather than the rule. In most other cases in which the functional properties of melanoma-specific T cells were studied, these T cells behaved as classical memory/effectors with *ex vivo* antigen-induced cytokine production and cytolytic activity. Collectively, these data indicate that as based on the observed antigen-specific T-cell expansions, the phenotype of these cells and their functional properties, tumor-induced T-cell deletion or anergy is unlikely to be a dominant factor in advanced stage melanoma.

Clinical significance of tetramer-identified melanoma-specific T-cell responses

It is currently unclear how *ex vivo* effector function of tumor-specific $CD8^+$ T cells relates to their *in vivo* function. To date, little clinical benefit from the presence of melanoma-specific T cells has been observed. On the contrary, our data on the presence of MDA-specific T-cell immunity in stage IV melanoma patients indicate a negative correlation between the presence of MDA-specific T cells and clinical outcome. However, there is mounting evidence that MDA-specific T cells can be functionally active *in vivo*.

First, as discussed above the presence of TIL could be correlated with a better clinical outcome for primary cutaneous melanomas and regional lymph node metastases. Furthermore, a small fraction of patients with melanoma develop vitiligo, an auto-immune skin disease that is caused by destruction of skin melanocytes. Although not proven formally, it is suggested that tumor-reactive T cells recognizing a self-antigen expressed by both melanocytes and melanoma are responsible for this phenomenon. In line with this, MART-1-specific $CD8^+$ T cells with homing capacity for the skin through expression of the skin homing receptor CLA were recently detected in the peripheral blood of vitiligo patients by MHC tetramer staining (Ogg *et al.*, 1998). Elegant work from Yee *et al.* (2000) has recently provided direct evidence for T-cell mediated vitiligo. In this study, MART-1-specific T cells were adoptively transferred to a stage IV melanoma patient in combination with IL-2. Adoptive transfer resulted in homing of infused T cells to both melanoma metastases and pigmented areas of the skin and subsequent destruction of MART-1-positive melanocytes. The frequencies of MART-1-specific T cells observed in peripheral blood of this patient post infusion were similar to the frequencies observed in some untreated melanoma patients. However, in the latter group of patients vitiligo remains uncommon. Collectively, these data indicate that although the presence of melanoma/melanocyte-specific T cells is an obvious prerequisite for tumor and melanocyte destruction by MDA-specific T cells, there are additional requirements. One of these factors may be the (lack of) expression of skin homing receptors. In addition, many of the cases of immunotherapy-induced vitiligo in melanoma patients are associated with the administration of IL-2 (Rosenberg and White, 1996). IL-2 may exert this effect either through an effect on T-cell activation state or by increasing vascular permeability. Clearly, in addition to simply counting T cells by MHC tetramer staining, it will be crucial to define the (patho-) physiological conditions and/or T-cell phenotype required for the *in vivo* function of these tumor/self-specific T cells.

Additional evidence for a role of melanoma-specific T cells comes from the observation that, in particular, in advanced stage disease, tumor cells have often become insensitive to T-cell attack. This has been shown to occur through loss of antigen expression, down-regulation of transporters associated with antigen processing (TAP) (Thor Straten *et al.*, 1997), and loss of HLA expression (Maeurer *et al.*, 1996; Cormier *et al.*, 1998; Hicklin *et al.*, 1998; Jimenez *et al.*, 1999). Recently, loss of expression of MART-1 antigen in primary cutaneous melanoma was shown to be a negative prognostic indicator for disease-free and overall survival (Berset *et al.*, 2001), perhaps suggesting that selection of immune escape variants is already an early phenomenon in the development to metastatic melanoma (Kageshita *et al.*, 1997; Geertsen *et al.*, 1998). In our own study on MDA-specific immunity in advanced stage melanoma patients loss of HLA-A expression was observed in about 50% of patients with rapid disease progression. Interestingly, in the majority of these patients MDA-specific T cells were detectable, suggesting that MHC class I-deficient tumors can induce or at least maintain tumor-specific T-cell responses, possibly through cross-presentation of tumor-derived antigens by professional APCs. In a mouse model in which T-cell responses against experimental tumors can be monitored directly, we have recently demonstrated that induction of tumor-specific $CD8^+$ T-cell immunity can indeed occur independent of MHC expression by the tumor, through cross-presentation of tumor-derived proteins (Wolkers *et al.*, 2001). This provides a somber view of the battle between the immune system and tumors, in which the body is mounting an increasingly strong T-cell response against a tumor that is completely insensitive to this attack.

Therapy-induced T-cell responses

A number of studies have used MHC tetramer technology to visualize the effects of immunotherapeutic interventions on T-cell responses against solid tumors in cancer patients (Lee *et al.*, 1999a; Fong *et al.*, 2001; Jonuleit *et al.*, 2001; Lau *et al.*, 2001; Pittet *et al.*, 2001). In a recently published paper by Fong *et al.* (2001) autologous DC pulsed with an APL of carcino-embryonic antigen (CEA) induced T cells that recognized tumor cells expressing endogenous CEA. In this study, the expansion of CEA-specific $CD8^+$ T cells correlated with clinical response. In melanoma, the currently available results have not revealed any clear correlation between increases in tumor-specific T cells in peripheral blood and clinical responses. In studies performed in the lab of S. Rosenberg and F. Marincola, vaccination of melanoma patients with an altered peptide ligand (APL) of a gp100 epitope resulted in an increase in the number of circulating gp100-specific T cells up to 2.5% (Lee *et al.*, 1999a). Although immunization was successful in 91% of patients, no clinical responses were observed. In contrast, patients that also received high dose IL-2 did show clinical responses, however, in this patient group an increase in gp100-specific T cells was found in only few cases. The authors suggest that in the latter patient group gp100-specific T cells may have extravasated to the tumor site. Although there are currently no data that directly support this hypothesis, it is becoming increasingly clear from mouse studies that T-cell migration patterns are highly regulated throughout T-cell activation and an important factor in T-cell function. In addition, we have observed that T-cell populations with different specificities may not be distributed equally over different sites. In a patient in which significant frequencies of gp100-, MART-1- and tyrosinase-specific T cells could be observed in peripheral blood, the latter cells were conspicuously absent from a tumor-infiltrated lymph node (Figure 7.2). Thereby these studies also point to one of the major limitations of MHC tetramer analysis of peripheral blood samples for T-cell responses against human solid tumors. Ideally the efficacy of vaccination protocols should be assessed not by determining the number of T cells induced, but rather by the number of T cells induced at the desired location. In particular the efficacy of vaccination protocols that affect T-cell distribution, for instance, through

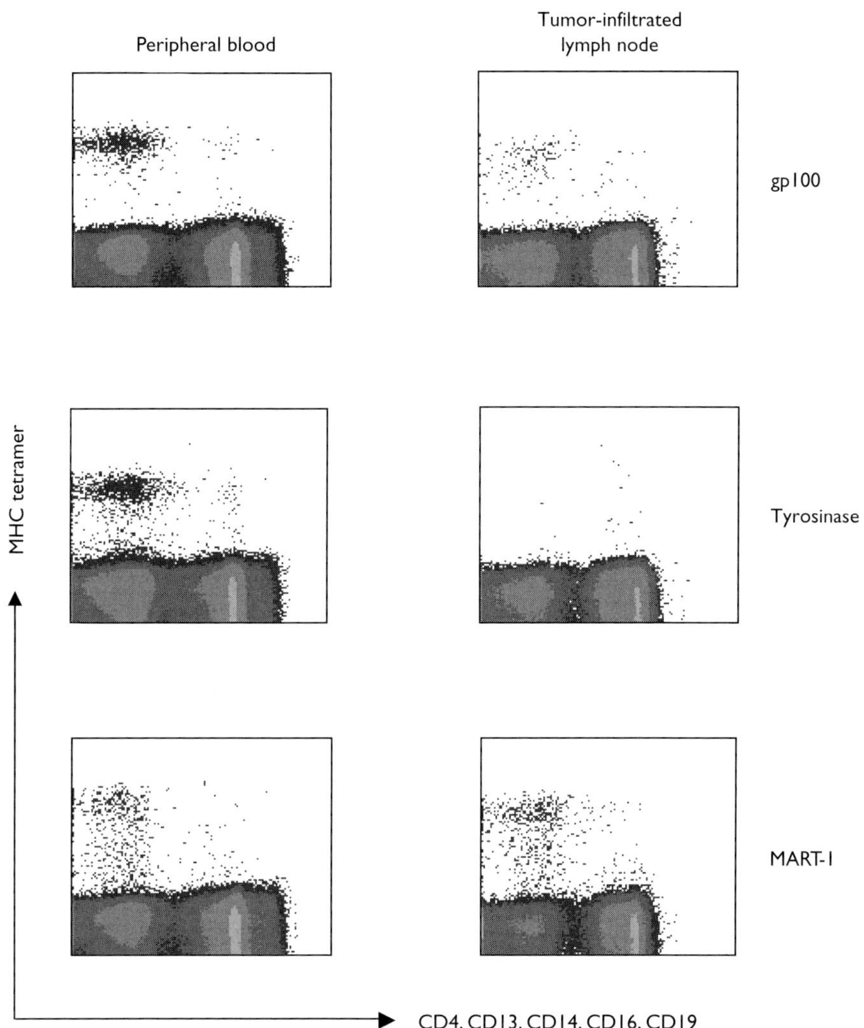

Figure 7.2 MHC tetramer staining of peripheral blood lymphocytes and tumor-infiltrated lymph node lymphocytes from a melanoma patient. Mononuclear cells isolated from peripheral blood and tumor-infiltrated lymph nodes were stained with HLA-A2.1 tetramers containing peptides gp100$_{280-288}$, tyrosinase$_{368-374}$ and MART-1$_{26-35}$. Cells were counterstained with a panel of monoclonal antibodies for lineage antigens CD4, CD13, CD14, CD16 and CD19, and thereafter analyzed by flow cytometry. MDA-specific CD8$^+$ T cells appear in the upper left quadrant. Note the absence of tyrosinase-specific T cells in the tumor-infiltrated lymph node.

inclusion of cytokines, may be misinterpreted by only monitoring the magnitude of T-cell responses in peripheral blood.

Allogeneic bone marrow transplantation (BMT) is increasingly used as an immunotherapeutic strategy for hematological malignancies and more recently also solid tumors. The anti-tumor effect of allogeneic transplants is mediated by minor histocompatibility antigen- and/or tumor antigen-specific T cells that are present in the transplant, and is strongly correlated with the development of Graft versus Host Disease (GvHD) (Appelbaum, 2001). MHC tetramer technology has recently been used to couple the development of antigen-specific T-cell responses to clinical course in BMT recipients, and these first studies have provided some very interesting insights.

Goulmy and coworkers reported a retrospective study of 17 allogeneic BMT recipients, using MHC tetramers complexed with minor histocompatibility antigens (Mutis *et al.*, 1999). This study shows that the development of mHag-specific T-cell responses that can be detected with MHC tetramers correlates with clinically evident GvHD. Furthermore, the effect of GvHD treatment on the magnitude of these T-cell responses could be correlated to the development of chronic GvHD.

Patients with chronic myelogenous leukemia (CML) that relapse after allogeneic bone marrow or peripheral stem cell transplantation have an 80% chance of complete remission (CR) upon donor lymphocyte infusion (DLI). Responses to DLI are often accompanied by GvHD and minor histocompatibility antigen-specific T cells are thought to mediate both GvHD and graft-versus-leukemia. Marijt *et al.* (2000) performed MHC tetramer analysis of blood samples prior to and weekly after DLI in a patient with relapsed CML after allogeneic stem cell transplantation. Their study revealed a transient but profound increase in both mHag HA-1 and HA-2-specific T-cell populations starting at week 5 after DLI and lasting about 3–4 weeks. This increase in circulating mHag-specific T-cell pool was associated with conversion to full donor chimerism and molecular CR of the disease.

In a separate study, Champlin and colleagues examined the development of T-cell immunity against the CML antigen proteinase 3 (Molldrem *et al.*, 2000). This antigen is a non-mutated self-protein that is modestly overexpressed in CML cells. No proteinase 3-specific T cells could be detected by MHC tetramer staining of PBL from CML patients that were treated with chemotherapy. In contrast, upon treatment of patients with interferon-alpha, and in particular upon allogeneic BMT, large numbers of MHC-tetramer-reactive T cells appeared. Interestingly, these expansions of proteinase 3-specific T cells correlated well with tumor response.

These studies reveal the value of MHC tetramer staining for our understanding of the clinical effects of immunotherapeutic interventions. MHC tetramers can be applied to couple expansions of specific T-cell populations to the development of either GvHD or tumor regression. It is noted though that these correlations do not formally prove a causal relationship between T-cell expansion and clinical manifestations. As an example, the presence of significant numbers of proteinase 3-specific T cells in patients with CML regression may in theory also be explained as a T-cell expansion that occurs as a consequence of increased antigen liberation upon tumor cell death. Definitive proof for a causal relationship may in this case come from clinical trials in which such T-cell responses are enhanced, for instance by vaccination of BMT donors.

Additional technologies

Detection of tumor-reactive T cells by tetramer-CLSM

Although MHC tetramer analysis of T cells by flow cytometry has been extremely valuable for the visualization of antigen-specific T-cell responses, it does not provide information on the spatial distribution of T-cell immunity in lymphoid organs or at effector sites.

Information on the distribution of antigen-specific T cells in tumor sites would be particularly interesting in view of the documented correlation between the nature of T-cell infiltrates and prognosis, as has been documented for primary melanoma (Clemente *et al.*, 1996). To extend MHC tetramer technology to the *in situ* visualization of polyclonal antigen-specific T-cell responses we have recently developed a strategy that permits the use of MHC tetramers in tissue sections. In a first set of experiments we tested a large number of conventional immunohistochemical techniques for the detection of TCR-transgenic cells in tissue sections by MHC tetramer staining. These experiments established that an optimized immunohistochemistry protocol using acetone-fixed cryosections showed low but detectable MHC tetramer staining of T cells in organs of TCR-transgenic mice. However, in spite of numerous attempts, this technique fails to reveal physiological T-cell responses, such as those following viral infections. The poor results obtained with conventional immunohistochemical techniques are not surprising. Binding of MHC tetramers requires TCRs in their native conformation and it is unclear to what extent this conformation is retained in conventional IHC techniques. In addition, MHC tetramer binding appears a dynamic process that depends on TCR reorganization (Drake and Braciale, 2001), a process that is undoubtedly lost in fixed tissue material. In an attempt to retain the TCR structure and mobility that seems to be required for MHC tetramer binding, we have examined the use of fresh, unfixed tissue sections. Noelle and coworkers previously showed that incubation of viable sections of secondary lymphoid organs with fluorochrome-labeled monoclonal antibody results in sufficient penetration to provide bona-fide staining when analyzed by confocal laser scanning microscopy (CLSM) (Gonzalez *et al.*, 1998). This type of fresh, non-manipulated sections can be prepared from most organs through the use of a vibrating blade microtome. MHC tetramer staining of organs of TCR-transgenic mice but also of non-transgenic mice infected with influenza A virus reveals an antigen-specific staining of T cells in spleen and lymph nodes. In addition, MHC tetramer CLSM can visualize polyclonal, antigen-specific tumor-infiltrating T cells in an experimental tumor model (Haanen *et al.*, 2000). In line with the proposed reorganization of TCRs upon MHC tetramer binding, we find that MHC tetramer-positive cells show a distinct, punctate staining pattern that is suggestive of TCR clustering. As compared to MHC tetramer flow cytometry the background of MHC tetramer CLSM is somewhat higher (approximately five-fold). In addition, because – unlike MHC tetramer flow cytometry – the number of events that can be analyzed in tetramer-stained CLSM sections is low, only T-cell responses of a sufficiently strong magnitude can be visualized.

Haase and colleagues have also obtained evidence for the feasibility of MHC tetramer staining of non-fixed tissue sections, using TCR-transgenic T cells (Skinner *et al.*, 2000). In these experiments, antigen-specific TCR-transgenic cells could also be visualized in lightly fixed and frozen tissues. The ability to stain antigen-specific T cells in previously archived material would clearly enhance the value of this method. Although our current understanding of the nature of the interaction between MHC tetramers and T cells (p. 114) might suggest limitations to the use of MHC tetramer CLSM for the detection of physiological T-cell responses, a recent murine study indicates that this technology can in fact be used to dissect *in situ* T cell responses (McGavern *et al.*, 2002).

We have not yet been able to apply this new tool for the *in situ* analysis of human tumor-specific T-cell responses. The reasons for this are two-fold. First of all, the technology requires fresh material. Second, the T-cell frequencies present within these sections must be significant, in part because the background of this technique is somewhat higher than for tetramer flow cytometry, but more importantly, because only a small number of events can be analyzed. Based on MHC tetramer flow cytometry data, the analysis of sections of lymphoid organs of patients with acute infectious mononucleosis may form a useful starting point (Callan *et al.*, 1998). In collaboration with Drs A. Dickinson and E. Goulmy we have however, used this technology for the *in situ* analysis of T-cell infiltration in a skin explant model (Dickinson *et al.*, 2002). In this assay, skin biopsies taken from transplant recipients prior to transplantation are incubated with CTL specific for minor histocompatibility antigens (mHag) and the degree of GvH reaction is later determined by histopathological analysis of skin sections. To gain insight into the migration patterns of CTLs, we have studied the infiltration of mHag-specific CTL in these human skin biopsies. These experiments reveal CTL infiltration in the dermis and epidermis of skin sections as evidenced by MHC tetramer CLSM and this infiltration is correlated to the degree of GvHD. These studies document that MHC tetramer CLSM can be applied to the study of clinically significant T-cell responses in a human setting, albeit if still for an *ex vivo* manipulated CTL response.

MHC class II tetramers

In the past years, mouse model systems have highlighted the role of tumor-specific $CD4^+$ T-cell responses both in $CD8^+$ T-cell activation and in the orchestration of immune responses mediated by cell types such as macrophages and eosinophils. In addition, in certain settings, tumor-specific $CD4^+$ T cells can mediate direct effector functions that suppress tumor growth (reviewed in Pardoll and Topalian, 1998; Toes *et al.*, 1999). These data have led to a renewed interest in the role of antigen-specific $CD4^+$ T cells in human tumor-specific immunity. However, as compared to the large number of MHC class I-restricted epitopes from tumor antigens that have been identified, only a small number of MHC class II-restricted epitopes has been defined to date (see Chapter 9 by Wang). Furthermore, part of these epitopes for $CD4^+$ T cells has been defined by "reverse immunology" and it is unclear to which of these epitopes T-cell responses occur in cancer patients.

The production of MHC class II tetramers is significantly more complex than that of MHC class I tetramers. Approaches for both the *in vitro* refolding of bacterially expressed MHC class II chains and expression of MHC class II heterodimers in eukaryotic cells have been developed (Crawford *et al.*, 1998; Gutgemann *et al.*, 1998; Cochran and Stern, 2000). Although the yield of both methods is low as compared to the production of MHC class I complexes, the resulting MHC class II tetramers stain antigen-specific $CD4^+$ T cells in specific fashion. Conditions of staining may be even more important than for MHC class I tetramers. As has been observed for MHC class I tetramers, staining of $CD4^+$ T cells with MHC class II tetramers is highly temperature-dependent. However, contrary to MHC class I tetramers, staining of $CD4^+$ T cells generally requires prolonged incubation of T cells with MHC class II tetramers. MHC class II tetramer-binding can be blocked by drugs that disrupt clustering of membrane proteins and may involve internalization of MHC complexes (Cameron *et al.*, 2001). Finally, $CD4^+$ T cells with lower antigen sensitivity do not always stain with the current generation MHC class II tetramers, and this could in some cases lead to an underestimate of T-cell frequencies (Rees *et al.*, 1999; Glaichenhaus, personal communication; Schepers, unpublished observations).

Notwithstanding these problems, several labs including ours have started the production of human MHC class II tetramers complexed with human tumor antigens. While no data are available yet on the analysis of human tumor-specific T-cell responses with these reagents, we have recently analyzed the antigen-specific $CD4^+$ T-cell response by MHC tetramer technology in a murine retrovirus-induced sarcoma model. Comparison of the antigen-specific $CD4^+$ and $CD8^+$ T-cell responses by MHC tetramer technology in this model reveals that the retrovirus-specific $CD4^+$ T-cell response is of a significantly smaller magnitude than the antigen-specific $CD8^+$ T-cell response (Schepers *et al.*, 2002). Interestingly, the wave of antigen-specific $CD4^+$ T cells appears to precede the antigen-specific $CD8^+$ response both in the lymphoid organs and at the effector site. These data suggest an important role of early $CD4^+$ T-cell responses in this murine sarcoma model and $CD4^+$ T-cell depletion studies support this notion. Although these experiments make an evaluation of human tumor-specific $CD4^+$ T-cell immunity with MHC class II tetramers all the more attractive, the observed low frequency of antigen-specific $CD4^+$ T cells in this highly immunogenic model suggests that this may be a difficult task. In line with this, available data with MHC class II tetramers in human models of viral disease and autoimmune disease show low frequencies of antigen-specific $CD4^+$ T cells. In human Lyme arthritis and autoimmune arthritis only very small numbers of tetramer-reactive $CD4^+$ T cells were found in synovial fluids (Kotzin *et al.*, 2000; Meyer *et al.*, 2000). Likewise, in humans previously infected with HSV or influenza A, MHC class II tetramer-reactive cells could be visualized, but only after *in vitro* restimulation (Novak *et al.*, 1999; Kwok *et al.*, 2000). Collectively, these data suggest that in many clinical settings, the applicability of current generation MHC class II tetramers in the study of human tumor-specific T-cell immunity may be limited as compared to MHC class I tetramers. However, with ongoing optimization of both the multimeric MHC technology and the equipment for detection, this may well change in the future.

Use of tetramers to select (high affinity) tumor-specific T cells

Apart from the use of MHC tetramers for the visualization of antigen-specific T-cell responses, these reagents are also becoming of significant use in the manipulation of tumor-specific T-cell responses. Although MHC tetramer binding appears to activate antigen-specific T cells (Smyth *et al.*, 1999), T-cell viability is not affected. Consequently, MHC tetramer staining can be used to isolate antigen-specific T cells from heterogeneous T-cell populations either by flow cytometry (Dunbar *et al.*, 1998, 1999; Molldrem *et al.*, 1999; Youde *et al.*, 2000) or by magnetic bead sorting (Bodinier *et al.*, 2000). Cerundolo and coworkers first demonstrated that MHC tetramer-based T-cell isolation can be used to rapidly obtain T cells with a desired antigen specificity that were present at low frequency in the starting population (Dunbar *et al.*, 1998). In addition, MHC tetramer-based sorting can be used to selectively isolate tumor-specific T-cell populations with higher activity. As discussed above, for at least some of the human tumor antigens, most of the T cells that can be detected by tetramer technology appear to have a low avidity for the relevant peptide–MHC complex. In an elegant series of experiments, Greenberg and colleagues showed that by sorting MART 1 or tyrosinase tetramer^{hi} T cells, tumor-specific T-cell populations could be obtained that recognized antigen-expressing melanoma cells. In contrast, MART-1 or tyrosinase tetramer^{lo} T cells isolated from these patients recognized peptide-loaded target cells but not tumor cells (Yee *et al.*, 1999). These experiments extend prior studies that showed that MHC tetramer binding correlates well with TCR affinity (Crawford *et al.*, 1998) and show that optimal tumor-reactive T cells can be isolated by MHC tetramer-based sorting. It seems likely that this type of approach in which tumor-reactive T cells are isolated with the aid of

MHC tetramers, are subsequently expanded *in vitro* and infused into patients will be tested in clinical trials in the forthcoming years.

Use of tetramers to select allo MHC-restricted or in vitro-modified TCRs

Although the tetramer-selection approach discussed above is valuable for the isolation of self-specific T cells for which small numbers of high affinity T cells are present naturally, it has become clear that for a significant fraction of self-antigens the high affinity repertoire has been deleted in its entirety (de Visser *et al.*, 2000; Hernandez *et al.*, 2000; Nugent *et al.*, 2000). A number of approaches have been put forward to circumvent the limitations of the naturally occurring T-cell repertoire in those cases, and MHC tetramer-based sorting is becoming an essential part of these strategies.

As first suggested by Stauss and colleagues, the effects of self-tolerance can be circumvented by using T-cell populations from MHC-mismatched individuals (Sadovnikova and Stauss, 1996). The induction of self-tolerance requires the presence of both the self-antigen and the relevant MHC molecule. By virtue of the fact that allogeneic T cells have not been exposed to the tolerizing peptide–MHC complex during thymic selection, the deletion of TCRs with the desired tumor-specificities has not taken place. To date, this approach for CTL generation has successfully been used to generate human and murine allo-MHC restricted T cells for a number of antigens (reviewed in Stauss, 1999). MHC tetramer-based T-cell sorting should be useful to selectively isolate *peptide-specific* allo-restricted T cells from T-cell cultures that also contain significant number of alloreactive T cells that show a slighter peptide dependence (Moris *et al.*, 2001). Furthermore, this selective isolation of T cells with a desired fine specificity can be facilitated by dual MHC tetramer staining (Haanen *et al.*, 1999).

An independent approach for the generation of TCRs with desirable tumor specificities is the selection of TCRs from *in vitro* generated libraries of TCRs. Expression systems that can be used to display libraries of TCRs have been developed in yeast and in T cells and have been used to select variant TCRs with either an increased affinity (Holler *et al.*, 2000), or altered specificity (Kessels *et al.*, 2000). It will be interesting to establish whether these approaches can be used to obtain improved tumor-specific TCRs, for instance by converting low affinity melanoma-specific TCRs into more potent TCRs.

The successful generation of a set of high affinity TCRs specific for human tumor antigens with any of the above-described methods may provide an effective way to circumvent the limitations of the physiological T-cell repertoire. Resulting high affinity TCRs could be used to redirect the specificity of T-cell populations by gene transfer of TCRs, thereby leading to strong and instantaneous immunity towards the antigen of choice. To test the feasibility of TCR gene therapy we have set up a murine model system for TCR gene transfer. A murine αβ TCR was introduced into peripheral T cells by retrovirus-mediated gene transfer and the effects of the resulting "redirected" T cells upon reintroduction into mice were established. T cells redirected by TCR gene transfer can be followed directly *ex vivo* by MHC tetramer technology. These T cells expand dramatically (approximately 3-log) upon *in vivo* antigen encounter and efficiently home to effector sites. In this model, TCR gene transfer is not associated with any significant autoimmune pathology. Furthermore, small numbers of TCR-transduced T cells can promote the rejection of antigen-expressing tumors *in vivo*. These data suggest the redirection of T cells by TCR gene transfer as a feasible strategy for the rapid induction of tumor-specific immunity (Kessels *et al.*, 2001). An important next step will be to establish the feasibility of TCR gene transfer as a therapeutic strategy in a human setting. It will be of particular interest to examine the value of this approach in conditions such as post-transplant lymphoproliferative disorder, in which infusion of antigen-specific T cells has previously been shown to be effective (Rooney *et al.*, 1997).

Acknowledgments

The authors would like to thank the Dutch Cancer Society (grants NKI 99–2036, NKI 2001–2417 and NKI 2001–2419) and the Netherlands Organization for Scientific Research (grant NWO Pioneer 00–03) for their financial support.

References

Altman, J. D., Moss, P. A., Goulder, P. J., Barouch, D. H., McHeyzer-Williams, M. G., Bell, J. I., McMichael, A. J. and Davis, M. M. (1996) *Science*, **274**, 94–6.

Appelbaum, F. R. (2001) *Nature*, **411**, 385–9.

Berset, M., Cerottini, J. P., Guggisberg, D., Romero, P., Burri, F., Rimoldi, D. and Panizzon, R. G. (2001) *Int. J. Cancer*, **95**, 73–7.

Bodinier, M., Peyrat, M. A., Tournay, C., Davodeau, F., Romagne, F., Bonneville, M. and Lang, F. (2000) *Nat. Med.*, **6**, 707–10.

Callan, M. F., Tan, L., Annels, N., Ogg, G. S., Wilson, J. D., O'Callaghan, C. A., Steven, N., McMichael, A. J. and Rickinson, A. B. (1998) *J. Exp. Med.*, **187**, 1395–402.

Cameron, T. O., Cochran, J. R., Yassine-Diab, B., Sekaly, R. P. and Stern, L. J. (2001) *J. Immunol.*, **166**, 741–5.

Chen, W., Anton, L. C., Bennink, J. R. and Yewdell, J. W. (2000) *Immunity*, **12**, 83–93.

Clark, W. H., Jr, Elder, D. E., Guerry, D. T., Braitman, L. E., Trock, B. J., Schultz, D., Synnestvedt, M. and Halpern, A. C. (1989) *J. Natl. Cancer Inst.*, **81**, 1893–904.

Clemente, C. G., Mihm, M. C., Jr, Bufalino, R., Zurrida, S., Collini, P. and Cascinelli, N. (1996) *Cancer*, **77**, 1303–10.

Cochran, J. R. and Stern, L. J. (2000) *Chem. Biol.*, **7**, 683–96.

Cormier, J. N., Hijazi, Y. M., Abati, A., Fetsch, P., Bettinotti, M., Steinberg, S. M., Rosenberg, S. A. and Marincola, F. M. (1998) *Int. J. Cancer*, **75**, 517–24.

Coulie, P. G., Van den Eynde, B. J., van der Bruggen, P., Van Pel, A., Boon, T. (1997) *Biochem. Soc. Trans.*, **25**, 544–8.

Crawford, F., Kozono, H., White, J., Marrack, P. and Kappler, J. (1998) *Immunity*, **8**, 675–82.

D'Souza, S., Rimoldi, D., Lienard, D., Lejeune, F., Cerottini, J. C. and Romero, P. (1998) *Int. J. Cancer*, **78**, 699–706.

Daniels, M. A. and Jameson, S. C. (2000) *J. Exp. Med.*, **191**, 335–46.

de Visser, K. E., Cordaro, T. A., Kioussis, D., Haanen, J. B., Schumacher, T. N. and Kruisbeek, A. M. (2000) *Eur. J. Immunol.*, **30**, 1458–68.

Denkberg, G., Cohen, C. J. and Reiter, Y. (2001) *J. Immunol.*, **167**, 270–6.

Dickinson, A. M., Wang, X. N., Sviland, L., Vyth-Dreese, F., Dunn, J., Jackson, G., Schumacher, T. N. M., Haanen, J. B. A. G., Mutis, T. and Goulmy, E. (2002) *Nat. Med.*, **8**, 410–14.

Drake, D. R., 3rd and Braciale, T. J. (2001) *J. Immunol.*, **166**, 7009–13.

Dunbar, P. R., Chen, J. L., Chao, D., Rust, N., Teisserenc, H., Ogg, G. S., Romero, P., Weynants, P. and Cerundolo, V. (1999) *J. Immunol.*, **162**, 6959–62.

Dunbar, P. R., Ogg, G. S., Chen, J., Rust, N., van der Bruggen, P. and Cerundolo, V. (1998) *Curr. Biol.*, **8**, 413–16.

Dunbar, P. R., Smith, C. L., Chao, D., Salio, M., Shepherd, D., Mirza, F., Lipp, M., Lanzavecchia, A., Sallusto, F., Evans, A., Russell-Jones, R., Harris, A. L. and Cerundolo, V. (2000) *J. Immunol.*, **165**, 6644–52.

Fong, L., Hou, Y., Rivas, A., Benike, C., Yuen, A., Fisher, G. A., Davis, M. M. and Engleman, E. G. (2001) *Proc. Natl. Acad. Sci. USA*, **26**, 26.

Gaugler, B., Van den Eynde, B., van der Bruggen, P., Romero, P., Gaforio, J. J., De Plaen, E., Lethe, B., Brasseur, F. and Boon, T. (1994) *J. Exp. Med.*, **179**, 921–30.

Geertsen, R. C., Hofbauer, G. F., Yue, F. Y., Manolio, S., Burg, G. and Dummer, R. (1998) *J. Invest. Dermatol.*, **111**, 497–502.

Gibbs, P., Hutchins, A. M., Dorian, K. T., Vaughan, H. A., Davis, I. D., Silvapulle, M. and Cebon, J. S. (2000) *Melanoma Res.*, **10**, 259–64.

Gonzalez, M., Mackay, F., Browning, J. L., Kosco-Vilbois, M. H. and Noelle, R. J. (1998) *J. Exp. Med.*, **187**, 997–1007.

Gutgemann, I., Fahrer, A. M., Altman, J. D., Davis, M. M. and Chien, Y. H. (1998) *Immunity*, **8**, 667–73.

Haanen, J. B., van Oijen, M. G., Tirion, F., Oomen, L. C., Kruisbeek, A. M., Vyth-Dreese, F. A. and Schumacher, T. N. (2000) *Nat. Med.*, **6**, 1056–60.

Haanen, J. B., Wolkers, M. C., Kruisbeek, A. M. and Schumacher, T. N. (1999) *J. Exp. Med.*, **190**, 1319–28.

Hamann, D., Baars, P. A., Rep, M. H., Hooibrink, B., Kerkhof-Garde, S. R., Klein, M. R. and van Lier, R. A. (1997) *J. Exp. Med.*, **186**, 1407–18.

Hernandez, J., Lee, P. P., Davis, M. M. and Sherman, L. A. (2000) *J. Immunol.*, **164**, 596–602.

Hicklin, D. J., Wang, Z., Arienti, F., Rivoltini, L., Parmiani, G. and Ferrone, S. (1998) *J. Clin. Invest.*, **101**, 2720–9.

Holler, P. D., Holman, P. O., Shusta, E. V., O'Herrin, S., Wittrup, K. D. and Kranz, D. M. (2000) *Proc. Natl. Acad. Sci. USA*, **97**, 5387–92.

Jager, E., Nagata, Y., Gnjatic, S., Wada, H., Stockert, E., Karbach, J., Dunbar, P. R., Lee, S. Y., Jungbluth, A., Jager, D., Arand, M., Ritter, G., Cerundolo, V., Dupont, B., Chen, Y. T., Old, L. J. and Knuth, A. (2000) *Proc. Natl. Acad. Sci. USA*, **97**, 4760–5.

Jimenez, P., Canton, J., Collado, A., Cabrera, T., Serrano, A., Real, L. M., Garcia, A., Ruiz-Cabello, F. and Garrido, F. (1999) *Int. J. Cancer*, **83**, 91–7

Jonuleit, H., Giesecke-Tuettenberg, A., Tuting, T., Thurner-Schuler, B., Stuge, T. B., Paragnik, L., Kandemir, A., Lee, P. P., Schuler, G., Knop, J. and Enk, A. H. (2001) *Int. J. Cancer*, **93**, 243–51.

Kageshita, T., Kawakami, Y., Hirai, S. and Ono, T. (1997) *J. Immunother.*, **20**, 460–5.

Kawakami, Y., Dang, N., Wang, X., Tupesis, J., Robbins, P. F., Wang, R. F., Wunderlich, J. R., Yannelli, J. R. and Rosenberg, S. A. (2000) *J. Immunother.*, **23**, 17–27.

Kessels, H. W., van Den Boom, M. D., Spits, H., Hooijberg, E. and Schumacher, T. N. (2000) *Proc. Natl. Acad. Sci. USA*, **97**, 14578–83.

Kessels, H. W., Wolkers, M. C., van den Boom, M. D., Van der Valk, M. and Schumacher, T. N. (2001) *Nat. Immunol.*, **2**, 957–61.

Kotzin, B. L., Falta, M. T., Crawford, F., Rosloniec, E. F., Bill, J., Marrack, P. and Kappler, J. (2000) *Proc. Natl. Acad. Sci. USA*, **97**, 291–6.

Kwok, W. W., Liu, A. W., Novak, E. J., Gebe, J. A., Ettinger, R. A., Nepom, G. T., Reymond, S. N. and Koelle, D. M. (2000) *J. Immunol.*, **164**, 4244–9.

Lau, R., Wang, F., Jeffery, G., Marty, V., Kuniyoshi, J., Bade, E., Ryback, M. E. and Weber, J. (2001) *J. Immunother.*, **24**, 66–78.

Lee, K. H., Wang, E., Nielsen, M. B., Wunderlich, J., Migueles, S., Connors, M., Steinberg, S. M., Rosenberg, S. A. and Marincola, F. M. (1999a) *J. Immunol.*, **163**, 6292–300.

Lee, P. P., Yee, C., Savage, P. A., Fong, L., Brockstedt, D., Weber, J. S., Johnson, D., Swetter, S., Thompson, J., Greenberg, P. D., Roederer, M. and Davis, M. M. (1999b) *Nat. Med.*, **5**, 677–85.

Maeurer, M. J., Gollin, S. M., Storkus, W. J., Swaney, W., Karbach, J., Martin, D., Castelli, C., Salter, R., Knuth, A. and Lotze, M. T. (1996) *Clin. Cancer Res.*, **2**, 641–52.

Malherbe, L., Filippi, C., Julia, V., Foucras, G., Moro, M., Appel, H., Wucherpfennig, K., Guery, J. C. and Glaichenhaus, N. (2000) *Immunity*, **13**, 771–82.

Marchand, M., van Baren, N., Weynants, P., Brichard, V., Dreno, B., Tessier, M. H., Rankin, E., Parmiani, G., Arienti, F., Humblet, Y., Bourlond, A., Vanwijck, R., Lienard, D., Beauduin, M., Dietrich, P. Y., Russo, V., Kerger, J., Masucci, G., Jager, E., De Greve, J., Atzpodien, J., Brasseur, F., Coulie, P. G., van der Bruggen, P. and Boon, T. (1999) *Int. J. Cancer*, **80**, 219–30.

Marijt, W. A. F., Kester, M. G. D., Goulmy, E., Mutis, T., Drijfhout, J. W., Willemze, R. and Falkenburg, J. H. F. (2000) *Blood*, **96**, 478A.

McGavern, D. B., Christen, U., Oldstone, M. B. (2002) *Nat. Immunol.*, **3**, 918–25.

McHeyzer-Williams, M. G. and Davis, M. M. (1995) *Science*, **268**, 106–11.

Meyer, A. L., Trollmo, C., Crawford, F., Marrack, P., Steere, A. C., Huber, B. T., Kappler, J. and Hafler, D. A. (2000) *Proc. Natl. Acad. Sci. USA*, **97**, 11433–8.

Molldrem, J. J., Lee, P. P., Wang, C., Champlin, R. E. and Davis, M. M. (1999) *Cancer Res.*, **59**, 2675–81.

Molldrem, J. J., Lee, P. P., Wang, C., Felio, K., Kantarjian, H. M., Champlin, R. E. and Davis, M. M. (2000) *Nat. Med.*, **6**, 1018–23.

Moris, A., Teichgraber, V., Gauthier, L., Buhring, H. J. and Rammensee, H. G. (2001) *J. Immunol.*, **166**, 4818–21.

Murali-Krishna, K., Altman, J. D., Suresh, M., Sourdive, D. J., Zajac, A. J., Miller, J. D., Slansky, J. and Ahmed, R. (1998) *Immunity*, **8**, 177–87.

Mutis, T., Gillespie, G., Schrama, E., Falkenburg, J. H., Moss, P. and Goulmy, E. (1999) *Nat. Med.*, **5**, 839–42.

Novak, E. J., Liu, A. W., Nepom, G. T. and Kwok, W. W. (1999) *J. Clin. Invest.*, **104**, R63–7.

Nugent, C. T., Morgan, D. J., Biggs, J. A., Ko, A., Pilip, I. M., Pamer, E. G. and Sherman, L. A. (2000) *J. Immunol.*, **164**, 191–200.

Ogg, G. S., Rod Dunbar, P., Romero, P., Chen, J. L. and Cerundolo, V. (1998) *J. Exp. Med.*, **188**, 1203–8.

Palermo, B., Campanelli, R., Mantovani, S., Lantelme, E., Manganoni, A. M., Carella, G., Da Prada, G., della Cuna, G. R., Romagne, F., Gauthier, L., Necker, A. and Giachino, C. (2001) *Eur. J. Immunol.*, **31**, 412–20.

Panelli, M. C., Bettinotti, M. P., Lally, K., Ohnmacht, G. A., Li, Y., Robbins, P., Riker, A., Rosenberg, S. A. and Marincola, F. M. (2000) *J. Immunol.*, **164**, 4382–92.

Pardoll, D. M. and Topalian, S. L. (1998) *Curr. Opin. Immunol.*, **10**, 588–94.

Pittet, M. J., Speiser, D. E., Lienard, D., Valmori, D., Guillaume, P., Dutoit, V., Rimoldi, D., Lejeune, F., Cerottini, J. C. and Romero, P. (2001) *Clin. Cancer Res.*, **7**, 796s–803s.

Pittet, M. J., Valmori, D., Dunbar, P. R., Speiser, D. E., Lienard, D., Lejeune, F., Fleischhauer, K., Cerundolo, V., Cerottini, J. C. and Romero, P. (1999) *J. Exp. Med.*, **190**, 705–15.

Rees, W., Bender, J., Teague, T. K., Kedl, R. M., Crawford, F., Marrack, P. and Kappler, J. (1999) *Proc. Natl. Acad. Sci. USA*, **96**, 9781–6.

Romero, P., Dunbar, P. R., Valmori, D., Pittet, M., Ogg, G. S., Rimoldi, D., Chen, J. L., Lienard, D., Cerottini, J. C. and Cerundolo, V. (1998) *J. Exp. Med.*, **188**, 1641–50.

Rooney, C. M., Smith, C. A. and Heslop, H. E. (1997) *Mol. Med. Today*, **3**, 24–30.

Rosenberg, S. A. and White, D. E. (1996) *J. Immunother. Emphasis. Tumor Immunol.*, **19**, 81–4.

Sadovnikova, E. and Stauss, H. J. (1996) *Proc. Natl. Acad. Sci. USA*, **93**, 13114–18.

Schepers, K., Toebes, M., Vyth-Dreese, F. A., Dellemijn, T. A. M., Melief, C. J. M., Ossendorp, F. and Schumacher, T. N. M. (2002) *J. Immunol.*, **169**, 3191–9.

Schneck, J. P. (2000) *Immunol. Invest.*, **29**, 163–9.

Skinner, P. J., Daniels, M. A., Schmidt, C. S., Jameson, S. C. and Haase, A. T. (2000) *J. Immunol.*, **165**, 613–17.

Smyth, L. A., Reynolds, L., Tyrrell, R., Williams, O., Norton, T., Schumacher, T., Tybulewicz, V., Ley, S. C. and Kioussis, D. (1999) *Cold Spring Harb. Symp. Quant. Biol.*, **64**, 275–81.

Stauss, H. J. (1999) *Immunol. Today*, **20**, 180–3.

Thor Straten, P., Kirkin, A. F., Seremet, T. and Zeuthen, J. (1997) *Int. J. Cancer*, **70**, 582–6.

Toes, R. E., Ossendorp, F., Offringa, R. and Melief, C. J. (1999) *J. Exp. Med.*, **189**, 753–6.

Traversari, C., van der Bruggen, P., Luescher, I. F., Lurquin, C., Chomez, P., Van Pel, A., De Plaen, E., Amar-Costesec, A. and Boon, T. (1992) *J. Exp. Med.*, **176**, 1453–7.

Valmori, D., Dutoit, V., Lienard, D., Rimoldi, D., Pittet, M. J., Champagne, P., Ellefsen, K., Sahin, U., Speiser, D., Lejeune, F., Cerottini, J. C. and Romero, P. (2000) *Cancer Res.*, **60**, 4499–506.

Valmori, D., Dutoit, V., Rubio-Godoy, V., Chambaz, C., Lienard, D., Guillaume, P., Romero, P., Cerottini, J. C. and Rimoldi, D. (2001) *Cancer Res.*, **61**, 509–12.

van der Bruggen, P., Bastin, J., Gajewski, T., Coulie, P. G., Boel, P., De Smet, C., Traversari, C., Townsend, A. and Boon, T. (1994) *Eur. J. Immunol.*, **24**, 3038–43.

van Oijen, M., Elias, S., Weder, P., de Gast, G., Gallee, M., Schumacher, T. N. M. and Haanen, J. Ms submitted.

Walter, J. B., Garboczi, D. N., Fan, Q. R., Zhou, X., Walker, B. D. and Eisen, H. N. (1998) *J. Immunol. Meth.*, **214**, 41–50.

Whelan, J. A., Dunbar, P. R., Price, D. A., Purbhoo, M. A., Lechner, F., Ogg, G. S., Griffiths, G., Phillips, R. E., Cerundolo, V. and Sewell, A. K. (1999) *J. Immunol.*, **163**, 4342–8.

Willcox, B. E., Gao, G. F., Wyer, J. R., Ladbury, J. E., Bell, J. I., Jakobsen, B. K. and van der Merwe, P. A. (1999) *Immunity*, **10**, 357–65.

Wolkers, M. C., Stoetter, G., Vyth-Dreese, F. A. and Schumacher, T. N. M. (2001) *J. Immunol.*, **167**, 3577–84.

Yee, C., Savage, P. A., Lee, P. P., Davis, M. M. and Greenberg, P. D. (1999) *J. Immunol.*, **162**, 2227–34.

Yee, C., Thompson, J. A., Roche, P., Byrd, D. R., Lee, P. P., Piepkorn, M., Kenyon, K., Davis, M. M., Riddell, S. R. and Greenberg, P. D. (2000) *J. Exp. Med.*, **192**, 1637–44.

Youde, S. J., Dunbar, P. R., Evans, E. M., Fiander, A. N., Borysiewicz, L. K., Cerundolo, V. and Man, S. (2000) *Cancer Res.*, **60**, 365–71.

Zorn, E. and Hercend, T. (1999) *Eur. J. Immunol.*, **29**, 602–7.

Part 3

Human tumor antigens recognized by class II HLA-restricted T cells

Chapter 8

Antigens of the MAGE family recognized by CD4+ T cells

Catia Traversari

Summary

Initial studies on the characterization of tumor-specific antigens have been focused on CD8-mediated immune responses. However, it is now established that CD4+ lymphocytes have a central role in the generation of effective immune response against tumors. Identification of tumor antigens recognized by CD4+ T cells in an HLA class II restricted fashion, are indeed the subject of several research studies. In particular, the identification of CD4+ epitopes encoded by genes of the MAGE family, will allow to perform clinical trials of vaccination with synthetic peptides corresponding to MAGE epitopes recognized by both CD8+ and CD4+ T cells. The additional use of peptides corresponding to the CD4+ epitopes would increase the effectiveness of the current peptide-based vaccination strategy.

Introduction

Several studies have demonstrated the existence of T lymphocytes recognizing autologous tumor cells. In the last years, many tumor associated antigens (TAAs) recognized by CD8 T cells have been identified in melanomas and in tumor of various histological types (Traversari, 1999). These findings have led to the development of clinical trials of vaccination based on the use of antigenic peptides derived by such TAAs, in patients affected by melanomas and other neoplastic diseases (Nestle *et al.*, 1998; Rosenberg *et al.*, 1998; Marchand *et al.*, 1999).

However, increasing evidence has suggested that optimal cancer vaccination strategies need the active role of both CD4+ and CD8+ T cells. Indeed, CD4+ T cells exert helper activity to enhance the induction and to extend the persistence of anti-tumor CD8+ T cells (Pardoll *et al.*, 1998). Recently, the target antigens of CD4-mediated responses have been identified at molecular level by the use of molecular (Wang *et al.*, 1999), biochemical (Pieper *et al.*, 1999) as well as reverse-immunology approaches (Chaux *et al.*, 1999; Manici *et al.*, 1999; Kobayashi *et al.*, 2001; Schultz *et al.*, 2000).

CD4+ specific epitopes have been identified in the sequence of antigens belonging to all the four classes of TAAs. The different classes, which have distinct degrees of tumor specificity and clinical relevance, have been defined according to the expression pattern in neoplastic and normal tissues (Traversari, 1999). The first class comprises antigens that are expressed in tumors of various histological origins but not in normal tissues other than testis and placenta; such antigens are now referred to as cancer–testis antigens. The second class contains differentiation antigens expressed by the tumor cells and by the normal tissues from which the tumors derive. The antigens belonging to the third class are generated by point mutations of genes that are ubiquitously expressed. The fourth class encodes ubiquitous genes that are overexpressed in certain tumors relative to normal cells.

This chapter describes the current knowledge on the CD4+ immune response against a gene family of the first class of tumor antigens.

The *MAGE* gene family

The MAGE-type gene family belongs to the class of antigens that has been designated as "cancer/testis antigens" because TAAs belonging to this group are expressed in variable proportions of a wide array of different cancers, but not in normal tissues except for testis and placenta. Four clusters of MAGE genes with a high degree of homology have been identified, namely MAGE-A, -B, -C and -D (Chomez *et al.*, 2001). Their function is yet unknown, however, a role in embryonic development has been suggested by the expression of some of the murine homologous genes (De Plaen *et al.*, 1999) in blastocysts and embryonal stem cells. The only normal adult cells that express *MAGE* genes are male germ-line and placental trophoblast cells. Both of the tissues, however, do not express HLA molecules and therefore cannot present the relevant antigens. The activation of the *MAGE* genes in tumor cells, usually results from demethylation of their own promoters (De Smet *et al.*, 1995), which correlates with the occurrence of an overall genomic demethylation in tumors. Indeed, the same mechanism seems to be involved in the activation of other members of the cancer–testis group. Among tumors, a frequent expression was detected in melanomas, lung carcinomas, head and neck, bladder and esophagus carcinomas.

The MAGE-A family is the most characterized, indeed, the first human tumor antigen identified, MAGE-A1, belongs to this family (van der Bruggen *et al.*, 1991; Traversari *et al.*, 1992). The MAGE-A family consists of 12 different genes (*MAGE-A1–A12*) located in the q28 region of chromosome X (De Plaen *et al.*, 1994). Among them, at least seven (i.e. *MAGE-A1, -A2, -A3, -A4, -A6, -A10* and *-A12*) encode T-cell defined epitopes that are presented by HLA class I and II molecules on the surface of a variety of tumors.

More than 20 peptides derived from MAGE-A antigens have been identified so far by the use of anti-tumor CTLs, or by reverse immunology approaches (http://www.istitutotumori.mi.it/menurisorse/listing/pdf/tables%201–7.pdf). The large majority is encoded by *MAGE-A1* and *MAGE-A3*; however, several antigenic peptides have also been isolated from *MAGE-A2, -A4, -A6, -A10* and *-A12*. Some of these peptides are currently utilized in experimental protocols of active immunotherapy (Nestle *et al.*, 1998; Marchand *et al.*, 1999; Thurner *et al.*, 1999). The existence of CD4+ T-cell epitopes derived from MAGE-A proteins, had been strongly suggested by the observation that some cancer patients produce anti-MAGE IgG antibodies *in vivo* (Sahin *et al.*, 1995) and indeed, peptides presented by HLA class II molecules have been recently identified.

Experimental approaches for the identification of HLA class II-restricted tumor antigen epitopes

Tumor-specific T cells recognize a trimolecular complex exposed on the cell surface of tumor cells. The association of an antigen-derived peptide with the invariant β2-microglobulin and the HLA class I heavy chains generates the class I-restricted "antigenic complex." Whereas, peptide association with HLA class II α and β chains yields class II-restricted TAAs. The peptides are produced inside the cell by an active mechanism of processing. Usually the class II pathway favors the presentation of exogenous proteins or endogenous antigens containing lysosomal targeting sequences (e.g. melanocyte-specific proteins), or a signal peptide (e.g. membrane associated and secretory proteins). However, there are clear indications that also cytosolic proteins can gain access to the class II presentation pathway (Lich *et al.*, 2000). After binding to HLA class I or class II molecules, the complexes are exposed on the cell surface of tumor cells or antigen presenting cells (APCs) where they are recognized by specific T-cell effectors.

All the experimental protocols utilized so far for the identification of class II-restricted MAGE epitopes rely on the "reverse immunology" strategy. In this approach, APCs (i.e. monocyte-derived dendritic cells or total PBMCs) are loaded with either synthetic peptides or recombinant proteins and then are utilized to elicit antigen-specific CD4+ T cells.

Reverse immunology based on synthetic peptides

First step in this strategy is the screening of the amino acid sequence of a known or potential TAA, for peptides carrying binding motifs for defined HLA class II molecules. General methods have been developed to facilitate the search for such peptides. Indeed, in spite of the very high degree of polymorphism in the human population, MHC molecules can be grouped in a relatively limited number of groups showing similar HLA-binding specificity. In particular, for HLA class II molecules the existence of peptides able to bind several different HLA molecules and to induce peptide-specific immune responses has been demonstrated since 10 years (O'Sullivan *et al.*, 1991; Alexander *et al.*, 1994; Malcherek *et al.*, 1995). The identification of binding motifs common to the worldwide predominant HLA-DR molecules have allowed the development of algorithms for the identification of broadly degenerated HLA class II binding peptides (Southwood *et al.*, 1998).

More recently, a T-cell epitope prediction software, TEPITOPE, has been developed, which is able to scan a defined protein sequence for the presence of promiscuous HLA-DR binding sites (Sturniolo *et al.*, 1999). The software incorporates both a large number of *in vitro* determined HLA-DR pocket profiles, and the derived virtual matrix data, which represent the majority of the HLA-DR binding specificities. While the ligand prediction algorithms usually assign a score to each peptide sequence present in a given protein, major advantage of TEPITOPE is the possibility of evaluating the predicted peptides in comparison with natural binding peptides, thus allowing the users to define a threshold for their analysis (Sturniolo *et al.*, 1999).

Following the identification and synthesis of the predicted peptides, CD4+ cells are elicited by multiple rounds of *in vitro* stimulation by appropriate peptide-pulsed APCs. Proliferation or cytokine release assays, in the presence of the synthetic peptide are usually utilized to evaluate the specificity of the induced effector cells. The ability of these CD4+ T cells to recognize the processed peptide on the surface of cells expressing the HLA restriction element and the antigen reveals whether or not the peptide is actually endogenously processed.

There are two major drawbacks in this approach. All the studies performed clearly showed that none of the tested donors or patients reacted with all the potential binding peptides. Moreover, some peptide-specific T cells do not recognize APCs and tumor cells expressing the protein endogenously. There are several explanations for this unexpected result, besides the obvious possibility that the predicted peptide is not actually processed and presented by the tumor cells. Since high amounts of peptide are used in the induction phase, low affinity effectors could be generated that are unable to recognize tumor cells expressing small amounts of peptide. In addition, differences in the access to the exogenous and endogenous presentation pathway could account for the lack of recognition of the antigen expressing cells (Schultz *et al.*, 2000).

Reverse immunology-based on recombinant proteins

Some of the limits of the previous approach can be overcome by using APCs loaded with the recombinant protein as stimulators (Chaux *et al.*, 1999; Schultz *et al.*, 2000). Indeed, uptake and processing of the entire protein, rather than pulsing with synthetic peptides, may allow APCs to present all the relevant peptides and thus to induce CD4+ populations that may, indeed, have a role in the tumor immune response. The stimulation schedule and the read-out systems are the same as utilized in the synthetic peptide approach (i.e. proliferation or cytokine release assays). Once the antigen-specific effectors have been isolated, the HLA restriction element is characterized by cross-recognition experiments with target cells sharing defined HLA class II molecules. To identify the antigenic epitope, small subfragments of the protein or sets of peptides covering the entire protein sequence (i.e. peptides of 16 aa that overlapped by 12 aa) are screened with the induced CD4+ T cells (Chaux *et al.*, 1999). The major disadvantage of this approach is related to the scarce purity of the recombinant proteins utilized for the stimulation. A large number of the CD4+ clones isolated were, indeed, directed against contaminants of the protein preparations (Chaux *et al.*, 1999). To avoid this limit, recombinant proteins from different sources or cells endogenously expressing the protein must be utilized during the induction phase and the final screening for specificity (Schultz *et al.*, 2000).

MAGE-A3 encoded epitopes recognized by CD4+ T cells

As already mentioned, the existence of MAGE-A epitopes recognized by CD4+ T cells has been strongly suggested by the detection of MAGE-specific IgG antibodies *in vivo*, in cancer patients (Sahin *et al.*, 1995).

The search for class II restricted MAGE-A3 epitopes has been focused on the MAGE-A3 antigen, because of the promising results obtained in clinical trials of cancer vaccination with tumor-peptides (Nestle *et al.*, 1998; Marchand *et al.*, 1999; Thurner *et al.*, 1999).

The presence of CD4+ epitope within MAGE-A3 was assessed by both of the experimental procedures described (i.e. synthetic peptides and recombinant protein). Consistent results were reported, which are summarized in Table 8.1.

Analysis of the 314 aa of the MAGE-A3 protein with the TEPITOPE software program (Manici *et al.*, 1999), allowed for the prediction of 11 promiscuous peptides with the threshold set at 5% (Protti, personal communication). The 11 predicted sequences have been synthesized and analyzed for their ability to actually bind seven of the most commonly found DR alleles (i.e. DR*0101, *0301, *0401, *0701, *0801, *1101 and *1401) in competition assays with allele-specific indicator peptides (Manici *et al.*, 1999; Consogno *et al.*, in press). Five peptides with the higher degree of promiscuity (i.e. $M3_{141-155}$, $M3_{146-160}$, $M3_{156-170}$, $M3_{171-185}$ and $M3_{281-295}$) were pooled and utilized to propagate CD4+ T cells from the blood of a healthy donor typed as DRB1*1101/*1104. After few rounds of stimulation with the peptides pool, the polyclonal T-cell line specifically proliferate in response to $M3_{281-295}$ and to a lesser extent to $M3_{141-155}$, and $M3_{146-160}$, (Table 8.1). When the analysis was extended to all the eleven peptides, four immunodominant regions (residues 111–125, 146–160, 191–205 and 281–295), encoding naturally processed epitopes presented in association with 3 to 4 different HLA-DR alleles were identified (Consogno *et al.*, in press).

Table 8.1 MAGE-A epitopes recognized by CD4+ T cells

HLA restriction	*Residues*	*Peptide sequence*	*Reference*
MAGE-A3			
DRB1*1301/1302	114–127	AELVHFLLLKYRAR	Chaux (1999)
DRB1*1301/1302	121–134	LLKYRAREPVTKAE	Chaux (1999)
DRB1*1101/1104	281–295	TSYVKVLHHMVKISG	Manici (1999)
DRB1*1101/1104	141–154	GNWQYFFPVIFSKA	Manici (1999)
DRB1*1101/1104	146–160	FFPVIFSKASSSLQL	Manici (1999)
DR4			Kobayashi (2001)
DR7			Kobayashi (2001)
DR4	22–36	ALGLVGAQAPATEEQ	Kobayashi (2001)
DPB1*0401	247–258	TQHFVQENYLEY	Schultz (2000)
MAGE-A1			
DR15	247–258	EYVIKVSARVRF	Chaux (2001)

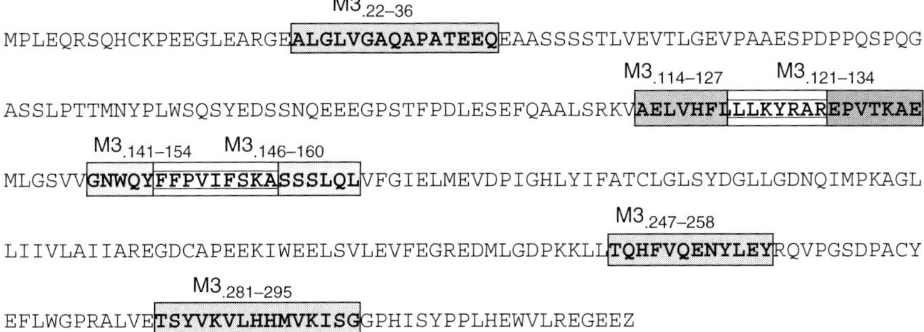

Figure 8.1 Localization of MAGE-A3 epitope recognized by CD4+ effectors. The amino acid sequence of MAGE-A3 is reported. Amino acid residues corresponding to the seven MAGE-A3 peptides studied are boxed. Two couples of peptides, namely $M3_{114-127}$/$M3_{121-134}$ and $M3_{141-154}$/$M3_{146-160}$, partially overlap.

The antigenicity and the stimulating properties of the peptide $M3_{146-160}$ have been recently confirmed by another study (Kobayashi *et al.*, 2001), which analyzed the response of three healthy individuals typed as DR1/13, DR4/15 and DR7/17. Along with $M3_{146-160}$, also the peptide $M3_{22-36}$ was analyzed. Both the peptides had been predicted to bind promiscuously to HLA-DR*0101, *0401 and *0701 (Southwood *et al.*, 1998), however, presentation by only the last two alleles was detected. Indeed, CD4+ T-cell clones specific for the peptide $M3_{146-160}$, presented in the context of HLA-DR4 and -DR7 were isolated, while peptide $M3_{22-36}$ was recognized only in association with HLA-DR4 molecules. Neither peptide was able to elicit T-cell responses in the HLA-DR1/DR13 donor.

Utilizing recombinant MAGE-A3 protein as source of antigens, and monocyte derived dendritic cells as stimulators, van der Bruggen has identified three peptides, namely $M3_{114-127}$, $M3_{121-134}$ and $M3_{247-258}$, which are presented by HLA-DR*1301/*1302 and DP*0401 molecules (Table 8.1). Noteworthy, the MAGE-$A3_{121-134}$ epitope was also encoded by MAGE-A1, -A2, and -A6. Localization of the seven antigenic peptides within the MAGE-A3 sequence is reported in Figure 8.1. Recently utilizing the same procedure a MAGE-A1 encoded epitope (i.e. $M1_{281-292}$) presented by HLA-DR15 has been identified (Chaux *et al.*, 2001).

In addition to the induction of peptide-specific CD4+ effectors, other factors such as actual antigen processing and transport on the cell surface, are fundamental to determining whether a predicted peptide can, indeed, function as an effective T-cell antigen. Apart from the three MAGE-A3 and the MAGE-A1 epitopes identified by the use of the recombinant protein (Chaux *et al.*, 1999; Schultz *et al.*, 2000), whose processing by dendritic cells is a prerequisite for the generation of CD4+ effector cells, clear indications on the natural processing of the other peptides have been reported only for $M3_{146-160}$ and $M3_{281-295}$ (Manici *et al.*, 1999; Kobayashi *et al.*, 2001) (Table 8.2).

Surprisingly, a disparity in the endogenous versus exogenous class II presentation pathway of MAGE-A3 epitopes was observed (Table 8.2). While peptides $M3_{247-258}$ and $M3_{281-295}$, were naturally processed and presented by MAGE-A3+ melanomas expressing the appropriate restriction alleles (i.e. DP*0401 and DR*1101), the DR13-restricted peptides (i.e. $M3_{114-127}$, $M3_{121-134}$) were exposed on the surface of melanoma cells only when the tumor was genetically modified to express an invariant chain-MAGE-A3 fusion protein (Ii-MAGE-A3) (Table 8.2). These results suggest that the endogenously synthesized MAGE-A3 protein should be at least partially digested, likely in the cytoplasm, before entering the endosomal compartment. Indeed, peptides $M3_{114-127}$ and $M3_{121-134}$, which are presented on the cell surface when MAGE-A3 is directly targeted to the endosomes, are absent when the cells express the natural cytoplasmic form of the MAGE-A3 protein. Even more complicated is the presentation pathway of $M3_{146-160}$, which is presented through the endogenous pathway of class II by HLA-DR7, but not by HLA-DR4 molecules (Kobayashi *et al.*, 2001). Presentation by DR4 seems to occur only via the exogenous pathway (i.e. by uptake of the recombinant protein).

Regardless, of the possible speculations on the molecular mechanisms responsible for the different processing of MAGE-A3 epitopes by the same cells, it appears evident that not only APCs but also tumor cells can directly present MAGE-A3 epitopes to CD4+ effector cells. This discovery has profound implications in the definition of the role that CD4+ cells have

Table 8.2 Actual processing of MAGE-A3 epitopes recognized by CD4+ T cells

Residues	*HLA-restriction*	*Class II presentation pathways*	
		Exogenous[a]	*Endogenous*
$M3_{22-36}$	DR4	?	?
$M3_{114-127}$	DR13	Yes	No
$M3_{121-134}$	DR13	Yes	No
$M3_{141-154}$	DR11	?	?
$M3_{146-160}$	DR7	Yes	Yes
$M3_{146-160}$	DR4	Yes	No
$M3_{281-295}$	DR13	Yes	Yes
$M3_{247-258}$	DP4	Yes	Yes

Note

a Access to the exogenous pathway of class II presentation was assessed by use of the recombinant MAGE-A3 protein (Chaux *et al.*, 1999; Manici *et al.*, 1999; Kobayashi *et al.*, 2001; Schultz *et al.*, 2000) and of the endosome-target fusion protein Ii-MAGE-A3 (Chaux *et al.*, 1999; Schultz *et al.*, 2000).

in controlling tumor growth. Indeed, it is now clear that CD4+ cells may act both in priming and effector phases of systemic anti-tumor immunity. During the priming phase, occurring within lymph nodes draining a tumor lesion, the major role of CD4+ cells is to activate dendritic cells through recognition of class II-restricted tumor-derived peptides and triggering of the CD40/CD40L system. In the effector phase, at the tumor site, CD4+ cells can be specifically activated by both resident macrophages and tumor cells expressing HLA class II molecules. Indeed, in the tumor mass inflammatory and immune reactions produce cytokine (i.e. IFN-γ) able to induce or to up-regulate class II expression on the tumor cells. Therefore, CD4+ cells may directly affect tumor growth by releasing cytokines, such as TNF, or by direct killing (Manici *et al.*, 1999; Schultz *et al.*, 2000).

Acknowledgments

We are grateful to M. P. Protti for helpful discussion. The author's work was supported by grants from the Italian Association for Cancer Research (AIRC), and from the Ministry of Health.

References

Alexander, J., Sidney, J., Southwood, S., Ruppert, J., Oseroff, C., Maewal, A. *et al.* (1994) Development of high potency universal DR-restricted helper epitopes by modification of high affinity DR-blocking peptides. *Immunity*, **1**, 751–761.

van der Bruggen, P., Traversari, C., Chomez, P., Lurquin, C., De Plaen, E., Van den Eynde, B. *et al.* (1991) A gene encoding an antigen recognized by cytolytic T lymphocytes on a human melanoma. *Science*, **254**, 1643–1647.

Chaux, P., Lethe, B., Van Snick, J., Corthals, J., Schultz, E. S., Cambiaso, C. L. *et al.* (2001) A MAGE-1 peptide recognized on HLA-DR15 by CD4(+) T cells. *Eur. J. Immunol.*, **31**, 1910–1916.

Chaux, P., Vantomme, V., Stroobant, V., Thielemans, K., Corthals, J., Luiten, R. *et al.* (1999) Identification of MAGE-3 epitopes presented by HLA-DR molecules to CD4(+) T lymphocytes. *J. Exp. Med.*, **189**, 767–778.

Chomez, P., De Backer, O., Bertrand, M., De Plaen, E., Boon, T. and Lucas, S. (2001) An overview of the mage gene family with the identification of all human members of the family. *Cancer Res.*, **61**, 5544–5551.

Consogno, G., Manici, S., Facchinetti, V., Bachi, A., Hammer, J. M., Conti-Fine, B. M., Rugarli, C., Traversari, C. and Protti, M. (2003) Identification of immunodominant regions among promiscuous HLA-DR restricted CD4+ T cell epitopes on the tumor antigen MAGE-3. *Blood*, first edition paper, published on line September 19, 2002; DOI 10.1182/blood-2002-03-0933.

De Plaen, E., Arden, K., Traversari, C., Gaforio, J. J., Szikora, J. P., De Smet, C. *et al.* (1994) Structure, chromosomal localization, and expression of 12 genes of the MAGE family. *Immunogenetics*, **40**, 360–369.

De Plaen, E., De Backer, O., Arnaud, D., Bonjean, B., Chomez, P., Martelange, V. *et al.* (1999) A new family of mouse genes homologous to the human MAGE genes. *Genomics*, **55**, 176–184.

De Smet, C., Courtois, S. J., Faraoni, I., Lurquin, C., Szikora, J. P., De Backer, O. *et al.* (1995) Involvement of two Ets binding sites in the transcriptional activation of the *MAGE-1* gene. *Immunogenetics*, **42**, 282–290.

Kobayashi, H., Song, Y., Hoon, D. S., Appella, E. and Celis, E. (2001) Tumor-reactive T helper lymphocytes recognize a promiscuous MAGE-A3 epitope presented by various major histocompatibility complex class II alleles. *Cancer Res.*, **61**, 4773–4778.

Lich, J. D., Elliott, J. F. and Blum, J. S. (2000) Cytoplasmic processing is a prerequisite for presentation of an endogenous antigen by major histocompatibility complex class II proteins. *J. Exp. Med.*, **191**, 1513–1524.

Malcherek, G., Gnau, V., Jung, G., Rammensee, H. G. and Melms, A. (1995) Supermotifs enable natural invariant chain-derived peptides to interact with many major histocompatibility complex-class II molecules. *J. Exp. Med.*, **181**, 527–536.

Manici, S., Sturniolo, T., Imro, M. A., Hammer, J., Sinigaglia, F., Noppen, C. *et al.* (1999) Melanoma cells present a MAGE-3 epitope to CD4(+) cytotoxic T cells in association with histocompatibility leukocyte antigen DR11. *J. Exp. Med.*, **189**, 871–876.

Marchand, M., van Baren, N., Weynants, P., Brichard, V., Dreno, B., Tessier, M. H. *et al.* (1999) Tumor regressions observed in patients with metastatic melanoma treated with an antigenic peptide encoded by gene *MAGE-3* and presented by HLA- A1. *Int. J. Cancer*, **80**, 219–230.

Nestle, F. O., Alijagic, S., Gilliet, M., Sun, Y., Grabbe, S., Dummer, R. *et al.* (1998) Vaccination of melanoma patients with peptide- or tumor lysate-pulsed dendritic cells. *Nat. Med.*, **4**, 328–332.

O'Sullivan, D., Sidney, J., Del Guercio, M. F., Colon, S. M. and Sette, A. (1991) Truncation analysis of several DR binding epitopes. *J. Immunol.*, **146**, 1240–1246.

Pardoll, D. M. and Topalian, S. L. (1998) The role of CD4+ T cell responses in antitumor immunity. *Curr. Opin. Immunol.*, **10**, 588–594.

Pieper, R., Christian, R. E., Gonzales, M. I., Nishimura, M. I., Gupta, G., Settlage, R. E. *et al.* (1999) Biochemical identification of a mutated human melanoma antigen recognized by CD4(+) T cells [see comments]. *J. Exp. Med.*, **189**, 757–766.

Rosenberg, S. A., Yang, J. C., Schwartzentruber, D. J., Hwu, P., Marincola, F. M., Topalian, S. L. *et al.* (1998) Immunologic and therapeutic evaluation of a synthetic peptide vaccine for the treatment of patients with metastatic melanoma. *Nat. Med.*, **4**, 321–327.

Sahin, U., Tureci, O., Schmitt, H., Cochlovius, B., Johannes, T., Schmits, R. *et al.* (1995) Human neoplasms elicit multiple specific immune responses in the autologous host. *Proc. Natl. Acad. Sci. USA*, **92**, 11810–11813.

Schultz, E. S., Lethe, B., Cambiaso, C. L., Van Snick, J., Chaux, P., Corthals, J. *et al.* (2000) A MAGE-A3 peptide presented by HLA-DP4 is recognized on tumor cells by CD4+ cytolytic T lymphocytes. *Cancer Res.*, **60**, 6272–6275.

Southwood, S., Sidney, J., Kondo, A., del Guercio, M. F., Appella, E., Hoffman, S. *et al.* (1998) Several common HLA-DR types share largely overlapping peptide binding repertoires. *J. Immunol.*, **160**, 3363–3373.

Sturniolo, T., Bono, E., Ding, J., Raddrizzani, L., Tuereci, O., Sahin, U. *et al.* (1999) Generation of tissue-specific and promiscuous HLA ligand databases using DNA microarrays and virtual HLA class II matrices. *Nat. Biotech.*, **17**, 555–561.

Thurner, B., Haendle, I., Roder, C., Dieckmann, D., Keikavoussi, P., Jonuleit, H. *et al.* (1999) Vaccination with mage-3A1 peptide-pulsed mature, monocyte-derived dendritic cells expands specific cytotoxic T cells and induces regression of some metastases in advanced stage IV melanoma. *J. Exp. Med.*, **190**, 1669–1678.

Traversari, C. (1999) Tumor-antigen recognized by T cells. *Minerva Biotech.*, **11**, 243–253.

Traversari, C., van der Bruggen, P., Luescher, I. F., Lurquin, C., Chomez, P., Van Pel, A. *et al.* (1992) A nonapeptide encoded by human gene *MAGE-1* is recognized on HLA-A1 by cytolytic T lymphocytes directed against tumor antigen MZ2-E. *J. Exp. Med.*, **176**, 1453–1457.

Wang, R. F., Wang, X., Atwood, A. C., Topalian, S. L. and Rosenberg, S. A. (1999) Cloning genes encoding MHC class II-restricted antigens: mutated CDC27 as a tumor antigen. *Science*, **284**, 1351–1354.

Chapter 9

Melanoma antigens recognized by $CD4^+$ T cells

Rong-Fu Wang

Summary

The identification of tumor antigens has generated a resurgence of interest in cancer immunotherapy. However, recent studies from both human and animal models indicate that therapeutic strategies mainly focusing on the exclusive use of $CD8^+$ T cells (and MHC class I-restricted tumor antigens) may not generate effective immunity against cancer cells. Thus, the identification of MHC class II-restricted tumor antigens, which are capable of stimulating $CD4^+$ T cells, might provide opportunities for developing effective cancer vaccines. This chapter will discuss the latest progress in identifying melanoma antigens recognized by $CD4^+$ T cells.

Introduction

Host immune system plays an essential role in immunosurveillance and destruction of cancer cells. The concept of immune surveillance of tumor was proposed three decades ago, but direct evidence to support the role of immune system in inhibiting spontaneous tumor growth was not provided until recently in animal systems (Shankaran *et al.*, 2001; Smyth *et al.*, 2001). More importantly, adoptive transfer of autologous tumor-infiltrating lymphocytes (TILs) along with interleukin 2 (IL-2) is associated with objective tumor regression in melanoma and renal carcinoma patients (Rosenberg *et al.*, 1988; Rosenberg, 2001). It is now widely accepted that immune targets on cancer cells can be recognized by both antibody and tumor-reactive T cells. Identification of immunogenic tumor antigens has significantly advanced our understanding of tumor immunity and provides unprecedented opportunity for the development of effective cancer therapy. This chapter will discuss the current progress in identification of MHC class II-restricted melanoma antigens recognized by $CD4^+$ T cells.

The need for identification of MHC class II-restricted tumor antigens

Since $CD8^+$ cytotoxic T lymphocytes (CTLs) are capable of lysing tumor cells directly and destroying large tumor masses *in vivo*, much attention has been given to the role of $CD8^+$ T cells in the immunotherapy of cancer in the last 10 years. As a result, a number of MHC class I-restricted tumor antigens have been defined using tumor-reactive $CD8^+$ T cells derived from peripheral blood mononuclear cells (PBMCs) or TILs that exhibit antitumor activity *in vivo* (Boon *et al.*, 1994; Rosenberg, 1999; Wang and Rosenberg, 1999). These antigens can be divided into several different classes based on their patterns of gene expression.

The tissue-specific differentiation antigens, which include tyrosinase, MART-1, gp100, TRP-1/gp75 and TRP-2, are expressed in melanoma as well as normal melanocytes (Brichard *et al.*, 1993; Kawakami *et al.*, 1994a, b; Wang *et al.*, 1995, 1996). Tumor-specific shared antigens such as MAGE-1 and NY-ESO-1 are expressed in a wide variety of tumors such as breast cancer and lung cancer (Van der Bruggen *et al.*, 1991; Jager *et al.*, 1998; Wang *et al.*, 1998). The expression of these products is limited to cancer cells and normal testis. Tumor-specific unique or mutated antigens such as CDK4, β-catenin and caspase 8 have also been described (Wolfel *et al.*, 1995; Robbins *et al.*, 1996; Mandruzzato *et al.*, 1997). Of particular interest is gp75/TRP-1 and NY-ESO-1 in that both proteins were first detected by serum antibodies (Mattes *et al.*, 1983; Chen *et al.*, 1997) and were found to encode two gene products from alternative open reading frames recognized by T cells (Wang *et al.*, 1996b, 1998). Recently, several additional examples such LAGE and M-CSF also encode two gene products, one of which functions as immune targets recognized by T cells (Aarnoudse *et al.*, 1999; Rimoldi *et al.*, 2000; Probst-Kepper *et al.*, 2001).

Identification of these antigens has set the stage for developing cancer vaccines. Since the majority of MHC class I-restricted tumor antigens identified to date are non-mutated self-proteins (Boon *et al.*, 1994; Rosenberg, 1999; Wang and Rosenberg, 1999), a potential consequence of active immunization with self-peptides is the development of autoimmune diseases. For instance, depigmentation or vitiligo has been found to be correlated with good prognosis or clinical responses to immunochemotherapy and TIL therapy in melanoma patients (Bystryn *et al.*, 1987; Richards *et al.*, 1992; Rosenberg and White, 1996). Animal studies using differentiation antigens suggest that induction of anti-self immunity may eradicate cancer cells, but at the same time it may cause normal tissue destruction as manifested by the development of vitiligo (Hara *et al.*, 1995; Yee *et al.*, 2000). Clinical studies conducted in several institutions have shown some evidence for a therapeutic effect on tumor growth inhibition and regression using peptides or peptide-pulsed dendritic cells (DCs) as vaccines (Jager *et al.*, 1996; Nestle *et al.*, 1998; Rosenberg *et al.*, 1998; Marchand *et al.*, 1999; Thurner *et al.*, 1999; Lau *et al.*, 2001), suggesting that immunization of patients with MHC class I-restricted self-antigens can generate antitumor immunity, leading to tumor regression. However, the objective complete clinical responses were sporadic, even though CTL reactivity was clearly evident after one round of stimulation *in vitro* of PBMC from the majority of vaccinated patients (Rosenberg *et al.*, 1998). Interestingly, transient CTL activity detected *in vitro* is not correlated with clinical responses observed in patients (Rosenberg *et al.*, 1998; Marchand *et al.*, 1999; Jager *et al.*, 2000a). Similar results were obtained when patients were immunized with DCs pulsed with MHC class I-restricted peptides (Dallal and Lotze, 2000). It is possible that tumor-reactive $CD8^+$ T cells traffic to tumor sites in clinically responding patients, while, in non-responding patients, these antigen-specific T cells remain in peripheral blood. Although the mechanism is unknown at the present time, T-cell trafficking to tumor lesions is an important issue that requires further investigation.

The studies described above demonstrate the potential and feasibility of immunotherapy of cancer using tumor antigens recognized by $CD8^+$ T cells, the overall immune responses, however, are weak and transient. One possible reason is that a lack of tumor-specific $CD4^+$ T-cell responses contributes to this failure. Indeed, a growing body of evidence suggests that $CD4^+$ T cells play a central role in initiating and maintaining immune responses against cancer (Toes *et al.*, 1999; Cohen *et al.*, 2000). Thus, optimal vaccination might require the participation of both $CD4^+$ and $CD8^+$ T cells to generate a strong and long-lasting antitumor immunity. However, a major obstacle for the development of optimal cancer vaccines is the lack of effective methods for identifying MHC class II-restricted tumor antigens, which are capable of stimulating CD4$^+$ T cells. Identification of such antigens would provide new opportunities for developing effective cancer vaccines and improve our understanding of the mechanisms by which CD4$^+$ T cells regulate the host immune system.

The role of CD4$^+$ T cells in immune response

Although the role of CD4$^+$ T cells in antitumor immune responses was reported in the 1980s, little attention has been given to CD4$^+$ T cells in the area of tumor immunology (Pardoll and Topalian, 1998; Cohen *et al.*, 2000). There are at least two reasons for this: first, CD4$^+$ T cells recognize peptides presented by MHC class II molecules, and most tumors express MHC class I, but not MHC class II molecules, and second, few MHC class II-restricted tumor antigens have been identified due to technical difficulties (see below). However, increasing evidence from animal studies and clinical trials indicates that CD4$^+$ T cells play a central role in orchestrating host immune responses against cancer and infectious diseases as well as in autoimmunity (O'Garra *et al.*, 1997; Kalams and Walker, 1998; Pardoll and Topalian, 1998; Zajac *et al.*, 1998). The role of CD4$^+$ T cells in modulating immune responses against cancer is described in the next section.

Initiation of immune responses

CD4$^+$ T cells can be divided into T helper 1 (Th1) and Th2 cells based on their cytokine secretion profile (Morel and Oriss, 1998). CD4$^+$ Th1 cells help prime CD8$^+$ T-cell responses (Toes *et al.*, 1999). For example, depletion of CD4$^+$ T cells by anti-CD4 antibodies or the use of CD4-knockout animals has demonstrated that CD4$^+$ T cells are essential in the induction of antigen-specific CD8$^+$ T cells (Hung *et al.*, 1998). It was proposed that CD4$^+$ Th cells and CD8$^+$ T cells must recognize an antigen on the same antigen-presenting cell (APC) such that CD4$^+$ T cells can provide help in the priming of CD8$^+$ T-cell responses. Although the mechanism(s) by which CD4$^+$ T cells provide help for priming naive CD8$^+$ T cells and activate memory CD8$^+$ T cells is not fully understood, several recent studies have demonstrated that CD40–CD40 ligand (L) interactions between DCs and CD4$^+$ T cells activate DCs for effective priming and activation of CD8$^+$ T cells (Bennett *et al.*, 1998; Ridge *et al.*, 1998; Schoenberger *et al.*, 1998). CD4$^+$ T cells recognize an antigen presented by professional APCs such as DCs and, in turn, activate antigen-bearing DCs (Banchereau and Steinman, 1998). Once activated, DCs become competent to prime CTLs that recognize an MHC class I-restricted determinant on the same APC. Thus, activation of APCs by CD4$^+$ T cells through antigen-specific recognition and CD40–CD40L engagement is essential to prime CD8$^+$ T cells (Bennett *et al.*, 1998; Ridge *et al.*, 1998; Schoenberger *et al.*, 1998). The lack of properly activated DCs in the absence of CD4$^+$ T helper cells can induce tolerance rather than activation of CD8$^+$ T cells.

Maintaining CD8$^+$ T-cell function and proliferation

In addition to their important role in priming CD8$^+$ T cells, CD4$^+$ T cells are also essential in the maintenance of CD8$^+$ T-cell effector functions by secreting cytokines such as IL-2 required for CD8$^+$ T-cell growth and proliferation (Greenberg, 1991; Rosenberg *et al.*, 1998). Progressive loss of CD8$^+$ T cells has also been observed in the CD4-deficient mice or in the absence of CD4$^+$ T cells, resulting in diminished resistance to subsequent virus challenge (Zajac *et al.*, 1998). These findings may have important implications in cancer therapy since complete tumor regression requires a prolonged antitumor immunity (Thurner *et al.*, 1999).

Inhibition of tumor growth in the absence of CD8$^+$ T cells

CD4$^+$ T cells are also able to mediate tumor regression in the absence of CD8$^+$ T cells as shown by adoptive transfer of tumor-specific CD4$^+$ T cells (Greenberg, 1991). Their role has been documented in many other tumor models (Cohen *et al.*, 2000), including CD4-knockout animals which fail to control tumor outgrowth (Hung *et al.*, 1998). Since most tumors do not express MHC class II molecules, CD4$^+$ T-cell-mediated antitumor immunity does not require direct contact between T cells and tumor. The mechanisms by which CD4$^+$ T cells mediate tumor eradication are not clear, but several studies suggest that cytokines such as interferon γ (INFγ) secreted by CD4$^+$ T cells might be involved in antitumor and anti-angiogenic activities (Mumberg *et al.*, 1999; Qin and Blankenstein, 2000). Other studies have proposed that CD4$^+$ T cells eliminate tumors through activation and recruitment of effector cells including macrophages and eosinophils (Greenberg, 1991; Hung *et al.*, 1998).

Providing help for B-cell activation

While Th1 provide help for cellular immunity, Th2 cells activate B cells to become antibody-secreting plasma cells. These activated B cells produce tumor-specific antibodies that might contribute to therapeutic antitumor immunity (Old, 1996; Glennie and Johnson, 2000). Serological screening of cDNA expression libraries with patient sera has identified many antibody-mediated antigens (Pfreundschuh, 2000).

Approach to identification of tumor antigens recognized by CD4$^+$ T cells

Although MHC class II-restricted tumor antigens are clearly important, little is known about them. A major obstacle is the lack of effective methods for identifying an unknown MHC class II tumor antigen recognized by tumor-specific CD4$^+$ T cells (Wang and Rosenberg, 1999). Several strategies have been developed for this purpose, including peptide elution from the cell surface of tumor cells (Halder *et al.*, 1997), biochemical purification of tumor cell lysates (Monach *et al.*, 1995; Pieper *et al.*, 1999), genetic targeting expression system (Wang *et al.*, 1999a; Wang and Rosenberg, 1999), and "reverse immunology" approach (Touloukian *et al.*, 2000; Zeng *et al.*, 2000). These methods and related rationales will be discussed in this section.

Peptide elution

T cells recognize a peptide bound to the MHC class II molecule. Tumor-specific peptides can be eluted with acid from either the tumor cell surface or purified peptide–MHC complexes, and subsequently separated by high pressure liquid chromatography (HPLC). The eluted peptide fractions are then tested for their ability to stimulate cytokine secretion from CD4$^+$ T cells when pulsed onto MHC matched APCs. A naturally processed tumor specific peptide recognized by CD4$^+$ T cells can be directly identified. The peptide sequence can then be used to search databases in order to find the gene encoding the antigenic peptide. This approach has been successfully used to identify several MHC class I-restricted peptides (Cox *et al.*, 1994; Skipper *et al.*, 1996; Meadows *et al.*, 1997), but it did not work well for the identification of MHC class II-restricted peptides. Only successful example is the gp100 peptide (Halder *et al.*, 1997). A major problem is the limiting amounts of MHC class II peptides and representation of antigenic peptides as nested sets with heterogeneous lengths (15–25 aa), which is different from MHC class I-restricted T-cell peptides with restricted peptide length (usually 9 or 10 aa).

Biochemical approach

Exogenous proteins are preferentially processed and presented to T cells through the MHC class II pathway. Thus, purified fractions or proteins from tumor cell lysates can be pulsed onto APCs such as B cells and DCs, and tested for the ability to stimulate $CD4^+$ T cells. In each step of purification, positive fractions are identified with antigen-specific $CD4^+$ T cells. After several purification steps, relatively pure protein fractions will be obtained, and analyzed by SDS-PAGE. Each protein band can be isolated for amino acid sequence determination. With this information at hand, one could search Databases to determine the identity of antigen, and design specific primers to amplify cDNA from tumor as well as normal cells for sequence analysis. This approach has been used to identify two murine antigens and one human melanoma antigens (Monach *et al.*, 1995; Pieper *et al.*, 1999; Matsutake and Srivastava, 2001). Interestingly, all these proteins are abundant ones, two of which are ribosomal proteins L9 and L11 containing unique point mutations. Thus, this method is useful to identify a tumor antigen with a high level of expression level in tumor cells. Similar to peptide elution, this method remains very laborious and slow compared with far more powerful targeting expression system. In addition, it has been difficult to identify tumor antigens with a low level of expression, and cannot be used to identify a tumor antigen when $CD4^+$ T cells fail to recognize B cell or DCs pulsed with tumor cell lysates.

A genetic targeting expression system

$CD4^+$ T cells recognize a peptide bound to MHC class II molecules on the cell surface of APCs. The formation of MHC class II–peptide complexes on the cell surface is a complicated, multistep process that favors presentation of antigens derived from exogenous proteins. This is distinct from antigen presentation by MHC class I, which favors endogenous proteins. Assembly of the MHC class II α and β chains, along with the associated invariant chain (Ii) begins in the endoplasmic reticulum (ER) (Germain, 1994; Cresswell, 1996). Ii association prevents an antigenic peptide from binding to the $\alpha\beta$ dimers and stablizes the $\alpha\beta$ complexes. Ii contains an endosome-targeting sequence at the N-terminus and CLIP peptide between amino acids 81–104. This targeting sequence in the cytoplasmic tail of Ii is responsible for the transport of nonameric $(\alpha\beta\ \mathrm{Ii})_3$ complexes from the ER to intracellular compartments with endosomal/lysosomal characteristics and ultimately to acidic endosomal and lysosomal-like structures called MHC class II compartments (MIIC) (Germain, 1994; Cresswell, 1996). HLA-DM molecules in this compartment facilitate disassociation of residual Ii peptide (CLIP) from the peptide-binding grooves of the MHC class II molecules and replacement with antigenic peptides (Cresswell, 1996). Thus, MHC class II antigen processing and presentation requires at least five genes (*DRα*, *DRβ*, *Ii*, *DMA* and *DMB*) and the specialized MIIC compartments.

Since Ii fusion has a critical role in enhancing presentation of antigens through the MHC class II pathway (Sanderson *et al.*, 1995b; van Bergen *et al.*, 1997; Fujii *et al.*, 1998), we developed a genetic targeting expression (GTE) approach that allows the screening of an Ii-fused cDNA library for the identification of MHC class II-restricted antigens based on the endogenous antigen presentation pathway (Wang *et al.*, 1999b). Since professional APCs are poorly transfectable, we generated HEK293IMDR cells by introducing cDNAs encoding DRα, DRβ, DMA, DMB and Ii into HEK293 cells, which have the same or similar capacity to process and present antigens as professional APCs (Wang *et al.*, 1999b). Hence, this system comprises of two essential components: generation of a highly transfectable HEK293IMDR cell line, and creation of an Ii fusion library such that the Ii fusion proteins are targeted to the endosomal/lysosomal compartment for efficient antigen processing and presentation (Wang *et al.*, 1999a,b). Three MHC class II-restricted tumor antigens including mutated triosephosphate isomerase (TPI), mutated CDC27 and a fusion protein derived from a chromosomal DNA rearrangement have been identified by this approach (see following section), and several others are under investigation (Wang *et al.*, 2002).

Alternative approach to the GTE system described above is to express recombinant proteins either in bacteria (Mougneau *et al.*, 1995) or mammalian cells (Scott *et al.*, 2000), and to allow recombinant proteins to be captured by macrophages or DCs for presentation to T cells restricted to MHC class II. Although these strategies have been used to clone genes encoding bacteria antigens and transplantation HY antigens (Mougneau *et al.*, 1995; Sanderson *et al.*, 1995a; Scott *et al.*, 2000), it has not been reported of successful use for the identification of tumor antigens.

Defining MHC class II-restricted peptides from candidate antigens

The strategies described in the previous section are useful for identifying new MHC class II-restricted tumor antigens with tumor-specific $CD4^+$ T cells. However, in many cases, tumor-specific $CD4^+$ T cells are not available, especially, for prostate and breast cancers. With advanced technologies, such as DNA microarrays, it is possible to identify cancer-specific antigens from weakly immunogenic tumors. In addition, serological analysis of recombinant cDNA expression libraries of human tumor with autologous serum (SEREX) has been used to isolate several putative human tumor antigens (Sahin *et al.*, 1997; Old and Chen, 1998). Among them are tyrosinase (Sahin *et al.*, 1995), MAGE (Sahin *et al.*, 1995), NY-ESO-1 (Chen *et al.*, 1997), SSX2 (Tureci *et al.*, 1996), SCP1 (Tureci *et al.*, 1998) and CT7 (Chen *et al.*, 1998). The key question is how to determine whether these antigens are immune targets recognized by T cells. DCs pulsed with peptide or protein for *in vitro* stimulation of PBMCs are useful in defining $CD4^+$ T-cell epitopes from a known putative tumor antigen (Chaux *et al.*, 1999; Manici *et al.*, 1999; Jager *et al.*, 2000b; Schultz *et al.*, 2000; Zarour *et al.*, 2000a).

We have recently used HLA-DR4-transgenic (Tg) mice to identify $CD4^+$ T-cell epitopes from candidate antigens (Touloukian *et al.*, 2000; Zeng *et al.*, 2000). This approach has been successfully used to define MHC class II-restricted T-cell peptides from autoantigens, which may trigger the immune system to cause autoimmune diseases (Gross *et al.*, 1998; Sonderstrup and McDevitt, 1998; Abraham and David, 2000). HLA-DR-Tg mice might have advantages for identifying putative peptides since they should have a high precursor frequency of specifically reactive T cells after immunization. Once candidate peptides are identified, CD4$^+$ T cells can be generated from human PBMCs stimulated with synthetic candidate peptides. Therefore, the combined use of Tg mice immunized with the intact protein antigen and stimulated with the peptides predicted by a computer-assisted algorithm may avoid the need to stimulate human PBMCs with a large number of peptides and several rounds of stimulation *in vitro*. Furthermore, candidate peptides identified using the immunized Tg mice are likely to be peptides that are naturally processed and presented on the cell surface. This may increase the likelihood that peptide-specific CD4$^+$ T cells also recognize the antigen-positive tumor cells.

MHC class II-restricted melanoma antigens

Class I-restricted antigens can contain class II-restricted T-cell epitopes

Data from studies of differentiation antigens indicate that MHC class I-restricted antigens also contain CD4$^+$ T cell-epitopes (Table 9.1). T-cell lines derived from a patient with melanoma has been shown to recognize T-cell peptides from tyrosinase in the context of HLA-DR4 and HLA-DR15, respectively (Topalian *et al.*, 1996; Kobayashi *et al.*, 1998b). Vogt–Koyanagi–Harada's disease is regarded as an autoimmune disorder of multiple organs containing melanocytes. A T-cell line from the peripheral blood of a patient with this disease was also reported to respond to the tyrosinase p193–203 peptide when presented by the HLA-DR4 molecule (Kobayashi *et al.*, 1998b). In addition, using *in vitro* stimulation, T-cell lines generated from PBMCs of melanoma patients recognized MART-1 in the context of HLA-DR4 (Zarour *et al.*, 2000a). The use of HLA-DR4-Tg mice identified an HLA-DR4-restricted T-cell peptide from gp100, that is identical to the peptide originally isolated by peptide elution of tumor cells (Li *et al.*, 1998; Touloukian *et al.*, 2000). This suggests that murine CD4$^+$ T cells generated from Tg mice after immunization can recognize the same peptides as human CD4$^+$ T cells (Touloukian *et al.*, 2000; Zeng *et al.*, 2000).

Similarly, MAGE-3 possesses CD4$^+$ T-cell peptides presented by either HLA-DR11, HLA-DR13 or DP4 (Chaux *et al.*, 1999; Manici *et al.*, 1999; Schultz *et al.*, 2000). More recently, several peptides from NY-ESO-1 were also shown to be recognized by CD4$^+$ T cells in the context of DRβ1*0401 and DRβ4*0101 (Zeng *et al.*, 2000; Jager *et al.*, 2000b; Zarour *et al.*, 2000b). NY-ESO-1 is a potent immunogen recognized by both antibody and T cells (Chen *et al.*, 1997; Jager *et al.*, 1998; Wang *et al.*, 1998). Of particular interest is that 10–13% of patients with advanced cancer developed a high titer of antibody

Table 9.1 Tumor antigens recognized by CD4$^+$ T cells

Antigens	*MHC class I restrictions*	*MHC class II restrictions*
Tyrosinase	A1, A2, A24, B44	DR4
MART-1/Melan-A	HLA-A2, B45	DR4
gp100	A2, A3	DR4
MAGE-3	A1, A2, B44	DR11, DR13, DP4
NY-ESO-1/CAG3	A2, A31, Cw	DR4, DP4
Eph		DR11

(Stockert *et al.*, 1998; Zeng *et al.*, 2000). This cannot be explained by the predicted frequency (3%) of antibody production in patients based on the frequencies of NY-ESO-1 (20%) in tumors and HLA-DR4 (15%) in the population. In the search for an explanation for this paradox, we identified a T-cell epitope presented by HLA-DP4, a predominant allele expressed in 40–70% of the population. Eleven out of twelve (92%) patients who developed high titers of antibody expressed the DP4 molecule, while only one out of seven who did not develop antibody against NY-ESO-1 expressed DP4. Furthermore, antigen-specific $CD4^+$ T cells could be generated from PBMCs of patients with NY-ESO-1-specific antibody, but not from PBMCs of patients without specific antibody (Zeng *et al.*, 2001). Identification of DP4-restricted T-cell peptides from MAGE-3 and NY-ESO-1 could be of great benefit for more than 50% of patients with cancer.

In addition, an antigenic peptide presented by DR11 was recently identified from Eph3, a member of the Eph family of tyrosine kinase receptors (Chiari *et al.*, 2000).

Taken together, these studies suggest that $CD4^+$ T-cell peptides are present in the known MHC class I-restricted tumor antigens, and thus might provide a critical help to antigen-specific $CD8^+$ T cells. Since these antigens as well as their T-cell peptides are non-mutated, an advantage is that both $CD4^+$ and $CD8^+$ T-cell peptides can be used for clinical application in a large population. On the other hand, however, these self-peptides might not be effective in breaking immune tolerance for generating potent antitumor immune responses. In any event, identification of these $CD4^+$ T-cell peptides has provided new opportunities for evaluating their usefulness in clinical trials.

Mutated or fusion proteins give rise to class II-restricted tumor antigens

Three MHC class II-restricted tumor antigens have been identified using the genetic targeting expression system described above (Table 9.2). One of these antigens is the fusion gene product LDFP recognized by HLA-DR1-restricted $CD4^+$TIL. DNA sequencing analysis indicated that LDFP was generated by fusing a low-density-lipid receptor (LDLR) gene at the 5′ end to a GDP-L-Fucose : β-D-Galactoside 2-α-L-Fucosyltransferase (FUT) in an antisense orientation at the 3′ end (Wang *et al.*, 1999b). Therefore, the fusion gene encodes the first five ligand-binding repeats of LDLR in the N-terminus followed by a new polypeptide translated in frame with LDLR from the FUT gene in an antisense direction. Two overlapping minimal peptides (PVIWRRAPA and WRRAPAPGA) were identified from the C-terminus of the fusion protein (Wang *et al.*, 1999b).

The second MHC class II-restricted tumor antigen is a mutated form of TPI identified by both biochemical and GTE approaches (Pieper *et al.*, 1999; Wang *et al.*, 1999a). The point

Table 9.2 MHC class II-restricted T-cell peptides resulting from fusion or mutated antigens

Antigens	*HLA restrictions*	*Peptides*
LDFP	HLA-DR1	PVIWRRAPA (312–323)
	HLA-DR1	WRRAPAPGA (315–323)
TPI	HLA-DR1	GELIGILNAAKVPAD (23–37)
CDC27	HLA-DR4	FSWAMDLDPKGA (760–771)

mutation, which results in an amino acid substitution of Ile for Thr, creats an HLA-DR1-restricted peptide recognized by CD4$^+$ T cells. The Thr to Ile conversion increases CD4$^+$ T-cell reactivity by 5 logs compared with the wild-type peptide (Pieper *et al.*, 1999).

The third antigen identified by this genetic approach is a mutated human CDC27, an important component of the anaphase promoting complex involved in cell cycle regulation. This mutated CDC27 gives rise to a melanoma target antigen recognized by CD4$^+$ HLA-DR4-restricted TIL1359 cells (Wang *et al.*, 1999a). Interestingly, in this case, the mutation itself does not constitute a T-cell epitope. Instead, the Ser to Leu mutation in a putative phosphorylation site allows a non-mutated peptide within CDC27 to be presented to T cells by MHC class II molecules. While wild-type CDC27 is predominately localized in the nucleus, the mutated form of the protein is found in the cytoplasm, and thereby gains the ability to access the MHC class II pathway. Therefore, the location or trafficking of protein, rather than its expression level, can determine whether it is processed and presented for T-cell recognition. It should be noted that CD4$^+$ TIL1359 cells are capable of recognizing the intact DR4-positive autologous tumor cells, but fail to respond to DR4-positive EBV B cells pusled with autologous tumor lysate. These findings suggest that although antigen "cross-priming" is an important mechanism for activating CD4$^+$ T cells through DCs that have captured antigens from tumor cells, direct antigen presentation by tumor cells to CD4$^+$ T cells occurs in tumor lesions when inflammatory responses and infiltrating immune cells produce cytokines such as IFN-γ that induce MHC class II molecules on tumor cells.

In summary, several mutated and fusion antigens have been identified as MHC class II-restricted tumor antigens recognized by CD4$^+$ T cells using the genetic targeting expression system. Of particular interest is that the mutation in CDC27 does not constitute a T-cell peptide, in stead, the mutation alters the ability of the mutated form of protein to enter MHC class II pathway. This represents a new mechanism of how a non-mutated MHC class II-restricted peptide can be processed and presented to CD4$^+$ T cells through the alteration of protein trafficking. To our surprise, the majority of MHC class II-restricted tumor antigens identified by tumor-reactive CD4$^+$ T cells are mutated or fusion antigens. Although their biological significance remains to be determined, one possibility is that these antigen-specific CD4$^+$ T cells might initiate cancer-specific immune responses since they are not tolerated in patients with cancer.

Development of cancer vaccines based on molecularly defined tumor antigens

Given the central role of CD4$^+$ T cells in initiating and maintaining immune responses against cancer, it is important to incorporate both MHC class I and II T-cell epitopes in tumor vaccines to generate potent antitumor immunity aimed at eradicating cancer cells. This notion is supported by studies showing that immunization with a virus-derived MHC class II T-cell peptide or adoptive transfer of tumor-reactive CD4$^+$ T cells results in antitumor immunity against MHC class II-negative tumors (Greenberg, 1991; Ossendorp *et al.*, 1998). Identification of MHC class II-restricted tumor antigens, as described above, opens new avenues for developing new vaccination strategies for cancer patients. Nonetheless, it is still necessary to address whether an MHC class II-restricted tumor antigen functions as a good tumor rejection antigen and how these antigens can be used effectively in cancer vaccines. With the availability of MHC class II-restricted tumor antigens, it is possible to develop specific immunotherapy based on attacking tumor cells bearing the identified antigens. A variety of clinical approaches utilizing these genes or gene products are possible. For example, peptide- and protein-based immunogens, DNA- and virus-based whole gene or minigene encoding a single or multiple T-cell epitopes and DC-based vaccines are being investigated in both animal and human clinical trials.

Vaccine strategies

Peptide-based vaccines Active immunotherapy involves the direct immunization of cancer patients with cancer antigens in an attempt to boost immune responses against the tumor. Vaccination of patients with peptides derived from MART-1, gp100, tyrosinase, MAGE-3 and NY-ESO-1 demonstrated significant induction of T-cell responses against melanoma (Cormier *et al.*, 1997; Rosenberg *et al.*, 1998; Marchand *et al.*, 1999; Jager *et al.*, 2000a,b). The immunodominant peptides derived from tumor antigens could readily be synthesized *in vitro* and used for immunization either alone or in a form intended to improve their immunogenicity, such as in combination with adjuvant, linkage to lipids/liposomes or helper peptides, or pulsed onto APC. Modification of the immunodominant peptides to improve binding efficiency to MHC antigens can potentially increase immunogenicity and induce stronger antitumor activity (Parkhurst *et al.*, 1996; Valmori *et al.*, 1998; Chen *et al.*, 2000). Clinical studies have indicated that peptide vaccines using a modified gp100 peptide combined with high doses of IL-2 result in a 42% of clinical response rate in HLA-A2$^+$ melanoma patients (Rosenberg *et al.*, 1998). Since multiple tumor antigens and multiple epitopes have been identified from melanoma, it is likely that the use of multiple epitope peptides including MHC class II-restricted T-cell peptides would enhance an antitumor activity.

Dendritic cell-based vaccines Dendritic cells are potent APC and play an important role in priming and maintenance of an antitumor immunity (Banchereau and Steinman, 1998). DCs pulsed with MHC class I-restricted peptides or proteins have been shown to induce potent antitumor immunity (Timmerman and Levy, 1999; Dallal and Lotze, 2000). It has recently been reported that patients treated with DCs pulsed peptides or tumor lysates resulted in an objective clinical regression of tumor in 5 of 16 patients evaluated (Nestle *et al.*, 1998). Vaccination of patients with DCs pulsed with MAGE-3 peptide showed tumor regressions of individual metastases (skin, lymph node, lung and liver) in 6/11 patients (Thurner *et al.*, 1999). Clinical and immunological responses were also observed when patients were treated with DCs pulsed with different peptides (Mackensen *et al.*, 2000). In a recent study, 16 melanoma patients were treated with the use of intravenous infusions of DCs derived by incubation of plastic-adherent peripheral blood mononuclear cells (PBMC) with IL-4 and GM-CSF for eight days in serum-free AIM-V medium, followed by overnight pulsing with peptides. Five of sixteen patients had an immune response to gp100 or tyrosinase in cytokine release assay; four of five were clinically stable or had tumor regression (Lau *et al.*, 2001). Although T helper peptides were used in some studies, these T helper peptides were not tumor-specific. Thus, tumor-specific T helper peptides should be tested in cancer vaccines in the future studies. Alternatively, genetically modified DCs could be used to stimulate both CD4$^+$ and CD8$^+$ T-cell responses, leading to potent antitumor immunity (Kirk and Mule, 2000).

Recombinant viruses/nucleotide acid-based vaccines One of the most effective cancer vaccines may involve the incorporation of genes encoding tumor antigens into recombinant plasmid or viruses such as vaccinia, foxlpox or adenovirus (Tang *et al.*, 1992; Ulmer *et al.*, 1993; Restifo, 1996; Gurunathan *et al.*, 2000). The major problem associated with the use of recombinant viruses is that patients develop strong antibody responses against virus, resulting in inefficient low infection activity (Rosenberg *et al.*, 1998b). The use of nucleic acid as anti-cancer vaccines (DNA or RNA that contain sequences coding for tumor antigens) has drawn much attention due to its obvious advantages such as simple, safe and cost effective. It has been demonstrated that nucleic acid vaccination can elicit a full range of immune responses, including antibodies, MHC class I-restricted CD8$^+$ CTL and class II-restricted helper T-cells response (Tang *et al.*, 1992; Ulmer *et al.*, 1993; Gurunathan *et al.*, 2000). Many methods have been used to improve nucleic acid vaccination, including the use of better promoters, intron and enhancer elements, polyadenylation sequences and immunostimulatory sequences (ISS), which are non-methylated, palindromic DNA-sequences containing CpG-oligodinucleotides (Krieg, 2000).

Recently, alphavirus self-replicons of Sindbis virus and Semliki Forest virus (SFV) have been used for improving antitumor immune responses (Kohno *et al.*, 1998; Ying *et al.*, 1999). To increase the stability of the construct and to facilitate the production and handling of the vaccine, the self-replicating RNA can be encoded by a DNA-plasmid where a CMV-promoter "jump-starts" the production of the self-replicating RNA. The alphavirus replicase functions in a broad range of host cells (mammalian, avian, reptilian, amphibian and insect cells), making it a very attractive delivery vehicle. Both RNA vaccine and DNA replicons encoding a model tumor antigen under the control of alphaviral RNA replicase have been demonstrated to be effective in the treatment of an experimental tumor (Ying *et al.*, 1999).

Monitoring CD4$^+$ T-cell response by MHC class II-peptide tetramers

The development of MHC class I-peptide tetramers has extended our ability to monitor CD8$^+$ T-cell response to immunization or virus infection and advanced our understanding of antigen-specific T cells (Altman *et al.*, 1996). To explore the role of CD4$^+$ T cells in anti-tumor immunity, MHC class II-peptide tetramers are badly needed. Two systems were recently developed for preparing MHC class II-peptide tetramers. The first is to link peptide epitopes to the amino terminus of the β chain using a short flexible linker to produce HLA-DR4 tetramers (Kotzin *et al.*, 2000). The second approach is to express HLA-DR4 in the empty form in insect cells, with the two chains linked at their carboxy termini by a leucine zipper. The purified HLA-DR4 were then incubated with a peptide to produce a stable molecule that could be biotinylated and used to stain peptide-specific CD4$^+$ T cells (Novak *et al.*, 1999). With the availability of such reagents, we are able to address the important questions regarding the role of CD4$^+$ T cells in tumor immunity. Thus, the combined use of MHC class I and II-restricted tumor antigens, co-stimulatory molecules and cytokines/chemokines that can be used to enhance immune responses and the powerful tools to monitor T-cell responses represent an unprecedented opportunity for the development of effective cancer vaccines.

References

Aarnoudse, C. A., van den Doel, P. B., Heemskerk, B. and Schrier, P. I. (1999) Interleukin-2-induced, melanoma-specific T cells recognize CAMEL, an unexpected translation product of LAGE-1. *Int. J. Cancer*, **82**, 442–8.

Abraham, R. S. and David, C. S. (2000) Identification of HLA-class-II-restricted epitopes of autoantigens in transgenic mice. *Curr. Opin. Immunol.*, **12**, 122–9.

Altman, J. D., Moss, P. A. H., Goulder, P. J. R., Barouch, D. H., McHeyzer-Williams, M. G., Bell, J. I., McMichael, A. J. and Davis, M. M. (1996) Phenotypic analysis of antigen-specific T lymphocytes [published erratum appears in *Science* 1998 June 19, **280** (5371), 1821], *Science*, **274**, 94–6.

Banchereau, J. and Steinman, R. M. (1998) Dendritic cells and the control of immunity. *Nature*, **392**, 245–52.

Bennett, S. R., Carbone, F. R., Karamalis, F., Flavell, R. A., Miller, J. F. and Heath, W. R. (1998) Help for cytotoxic-T-cell responses is mediated by CD40 signalling. *Nature*, **393**, 478–80.

van Bergen, J., Schoenberger, S. P., Verreck, F., Amons, R., Offringa, R. and Koning, F. (1997) Efficient loading of HLA-DR with a T helper epitope by genetic exchange of CLIP. *Proc. Natl. Acad. Sci. USA*, **94**, 7499–502.

Boon, T., Cerottini, J.-C., Van Den Eynde, B., Van der Bruggen, P. and Van Pel, A. (1994) Tumor antigens recognized by T lymphocytes. *Annu. Rew. Immunol.*, **12**, 337–65.

Brichard, V., Van Pel, A., Wölfel, T., Wölfel, C., De Plaen, E., Lethë, B., Coulie, P. and Boon, T. (1993) The tyrosinase gene codes for an antigen recognized by autologous cytolytic T lymphocytes on HLA-A2 melanomas. *J. Exp. Med.*, **178**, 489–95.

Bystryn, J.-C., Rigel, D., Friedman, R. J. and Kopf, A. (1987) Prognostic significance of hypopigmentation in malignant melanoma. *Arch. Dermatol.*, **123**, 1053–5.

Chaux, P., Vantomme, V., Stroobant, V., Thielemans, K., Corthals, J., Luiten, R., Eggermont, A. M., Boon, T. and van der Bruggen, P. (1999) Identification of MAGE-3 epitopes presented by HLA-DR molecules to CD4(+) T lymphocytes. *J. Exp. Med.*, **189**, 767–78.

Chen, J. L., Dunbar, P. R., Gileadi, U., Jager, E., Gnjatic, S., Nagata, Y., Stockert, E., Panicali, D. L., Chen, Y. T., Knuth, A., Old, L. J. and Cerundolo, V. (2000) Identification of NY-ESO-1 peptide analogues capable of improved stimulation of tumor-reactive CTL. *J. Immunol.*, **165**, 948–55.

Chen, Y. T., Gure, A. O., Tsang, S., Stockert, E., Jager, E., Knuth, A. and Old, L. J. (1998) Identification of multiple cancer/testis antigens by allogeneic antibody screening of a melanoma cell line library. *Proc. Natl. Acad. Sci. USA*, **95**, 6919–23.

Chen, Y. T., Scanlan, M. J., Sahin, U., Tureci, O., Gure, A. O., Tsang, S., Williamson, B., Stockert, E., Pfreundschuh, M. and Old, L. J. (1997) A testicular antigen aberrantly expressed in human cancers detected by autologous antibody screening. *Proc. Natl. Acad. Sci. USA*, **94**, 1914–18.

Chiari, R., Hames, G., Stroobant, V., Texier, C., Maillere, B., Boon, T. and Coulie, P. G. (2000) Identification of a tumor-specific shared antigen derived from an Eph receptor and presented to CD4 T cells on HLA class II molecules. *Cancer Res.*, **60**, 4855–63.

Cohen, P. A., Peng, L., Plautz, G. E., Kim, J. A., Weng, D. E. and Shu, S. (2000) CD4+ T cells in adoptive immunotherapy and the indirect mechanism of tumor rejection. *Crit. Rev. Immunol.*, **20**, 17–56.

Cormier, J. N., Salgaller, M. L., Prevette, T., Barracchini, K. C., Rivoltini, L., Restifo, N. P., Rosenberg, S. A. and Marincola, F. M. (1997) Enhancement of cellular immunity in melanoma patients immunized with a peptide from MART-1/Melan A. *Cancer J. Sci. Am.*, **3**, 37–44.

Cox, A. L., Skipper, J., Cehn, Y., Henderson, R. A., Darrow, T. L., Shabanowitz, J., Engelhard, V. H., Hunt, D. F. and Slingluff, C. L. (1994) Identification of a peptide recognized by five melanoma-specific human cytotoxic T cell lines. *Science*, **264**, 716–19.

Cresswell, P. (1996) Invariant chain structure and MHC class II function. *Cell*, **84**, 505–7.

Dallal, R. M. and Lotze, M. T. (2000) The dendritic cell and human cancer vaccines. *Curr. Opin. Immunol.*, **12**, 583–8.

Fujii, S., Senju, S., Chen, Y. Z., Ando, M., Matsushita, S. and Nishimura, Y. (1998) The CLIP-substituted invariant chain efficiently targets an antigenic peptide to HLA class II pathway in L cells. *Hum. Immunol.*, **59**, 607–14.

Germain, R. N. (1994) MHC-dependent antigen processing and peptide presentation: providing ligands for T lymphocyte activation. *Cell*, **76**, 287–99.

Glennie, M. J. and Johnson, P. W. (2000) Clinical trials of antibody therapy. *Immunol. Today*, **21**, 403–10.

Greenberg, P. D. (1991) Adoptive T cell therapy of tumors: mechanisms operative in the recognition and elimination of tumor cells. *Adv. Immunol.*, **49**, 281–355.

Gross, D. M., Forsthuber, T., Tary-Lehmann, M., Etling, C., Ito, K., Nagy, Z. A., Field, J. A., Steere, A. C. and Huber, B. T. (1998) Identification of LFA-1 as a candidate autoantigen in treatment-resistant Lyme arthritis. *Science*, **281**, 703–6.

Gurunathan, S., Klinman, D. M. and Seder, R. A. (2000) DNA vaccines: immunology, application, and optimization. *Annu. Rev. Immunol.*, **18**, 927–74.

Halder, T., Pawelec, G., Kirkin, A. F., Zeuthen, J., Meyer, H. E., Kun, L. and Kalbacher, H. (1997) Isolation of novel HLA-DR restricted potential tumor-associated antigens from the melanoma cell line FM3. *Cancer Res.*, **57**, 3238–44.

Hara, I., Takechi, Y. and Houghton, A. N. (1995) Implicating a role for immune recognition of self in tumor rejection: passive immunization against the *Brown* locus protein. *J. Exp. Med.*, **182**, 1609–14.

Hung, K., Hayashi, R., Lafond-Walker, A., Lowenstein, C., Pardoll, D. and Levitsky, H. (1998) The central role of CD4(+) T cells in the antitumor immune response. *J. Exp. Med.*, **188**, 2357–68.

Jager, E., Bernhard, H., Romero, P., Ringhoffer, M., Arand, M., Karbach, J., Ilsemann, C., Hagedorn, M. and Knuth, A. (1996) Generation of cytotoxic T cell responses with synthetic melanoma-associated peptides *in vivo*: implication for tumor vaccines with melanoma-associated antigens. *Int. J. Cancer*, **66**, 162–9.

Jager, E., Chen, Y. T., Drijfhout, J. W., Karbach, J., Ringhoffer, M., Jager, D., Arand, M., Wada, H., Noguchi, Y., Stockert, E., Old, L. J. and Knuth, A. (1998) Simultaneous humoral and cellular immune response against cancer–testis antigen NY-ESO-1: definition of human histocompatibility leukocyte antigen (HLA)-A2-binding peptide epitopes. *J. Exp. Med.*, **187**, 265–70.

Jager, E., Gnjatic, S., Nagata, Y., Stockert, E., Jager, D., Karbach, J., Neumann, A., Rieckenberg, J., Chen, Y. T., Ritter, G., Hoffman, E., Arand, M., Old, L. J. and Knuth, A. (2000a) Induction of primary NY-ESO-1 immunity: CD8+ T lymphocyte and antibody responses in peptide-vaccinated patients with NY-ESO-1+ cancers. *Proc. Natl. Acad. Sci. USA*, **97**, 12198–203.

Jager, E., Jager, D., Karbach, J., Chen, Y. T., Ritter, G., Nagata, Y., Gnjatic, S., Stockert, E., Arand, M., Old, L. J. and Knuth, A. (2000b) Identification of NY-ESO-1 epitopes presented by human histocompatibility antigen (HLA)-DRB4*0101-0103 and recognized by CD4(+) T lymphocytes of patients with NY-ESO-1-expressing melanoma. *J. Exp. Med.*, **191**, 625–30.

Jager, E., Maeurer, M., Hohn, H., Karbach, J., Jager, D., Zidianakis, Z., Bakhshandeh-Bath, A., Orth, J., Neukirch, C., Necker, A., Reichert, T. E. and Knuth, A. (2000c) Clonal expansion of Melan A-specific cytotoxic T lymphocytes in a melanoma patient responding to continued immunization with melanoma-associated peptides. *Int. J. Cancer*, **86**, 538–47.

Jager, E., Nagata, Y., Gnjatic, S., Wada, H., Stockert, E., Karbach, J., Dunbar, P. R., Lee, S. Y., Jungbluth, A., Jager, D., Arand, M., Ritter, G., Cerundolo, V., Dupont, B., Chen, Y. T., Old, L. J. and Knuth, A. (2000d) Monitoring CD8 T cell responses to NY-ESO-1: correlation of humoral and cellular immune responses. *Proc. Natl. Acad. Sci. USA*, **97**, 4760–5.

Kalams, S. A. and Walker, B. D. (1998) The critical need for CD4 help in maintaining effective cytotoxic T lymphocyte responses. *J. Exp. Med.*, **188**, 2199–204.

Kawakami, Y., Eliyahu, S., Delgado, C. H., Robbins, P. F., Sakaguchi, K., Appella, E., Yannelli, J. R., Adema, G. J., Miki, T. and Rosenberg, S. A. (1994a) Identification of a human melanoma antigen recognized by tumor infiltrating lymphocytes associated with *in vivo* tumor rejection. *Proc. Natl. Acad. Sci. USA*, **91**, 6458–62.

Kawakami, Y., Eliyahu, S., Delgaldo, C. H., Robbins, P. F., Rivoltini, L., Topalian, S. L., Miki, T. and Rosenberg, S. A. (1994b) Cloning of the gene coding for a shared human melanoma antigen recognized by autologous T cells infiltrating into tumor. *Proc. Natl. Acad. Sci. USA*, **91**, 3515–19.

Kirk, C. J. and Mule, J. J. (2000) Gene-modified dendritic cells for use in tumor vaccines. *Hum. Gene. Ther.*, **11**, 797–806.

Kobayashi, H., Kokubo, T., Sato, K., Kimura, S., Asano, K., Takahashi, H., Iizuka, H., Miyokawa, N. and Katagiri, M. (1998a) CD4+ T cells from peripheral blood of a melanoma patient recognize peptides derived from nonmutated tyrosinase. *Cancer Res.*, **58**, 296–301.

Kobayashi, H., Kokubo, T., Takahashi, M., Sato, K., Miyokawa, N., Kimura, S., Kinouchi, R. and Katagiri, M. (1998b) Tyrosinase epitope recognized by an HLA-DR-restricted T-cell line from a Vogt–Koyanagi–Harada disease patient. *Immunogenetics*, **47**, 398–403.

Kohno, A., Emi, N., Kasai, M., Tanimoto, M. and Saito, H. (1998) Semliki Forest virus-based DNA expression vector: transient protein production followed by cell death. *Gene Ther.*, **5**, 415–18.

Kotzin, B. L., Falta, M. T., Crawford, F., Rosloniec, E. F., Bill, J., Marrack, P. and Kappler, J. (2000) Use of soluble peptide-DR4 tetramers to detect synovial T cells specific for cartilage antigens in patients with rheumatoid arthritis. *Proc. Natl. Acad. Sci. USA*, **97**, 291–6.

Krieg, A. M. (2000) The role of CpG motifs in innate immunity. *Curr. Opin. Immunol.*, **12**, 35–43.

Lau, R., Wang, F., Jeffery, G., Marty, V., Kuniyoshi, J., Bade, E., Ryback, M. E. and Weber, J. (2001) Phase I trial of intravenous peptide-pulsed dendritic cells in patients with metastatic melanoma. *J. Immunother.*, **24**, 66–78.

Li, K., Adibzadeh, M., Halder, T., Kalbacher, H., Heinzel, S., Muller, C., Zeuthen, J. and Pawelec, G. (1998) Tumour-specific MHC-class-II-restricted responses after *in vitro* sensitization to synthetic peptides corresponding to gp100 and Annexin II eluted from melanoma cells. *Cancer Immunol Immunother.*, **47**, 32–8.

Mackensen, A., Herbst, B., Chen, J. L., Kohler, G., Noppen, C., Herr, W., Spagnoli, G. C., Cerundolo, V. and Lindemann, A. (2000) Phase I study in melanoma patients of a vaccine with peptide-pulsed dendritic cells generated *in vitro* from CD34(+) hematopoietic progenitor cells. *Int. J. Cancer*, **86**, 385–92.

Mandruzzato, S., Brasseur, F., Andry, G., Boon, T. and van der Bruggen, P. (1997) A CASP-8 mutation recognized by cytolytic T lymphocytes on a human head and neck carcinoma. *J. Exp. Med.*, **186**, 785–93.

Manici, S., Sturniolo, T., Imro, M. A., Hammer, J., Sinigaglia, F., Noppen, C., Spagnoli, G., Mazzi, B., Bellone, M., Dellabona, P. and Protti, M. P. (1999) Melanoma cells present a MAGE-3 epitope to CD4(+) cytotoxic T cells in association with histocompatibility leukocyte antigen DR11. *J. Exp. Med.*, **189**, 871–6.

Marchand, M., van Baren, N., Weynants, P., Brichard, V., Dreno, B., Tessier, M. H., Rankin, E., Parmiani, G., Arienti, F., Humblet, Y., Bourlond, A., Vanwijck, R., Lienard, D., Beauduin, M., Dietrich, P. Y., Russo, V., Kerger, J., Masucci, G., Jager, E., De Greve, J., Atzpodien, J., Brasseur, F., Coulie, P. G., van der Bruggen, P. and Boon, T. (1999) Tumor regressions observed in patients with metastatic melanoma treated with an antigenic peptide encoded by gene *MAGE-3* and presented by HLA- A1. *Int. J. Cancer*, **80**, 219–30.

Matsutake, T. and Srivastava, P. K. (2001) The immunoprotective MHC II epitope of a chemically induced tumor harbors a unique mutation in a ribosomal protein. *Proc. Natl. Acad. Sci. USA*, **98**, 3992–7.

Mattes, M. J., Thomson, T. M., Old, L. J. and Lloyd, K. O. (1983) A pigmentation-associated, differentiation antigen of human melanoma defined by a precipitating antibody in human serum. *Int. J. Cancer*, **32**, 717–21.

Meadows, L., Wang, W., den Haan, J. M., Blokland, E., Reinhardus, C., Drijfhout, J. W., Shabanowitz, J., Pierce, R., Agulnik, A. L., Bishop, C. E. and Hunt, D. (1997) The HLA-A2*0201-restricted H-Y antigen contains a postranslationally modified cysteine that significantly affects T cell recognition. *Immunity*, **6**, 273–81.

Monach, P. A., Meredith, S. C., Siegel, C. T. and Schreiber, H. (1995) A unique tumor antigen produced by a single amino acid substitution. *Immunity*, **2**, 45–59.

Morel, P. A. and Oriss, T. B. (1998) Crossregulation between Th1 and Th2 cells. *Crit. Rev. Immunol.*, **18**, 275–303.

Mougneau, E., Altare, F., Wakil, A. E., Zheng, S., Coppola, T., Wang, Z. E., Waldmann, R., Locksley, R. M. and Glaichenhaus, N. (1995) Expression cloning of a protective Leishmania antigen. *Science*, **268**, 563–6.

Mumberg, D., Monach, P. A., Wanderling, S., Philip, M., Toledano, A. Y., Schreiber, R. D. and Schreiber, H. (1999) CD4(+) T cells eliminate MHC class II-negative cancer cells *in vivo* by indirect effects of IFN-gamma. *Proc. Natl. Acad. Sci. USA*, **96**, 8633–8.

Nestle, F. O., Alijagic, S., Gilliet, M., Sun, Y., Grabbe, S., Dummer, R., Burg, G. and Schadendorf, D. (1998) Vaccination of melanoma patients with peptide- or tumor lysate-pulsed dendritic cells. *Nat. Med.*, **4**, 328–32.

Novak, E. J., Liu, A. W., Nepom, G. T. and Kwok, W. W. (1999) MHC class II tetramers identify peptide-specific human CD4(+) T cells proliferating in response to influenza A antigen. *J. Clin. Invest.*, **104**, R63–7.

O'Garra, A., Steinman, L. and Gijbels, K. (1997) CD4+ T-cell subsets in autoimmunity. *Curr. Opin. Immunol.*, **9**, 872–83.

Old, L. J. (1996) Immunotherapy for cancer. *Sci. Am.*, **275**, 136–43.

Old, L. J. and Chen, Y. T. (1998) New paths in human cancer serology. *J. Exp. Med.*, **187**, 1163–7.

Ossendorp, F., Mengede, E., Camps, M., Filius, R. and Melief, C. J. (1998) Specific T helper cell requirement for optimal induction of cytotoxic T lymphocytes against major histocompatibility complex class II negative tumors. *J. Exp. Med.*, **187**, 693–702.

Pardoll, D. M. and Topalian, S. L. (1998) The role of CD4(+) T cell responses in antitumor immunity. *Curr. Opin. Immunol.*, **10**, 588–94.

Parkhurst, M. R., Salgaller, M., Southwood, S., Robbins, P., Sette, A., Rosenberg, S. A. and Kawakami, Y. (1996) Improved induction of melanoma reactive CTL with peptides from the melanoma antigen gp100 modified at HLA-A0201 binding residues. *J. Immunol.*, **157**, 2539–48.

Pfreundschuh, M. (2000) Exploitation of the B cell repertoire for the identification of human tumor antigens. *Cancer Chemother. Pharmacol.*, **46**, S3–7.

Pieper, R., Christian, R. E., Gonzales, M. I., Nishimura, M. I., Gupta, G., Settlage, R. E., Shabanowitz, J., Rosenberg, S. A., Hunt, D. F. and Topalian, S. L. (1999) Biochemical identification of a mutated human melanoma antigen recognized by CD4(+) T cells. *J. Exp. Med.*, **189**, 757–66.

Probst-Kepper, M., Stroobant, V., Kridel, R., Gaugler, B., Landry, C., Brasseur, F., Cosyns, J. P., Weynand, B., Boon, T. and Van Den Eynde, B. J. (2001) An alternative open reading frame of the human macrophage colony-stimulating factor gene is independently translated and codes for an antigenic peptide of 14 amino acids recognized by tumor-infiltrating CD8 T lymphocytes. *J. Exp. Med.*, **193**, 1189–98.

Qin, Z. and Blankenstein, T. (2000) CD4+ T cell-mediated tumor rejection involves inhibition of angiogenesis that is dependent on IFNgamma receptor expression by nonhematopoietic cells. *Immunity*, **12**, 677–86.

Restifo, N. P. (1996) The new vaccines: building viruses that elicit antitumor immunity. *Curr. Opin. Immunol.*, **8**, 658–63.

Richards, J. M., Mehta, N., Ramming, K. and Skosey, P. (1992) Sequential chemoimmunotherapy in the treatment of metastatic melanoma. *J. Clin. Oncol.*, **10**, 1338–43.

Ridge, J. P., Di Rosa, F. and Matzinger, P. (1998) A conditioned dendritic cell can be a temporal bridge between a CD4+ T-helper and a T-killer cell. *Nature*, **393**, 474–8.

Rimoldi, D., Rubio-Godoy, V., Dutoit, V., Lienard, D., Salvi, S., Guillaume, P., Speiser, D., Stockert, E., Spagnoli, G., Servis, C., Cerottini, J. C., Lejeune, F., Romero, P. and Valmori, D. (2000) Efficient simultaneous presentation of NY-ESO-1/LAGE-1 primary and nonprimary open reading frame-derived CTL epitopes in melanoma. *J. Immunol.*, **165**, 7253–61.

Robbins, P. F., El Gamil, M., Li, Y. F., Kawakami, Y., Loftus, D., Appella, E. and Rosenberg, S. A. (1996) A mutated β-catenin gene encodes a melanoma-specific antigen recognized by tumor infiltrating lymphocytes. *J. Exp. Med.*, **183**, 1185–92.

Rosenberg, S. A. (1999) A new era for cancer immunotherapy based on the genes that encode cancer antigens. *Immunity*, **10**, 281–7.

Rosenberg, S. A. (2001) Progress in human tumour immunology and immunotherapy. *Nature*, **411**, 380–4.

Rosenberg, S. A. and White, D. E. (1996) Vitiligo in patients with melanoma: normal tissue antigens can be targeted for cancer immunotherapy. *J. Immunother.*, **19**, 81–4.

Rosenberg, S. A., Packard, B. S., Aebersold, P. M., Solomon, D., Topalian, S. L., Toy, S. T., Simon, P., Lotze, M. T., Yang, J. C., Seipp, C. A., Simpson, C., Carter, C., Bock, S., Schwartzentruber, D., Wei, J. P. and White, D. E. (1988) Use of tumor infiltrating lymphocytes and interleukin-2 in the immunotherapy of patients with metastatic melanoma. Preliminary report. *N. Engl. J. Med.*, **319**, 1676–80.

Rosenberg, S. A., Yang, J. C., Schwartzentruber, D. J., Hwu, P., Marincola, F. M., Topalian, S. L., Restifo, N. P., Dudley, M. E., Schwarz, S. L., Spiess, P. J., Wunderlich, J. R., Parkhurst, M. R., Kawakami, Y., Seipp, C. A., Jan, R. N., Einhorn, R. N. and White, D. E. (1998a) Immunologic and therapeutic evaluation of a synthetic tumor-associated peptide vaccine for the treatment of patients with metastatic melanoma. *Nat. Med.*, **4**, 321–7.

Rosenberg, S. A., Zhai, Y., Yang, J. C., Schwartzentruber, D. J., Hwu, P., Marincola, F. M., Topalian, S. L., Restifo, N. P., Seipp, C. A., Einhorn, J. H., Roberts, B. and White, D. E. (1998b) Immunizing patients with metastatic melanoma using recombinant adenoviruses encoding MART-1 or gp100 melanoma antigens. *J. Natl. Cancer Inst.*, **90**, 1894–900.

Sahin, U., Tureci, O. and Pfreundschuh, M. (1997) Serological identification of human tumor antigens. *Curr. Opin. Immunol.*, **9**, 709–16.

Sahin, U., Tureci, O., Schmitt, H., Cochlovius, B., Johannes, T., Schmits, R., Stenner, F., Luo, G., Schobert, I. and Pfreundschuh, M. (1995) Human neoplasms elicit multiple immune responses in the autologous host. *Proc. Natl. Acad. Sci. USA*, **92**, 11810–13.

Sanderson, S., Campbell, D. J. and Shastri, N. (1995a) Identification of a CD4+ T cell-stimulating antigen of pathogenic bacteria by expression cloning. *J. Exp. Med.*, **182**, 1751–7.

Sanderson, S., Frauwirth, K. and Shastri, N. (1995b) Expression of endogeous peptide–major histocompatibility complex class II complexes derived from invariant chain–antigen fusion proteins. *Proc. Natl. Acad. Sci. USA*, **92**, 7217–21.

Schoenberger, S. P., Toes, R. E., E.L., v. d. V., Offringa, R. and Melief, C. J. (1998) T-cell help for cytotoxic T lymphocytes is mediated by CD40–CD40L interactions. *Nature*, **393**, 473–4.

Schultz, E. S., Lethe, B., Cambiaso, C. L., Van Snick, J., Chaux, P., Corthals, J., Heirman, C., Thielemans, K., Boon, T. and van der Bruggen, P. (2000) A MAGE-A3 peptide presented by HLA-DP4 is recognized on tumor cells by CD4+ cytolytic T lymphocytes. *Cancer Res.*, **60**, 6272–5.

Scott, D., Addey, C., Ellis, P., James, E., Mitchell, M. J., Saut, N., Jurcevic, S. and Simpson, E. (2000) Dendritic cells permit identification of genes encoding MHC class II-restricted epitopes of transplantation antigens. *Immunity*, **12**, 711–20.

Shankaran, V., Ikeda, H., Bruce, A. T., White, J. M., Swanson, P. E., Old, L. J. and Schreiber, R. D. (2001) IFNgamma and lymphocytes prevent primary tumour development and shape tumour immunogenicity. *Nature*, **410**, 1107–11.

Skipper, J. C., Hendrickson, R. C., Gulden, P. H., Brichard, V., Van Pel, A., Chen, Y., Shabanowitz, J., Wolfel, T., Slingluff, C. L., Boon, T. and Hunt, D. (1996) An HLA-A2-restricted tyrosinase antigen on melanoma cells results from posttranslational modification and suggests a novel pathway for processing of membrane proteins. *J. Exp. Med.*, **183**, 527–34.

Smyth, M. J., Godfrey, D. I. and Trapani, J. A. (2001) A fresh look at tumor immunosurveillance and immunotherapy. *Nat. Immunol.*, **2**, 293–9.

Sonderstrup, G. and McDevitt, H. (1998) Identification of autoantigen epitopes in MHC class II transgenic mice. *Immunol. Rev.*, **164**, 129–38.

Stockert, E., Jager, E., Chen, Y. T., Scanlan, M. J., Gout, I., Karbach, J., Arand, M., Knuth, A. and Old, L. J. (1998) A survey of the humoral immune response of cancer patients to a panel of human tumor antigens. *J. Exp. Med.*, **187**, 1349–54.

Tang, D., DeVit, M. and Johnston, A. (1992) Genetic immunization is a simple method for eliciting an immune response. *Nature*, **356**, 152–3.

Thurner, B., Haendle, I., Roder, C., Dieckmann, D., Keikavoussi, P., Jonuleit, H., Bender, A., Maczck, C., Schreiner, D., von den Driesch, P., Brocker, E. B., Steinman, R. M., Enk, A., Kampgen, E. and Schuler, G. (1999) Vaccination with mage-3A1 peptide-pulsed mature, monocyte-derived dendritic cells expands specific cytotoxic T cells and induces regression of some metastases in advanced stage IV melanoma. *J. Exp. Med.*, **190**, 1669–78.

Timmerman, J. M. and Levy, R. (1999) Dendritic cell vaccines for cancer immunotherapy. *Annu. Rev. Med.*, **50**, 507–29.

Toes, R. E., Ossendorp, F., Offringa, R. and Melief, C. J. (1999) CD4 T Cells and their role in antitumor immune responses. *J. Exp. Med.*, **189**, 753–6.

Topalian, S. L., Gonzales, M. I., Parkhurst, M., Li, Y. F., Southwood, S., Sette, A., Rosenberg, S. A. and Robbins, P. F. (1996) Melanoma-specific CD4+ T cells recognize nonmutated HLA-DR-restricted tyrosinase epitopes. *J. Exp. Med.*, **183**, 1965–71.

Touloukian, C. E., Leitner, W. W., Topalian, S. L., Li, Y. F., Robbins, P. F., Rosenberg, S. A. and Restifo, N. P. (2000) Identification of a MHC class II-restricted human gp100 epitope using DR4-IE transgenic mice. *J. Immunol.*, **164**, 3535–42.

Tureci, O., Sahin, U., Zwick, C., Koslowski, M., Seitz, G. and Pfreundschuh, M. (1998) Identification of a meiosis-specific protein as a member of the class of cancer–testis antigens. *Proc. Natl. Acad. Sci. USA*, **95**, 5211–16.

Tureci, O., Sahin, U., Schobert, I., Koslowski, M., Scmitt, H., Schild, H. J., Stenner, F., Seitz, G., Rammensee, H. G. and Pfreundschuh, M. (1996) The *SSX-2* gene, which is involved in the t(X;18) translocation of synovial sarcomas, codes for the human tumor antigen HOM-MEL-40. *Cancer Res.*, **56**, 4766–72.

Ulmer, J. B., Donnelly, J. J., Parker, S. E., Rhodes, G. H., Felgner, P. L., Dwarki, V. J., Gromkowski, S. H., Deck, R. R., DeWitt, C. M., Friedman, A. *et al.* (1993) Heterologous protection against influenza by injection of DNA encoding a viral protein. *Science*, **259**, 1745–9.

Valmori, D., Fonteneau, J. F., Lizana, C. M., Gervois, N., Lienard, D., Rimoldi, D., Jongeneel, V., Jotereau, F., Cerottini, J. C. and Romero, P. (1998) Enhanced generation of specific tumor-reactive CTL *in vitro* by selected Melan-A/MART-1 immunodominant peptide analogues. *J. Immunol.*, **160**, 1750–8.

Van der Bruggen, P., Traversari, C., Chomez, P., Lurquin, C., DePlaen, E., Van Den Eynde, B., Knuth, A. and Boon, T. (1991) A gene encoding an antigen recognized by cytolytic T lymphocytes on a human melanoma. *Science*, **254**, 1643–7.

Wang, H. Y., Zhou, J., Zhou, K., Marincola, F. M. and Wang, R. F. (2002) Identification of a mutated fibronectin as a tumor antigen recognized by CD4+ T cells: its role in extracellular matrix formation and tumor metastasis. *J. Exp. Med.*, **195**, 1397–406.

Wang, R.-F. and Rosenberg, S. A. (1999) Human tumor antigens for cancer vaccine development. *Immunol. Rev.*, **170**, 85–100.

Wang, R.-F., Johnston, S. L., Zeng, G., Schwartzentruber, D. J. and Rosenberg, S. A. (1998) A breast and melanoma-shared tumor antigen: T cell responses to antigenic peptides translated from different open reading frames. *J. Immunol.*, **161**, 3596–606.

Wang, R.-F., Wang, X., Atwood, A. C., Topalian, S. L. and Rosenberg, S. A. (1999a) Cloning genes encoding MHC class II-restricted antigens: mutated CDC27 as a tumor antigen. *Science*, **284**, 1351–4.

Wang, R.-F., Wang, X. and Rosenberg, S. A. (1999b) Identification of a novel MHC class II-restricted tumor antigen resulting from a chromosomal rearrangement recognized by CD4+ T cells. *J. Exp. Med.*, **189**, 1659–67.

Wang, R.-F., Appella, E., Kawakami, Y., Kang, X. and Rosenberg, S. A. (1996a) Identification of TRP-2 as a human tumor antigen recognized by cytotoxic T lymphocytes. *J. Exp. Med.*, **184**, 2207–16.

Wang, R.-F., Parkhurst, M. R., Kawakami, Y., Robbins, P. F. and Rosenberg, S. A. (1996b) Utilization of an alternative open reading frame of a normal gene in generating a novel human cancer antigen. *J. Exp. Med.*, **183**, 1131–40.

Wang, R.-F., Robbins, P. F., Kawakami, Y., Kang, X. Q. and Rosenberg, S. A. (1995) Identification of a gene encoding a melanoma tumor antigen recognized by HLA-A31-restricted tumor-infiltrating lymphocytes. *J. Exp. Med.*, **181**, 799–804.

Wolfel, T., Hauer, M., Schneider, J., Serrano, M., Wolfel, C., Klehmann-Hieb, E., De Plaen, E., Hankeln, T., Meyer Zum Buschenfelde, K.-H. and Beach, D. (1995) A p16INK4a-insensitive CDK4 mutant targeted by cytolytic T lymphocytes in a human melanoma. *Science*, **269**, 1281–4.

Yee, C., Thompson, J. A., Roche, P., Byrd, D. R., Lee, P. P., Piepkorn, M., Kenyon, K., Davis, M. M., Riddell, S. R. and Greenberg, P. D. (2000) Melanocyte destruction after antigen-specific immunotherapy of melanoma. Direct evidence of t cell-mediated vitiligo. *J. Exp. Med.*, **192**, 1637–44.

Ying, H., Zaks, T. Z., Wang, R. F., Irvine, K. R., Kammula, U. S., Marincola, F. M., Leitner, W. W. and Restifo, N. P. (1999) Cancer therapy using a self-replicating RNA vaccine. *Nat. Med.*, **5**, 823–7.

Zajac, A. J., Murali-Krishna, K., Blattman, J. N. and Ahmed, R. (1998) Therapeutic vaccination against chronic viral infection: the importance of cooperation between CD4+ and CD8+ T cells. *Curr. Opin. Immunol.*, **10**, 444–9.

Zarour, H. M., Kirkwood, J. M., Kierstead, L. S., Herr, W., Brusic, V., Slingluff, C. L., Jr., Sidney, J., Sette, A. and Storkus, W. J. (2000a) Melan-A/MART-1(51-73) represents an immunogenic HLA-DR4-restricted epitope recognized by melanoma-reactive CD4(+) T cells. *Proc. Natl. Acad. Sci. USA*, **97**, 400–5.

Zarour, H. M., Storkus, W. J., Brusic, V., Williams, E. and Kirkwood, J. M. (2000b) NY-ESO-1 encodes DRB1*0401-restricted epitopes recognized by melanoma-reactive CD4+ T cells. *Cancer Res.*, **60**, 4946–52.

Zeng, G., Touloukian, C. E., Wang, X., Restifo, N. P., Rosenberg, S. A. and Wang, R.-F. (2000) Identification of CD4+ T cell epitopes from NY-ESO-1 presented by HLA-DR molecules. *J. Immunol.*, **165**, 1153–59.

Zeng, G., Wang, X., Robbins, P. F., Rosenberg, S. A. and Wang, R.-F. (2001) CD4+ T cell recognition of MHC class II-restricted epitopes from NY-ESO-1 presented by a prevalent HLA-DP4 allele: association with NY-ESO-1 antibody production. *Proc. Natl. Acad. Sci. USA*, **98**, 3964–9.

Part 4

Human tumor antigens recognized by antibodies

Chapter 10

Human tumor antigens recognized by antibodies (SEREX)

Michael Pfreundschuh, Klaus-Dieter Preuss, Carsten Zwick, Claudia Bormann and Frank Neumann

Summary

The analysis of the B-cell repertoire against tumors using tumor cDNA expression cloning and autologous serum by *se*rological analysis of antigens by *r*ecombinant *ex*pression cloning (SEREX) revealed that many, if not all, human tumors express multiple antigens which are recognized by the patient's immune system. There are different antigen specificities: (1) shared tumor antigens, (2) differentiation antigens; products of (3) mutated, (4) viral, (5) overexpressed and (6) amplified genes, as well as (7) splice variants of normal genes, (8) widely expressed autoantigens, the immunogenicity of which is restricted to cancer patients, (9) common autoantigens, to which antibodies are found in the sera from patients with other than malignant diseases; and finally (10) products of genes which are underexpressed in the autologous tumor compared to normal tissues. Our results indicate that the context of presentation of a molecule is more important for its immunogenicity than its more or less restricted expression in tumors. This implies that only immunotherapeutic approaches based on specific antigens are likely to induce tumor-specific reactions, whereas whole-tumor cell approaches would rather induce tolerance and/or autoimmune disease. CD8+ as well as CD4+ responses could be demonstrated against SEREX antigens and the relevant antigenic peptides have been defined. With the availability of specific antigens for the majority of human cancers, carefully designed trials have to be performed to determine the value of our knowledge about these antigens for immuno- and genetherapeutic approaches in patients with malignant disease.

Introduction

The identification and molecular characterization of tumor antigens, which are able to elicit specific immune responses in the tumor-bearing host, has become a major effort in tumor immunology. In the 70s and 80s of the 20th century, the hybridoma technology was exploited for the identification of molecules on tumor cells which could be used as diagnostic markers or as target structures for immunotherapeutic approaches with monoclonal antibodies. While some of these efforts have yielded new therapeutic tools, such as the anti-CD20 antibody rituximab which shows considerable activity and has been licensed for the treatment of follicular lymphomas (Maloney *et al.*, 1997), approaches of *active* immunotherapy require the identification of target structures which are immunogenic in the autologous tumor-bearing host.

The analysis of humoral and cellular immune responses against such antigens in cancer patients had indicated for a long time that cancer specific antigens do indeed exist which are recognized by the patient's immune system (Old, 1981). To disclose the molecular nature of these antigens, cloning techniques were developed that used established CTL clones (van der Bruggen *et al.*, 1991) or circulating antibodies (Sahin *et al.*, 1995) as probes for screening tumor-derived expression libraries. While the molecular characterization of the first human tumor antigens was accomplished with cloning techniques that used established CTL clones (van der Bruggen *et al.*, 1991), it is now commonly accepted that immune recognition of tumors is a concerted action. Thus, high-titered circulating tumor-associated antibodies of the IgG class may reflect a significant host–tumor interaction and may identify such gene products to which at least cognate T cell help, but also specific cytotoxic T cells should exist. This rationale made us design a novel strategy using the antibody repertoire of cancer patients for the molecular definition of antigens. Serologically defined antigens could then be subjected to procedures of "reverse" T-cell immunology for the definition of epitopes which are presented by MHC class I or II molecules, respectively, and are recognized by T-lymphocytes. The serological cloning approach for the identification of tumor antigens recognized by antibodies in the serum of cancer patients was designated SEREX. SEREX allows a systematic and unbiased search for antibody responses against proteins and the direct molecular definition of the respective tumor antigens based on their reactivity with autologous patient serum (Sahin *et al.*, 1997; Türeci *et al.*, 1997).

Material and methods

The study had been approved of by the local ethical review board ("Ethikkommission der Ärztekammer des Saarlandes"). Recombinant DNA work was done with the official permission and according to the rules of the State Government of Saarland. All healthy donors and cancer patients gave informed consent.

Sera and Tissues. Sera and tumor tissues were obtained during routine diagnostic or therapeutic procedures. Normal tissues were collected from autopsies of tumor-free patients.

Construction of cDNA expression libraries. Out of total RNA isolated from fresh tumor biopsies, 5–8 μg poly(A) + RNA was prepared with an mRNA isolation kit (Stratagene). cDNA expression libraries were directionally cloned into λ-ZAP Express vector using a commercially available adaptor ligation system according to the manufacturer's instructions (Stratagene, Heidelberg, Germany). After packaging into phage particles these cDNA-expression libraries were transfected into XLI MRF′ bacteria for one round of amplification. In other experiments, testis-derived expression libraries were used.

Immunoscreening for antigens expressed by human tumors. XLI MRF′ bacteria transfected with recombinant λ-ZAP Express phages were plated onto Luria–Bertani agar plates. Expression of recombinant proteins in lytic phage plaques on the bacterial lawn was induced with IPTG. Plates were incubated at 37°C until plaques were visible and then blotted onto nitrocellulose membranes. The membranes were blocked with 5% low-fat milk in Tris-buffered saline and incubated with a 1 : 500 dilution of the patient's serum, which had been preabsorbed with transfected *Escherechia coli.* Serum antibodies binding to recombinant proteins expressed in lytic plaques were detected by incubation with alkaline phosphatase-conjugated goat anti-human IgG and visualization by staining with 5-bromo-4-chloro-3-indolyl phosphate and nitroblue tetrazolium. For screening of testis-derived libraries allogeneic sera of patients with different tumor types were used. Positive clones were isolated and monoclonalized by repeated rounds of immunoscreening followed by plaque elution.

Sequence analysis. Positive clones were subcloned to monoclonality and submitted to *in vivo* excision of pBK-CMV phagemids. The nucleotide sequence of cDNA inserts was determined using Excel cycle sequencing kit (Epicentre) on a LICOR automatic sequencer. Sequencing was performed according to the manufacturers' instructions starting with the vector specific primers. Insert-specific primers were designed as the sequencing proceeded. Sequence alignments were performed with DNASIS (Pharmacia Biotech) and BLAST software on EMBL, Genbank and PROSITE databases.

Northern blot analysis. Northern blots were performed with RNA extracted from tumors and normal tissues. Integrity of RNA was checked by electrophoresis in formalin/MOPS-gels. Gels containing 10 μg RNA per lane were blotted onto nylon membranes. After pre-hybridization the membranes were incubated with the specific 32P-labeled insert-specific full-length cDNA probes overnight at 42°C in hybridization solution (50% formamide, 6 × SSC, 5 × Denhardt's, 0.2% SDS). The membranes were then washed at progressively higher stringency, with the final wash in 1 × SSC and 0.2% SDS at 65°C. Autoradiography was conducted at 70°C for upto seven days using Kodak X-OMAT-AR film and intensifying screen. Northern blots were used to assess expression levels of transcript in a panel of normal tissues and tumors and particularly to compare the tumors with their normal counterpart tissue from the same patient.

Reverse transcription PCR. Total cellular RNA was extracted and primed with a mixture of dT(18) oligonucleotide and random hexamer primers and reverse-transcribed with Superscript RT (Gibco, Eggenstein, Germany). cDNA thus obtained was tested for integrity by amplification of ß-actin transcripts in a 30-cycle PCR reaction. For each clone positive in the SEREX screening specific primers located in different exons were designed and a panel of normal tissues was screened for expression.

Detection of antigen-specific antibodies in allogeneic sera. Monoclonalized phages from positive clones were mixed with non-reactive phages of the cDNA-library as internal negative controls at a ratio of 1 : 10 and used to transfect bacteria. IgG antibodies in the 1 : 200 diluted *E. coli*-preabsorbed sera from allogeneic patients and healthy controls were tested with the above-described immunoscreening assay to assess for tumor-associated antibody-responses.

Results

Immunoscreening for human tumor antigens. Expression libraries were constructed and analyzed by SEREX from a variety of different neoplasms, including three different renal cell carcinomas of the clear cell type, two melanomas, one ovarian carcinoma, one hepatocarcinoma, three gliomas, two colorectal cancers, one pancreatic cancer, two breast cancers, two Hodgkin's lymphomas, two acute T-cell leukemias and two acute myelogenous leukemias. Primary libraries with at least 1×10^6 independent clones were established. The screening of at least 1×10^6 clones per library revealed multiple reactive clones in each library. Some transcripts were detected repeatedly, indicating that they were multiply represented in the library. In order to bias for the detection of antigens of the cancer testis class, libraries were constructed from normal testis tissue and screened with allogeneic tumor patients' sera.

Molecular characterization of SEREX antigens. To select clones of potential biological or clinical interest for more in-depth analysis, a three-step strategy was pursued. This included: (1) the comparison of the sequence data with databases to reveal identity or homologies with known genes and to identify domains or motifs informative for a putative function or cellular localization; (2) the analysis of the expression pattern of the respective antigen in normal tissues and tumors by reverse transcription polymerase chain reaction (RT-PCR), by Northern blot hybridization with specific probes and by analysis in expressed sequence tag containing databases; and (3) an initial survey for antibodies in the sera from healthy controls and allogeneic tumor patients to evaluate the incidence of serum antibodies to the respective antigen.

Four different groups of genes coding for antigens were identified. The first group codes for known tumor antigens such as the melanoma antigens MAGE-1, MAGE-4a and tyrosinase. A second group encodes known classical autoantigens for which immunogenicity is associated with autoimmune diseases, for example, anti-mitochondrial antibodies or antibodies to U1-snRNP. When patients known to have autoimmune or rheumatic disorders are excluded from SEREX analysis, the incidence of such antigens is not higher than 1%. A third group codes for transcripts that are either identical or highly homologous to known genes, but have not been known to elicit immune responses in humans. Examples are restin which had originally been identified by a murine monoclonal antibody specific for Hodgkin and Reed–Sternberg cells and lactate dehydrogenase, an enzyme overexpressed in many human tumors. The fourth group of serologically defined antigens represents products of previously unknown genes.

Specificity of SEREX antigens. According to their expression pattern in normal and malignant tissues, several classes of tumor antigens can be distinguished (Table 10.1): (1) shared tumor antigens, (2) differentiation antigens; products of (3) mutated, (4) viral, (5) overexpressed, and (6) amplified genes as well as (7) splice variants, (8) widely expressed, but cancer-associated autoantigens, the immunogenicity of which is restricted to cancer patients, (9) common autoantigens to which antibodies are found in sera from patients with other than malignant diseases; and finally (10) products of genes which are *underexpressed* in the autologous tumor compared to normal tissues.

Shared tumor antigens. A variable proportion of human tumors, ranging from 10% to 70% depending on the type of tumor, express a given shared tumor antigen. Interestingly, all human shared tumor antigens identified to date are expressed in a variety of human cancers, but not in normal tissues, except for testis; therefore, the term *cancer–testis antigens* has been coined for them. The prototypes of this category, MAGE (van der Bruggen *et al.*, 1991), BAGE (Boel *et al.*, 1995), and GAGE (van den Eynde *et al.*, 1995), were initially identified as targets for cytotoxic T cells. The HOM-MEL-40 antigen, which was detected in

Table 10.1 Specificity of tumor antigens detected by SEREX

Specificity	*Example*	*Source*
Shared tumor antigens	HOM-MEL-40	Melanoma
Differentiation antigens	HOM-MEL-55 (tyrosinase)	Melanoma
Mutated genes	NY-COL-2 (p53)	Colorectal carcinoma
Splice variants	HOM-HD-397 (restin)	Hodgkin's disease
Viral antigens	HOM-RCC-1.14 (HERV-K10)	Renal cell cancer
Overexpression	HOM-HD-21 (galectin-9)	Hodgkin's disease
Gene amplifications	HOM-NSCLC-11 (eIF-4γ)	Lung cancer
Cancer-related autoantigens	HOM-MEL-2.4 (CEBP)	Melanoma
Cancer-independent autoantigens	NY-ESO-2 (U1-snRNP)	Esophageal carcinoma
Underexpressed genes	HOM-HCC-8	Hepatocellular carcinoma

a melanoma library, is the first cancer/testis antigen identified by SEREX. It is encoded by the *SSX-2* gene (Türeci *et al.*, 1996). Members of the *SSX* gene family, SSX1 and SSX2, have been shown to be involved in the t(X;18)(p11.2; q11.2) translocation which is found in the majority of human synovial sarcomas (Clark *et al.*, 1994) and fuses the respective *SSX* gene with the *SYT* gene from chromosome 18. Using homology cloning, additional members of the SSX-family were identified (Güre *et al.*, 1995) revealing at least five genes, of which four (*SSX-1, 2, 4* and *5*) demonstrate a CT antigen-like expression (Türeci *et al.*, 1998a). Using SEREX, Chen *et al.* (1997) identified NY-ESO-1 as a new CT antigen. NY-ESO-1 mRNA expression is detectable in a wide array of human cancers, including melanomas, breast cancer, bladder cancer and prostate cancer. A homologous gene, named *LAGE-1* was subsequently isolated by a subtractive cloning approach (Lethe *et al.*, 1998) demonstrating that NY-ESO-1 belongs to a gene family with at least two members. NY-ESO-1 as well as its homologue LAGE-1 were discovered by independent groups using tumor specific CTL or tumor-infiltrating lymphocytes (TIL) derived from melanoma patients as probes, thus disclosing several HLA-A0201 and HLA-A31 restricted epitopes (Jäger *et al.*, 1998) and demonstrating that NY-ESO-1 is a target for both antibody and CTL responses in the same patient (Jäger *et al.*, 1998, 1999). IgG antibody responses directed against NY-ESO-1 are present in upto 50% of antigen-expressing patients indicating that this antigen may also be a frequent target for CD4+ T-lymphocytes (Stockert *et al.*, 1998). Another new CT antigen is HOM-TES-14 (Türeci *et al.*, 1998b) which is encoded by the gene coding for the synaptonemal complex protein-1 (SCP-1).

Differentiation antigens. They demonstrate a lineage-specific expression in tumors, but also in normal cells of the same origin; examples are tyrosinase and glial fibrillary acidic protein (GFAP) which are antigenic in malignant melanoma and glioma, but are also expressed in melanocytes or brain cells, respectively.

Antigens encoded by mutated genes. They have been demonstrated only rarely by the serological approach, with mutated p53 being one example (Scanlan *et al.*, 1997). For SEREX-detected antigens proof of an underlying mutation is technically challenging since antibody responses induced by a mutation may be directed to the wild-type backbone of the molecule and thus the wild-type allele may be picked up during the immunoscreening, so that sequencing of several independent clones from the same library as well as exclusion of polymorphisms is mandatory.

Viral genes. A virus-encoded antigen that elicits an autologous antibody response is the env protein of the human endogenous retrovirus HERV-K10, which was found in a renal cell cancer and in a seminoma.

Overexpressed genes. These code for many tumor antigens identified by SEREX, which has an inherent methodological bias for the detection of abundant transcripts. The members of this class are expressed at low levels in normal tissues (usually detectable by RT-PCR), but are upto 100-fold overexpressed in tumors. An example is HOM-RCC-3.1.3, a new carbonic anhydrase which is overexpressed in a fraction of renal cell cancers (Türeci *et al.*, 1998c).

Amplified genes. They may also code for tumor antigens. The overexpression of a transcript resulting from a gene amplification has been demonstrated for the translation initiation factor eIF-4γ in a squamous cell lung cancer (Brass *et al.*, 1997).

Splice variants of known genes. They were also found to be immunogenic in cancer patients. Examples are NY-COL-38 and restin, which represents a splice variant of the formerly described cytoplasmic linker protein CLIP-170 (Pierre *et al.*, 1992).

Cancer-related autoantigens. These are expressed ubiquitously and at a similar level in healthy as well as malignant tissues. The encoding genes are not altered in tumor samples. However, they elicit antibody responses only in cancer patients, but not in healthy individuals. This might result from tumor-associated post-translational modifications or changes in the antigen processing and/or presentation in tumor cells. An example is HOM-MEL-2.4, which represents the CCAAT enhancer binding protein.

Non-cancer-related autoantigens. These are also expressed ubiquitously in most human tissues; in contrast to cancer-related autoantigens antibodies against these antigens are found in non-tumor bearing controls at a similar frequency as in tumor patients. An example is HOM-RCC-10, which represent mitochondrial DNA, and HOM-TES-11, which is identical to pericentriol material-1 (PCM-1).

Products of underexpressed genes. An antigen was detected during the SEREX analysis of a hepatocellular carcinoma, which was underexpressed in the malignant tissue when compared to normal liver (Stenner-Lieven *et al.*, 2000).

Incidence of antibodies to SEREX antigens and clinical significance. The analysis of sera from patients with various malignant diseases and from healthy controls showed that different patterns of seroresponses against SEREX antigens exist. Clinically most interesting is the group of antibodies which occurs exclusively in patients with cancer. Such strictly tumor-associated antibody responses are detected with varying frequencies only in the sera of patients with tumors that express the respective antigen. Examples are antibodies against HOM-TES-14/SCP-1, HOM-HD-21/galectin-9 (Sahin *et al.*, 1995; Türeci *et al.*, 1997), NY-ESO-I (Stockert *et al.*, 1998), and against several antigens cloned from colon cancer (Scanlan *et al.*, 1997). The incidence of tumor-associated antibodies in unselected tumor patients ranges between 5% and 50% depending on the tumor-type and the respective antigen. For many antigens identified by SEREX, antibodies are only detected in the patient whose serum was used for the SEREX assay. A third group of antibodies occurs in cancer patients and healthy controls at a similar rate. While most of these antibodies are directed against non-cancer-associated, widely expressed autoantigens, for example, poly-adenosyl-ribosyl transferase, antibodies of this category are also found to be directed against antigens with a very restricted expression pattern, for example, restin. Restin represents a differentiation antigen, since its expression is limited to Hodgkin- and Reed–Sternberg cells and immature dendritic cells.

There is little information as to the significance of antibody responses and their correlation with the clinical course of the malignant disease. Well-designed prospective studies are necessary to identify those antibodies that might be (alone or in combination with others) a valuable tool for the diagnosis and therapeutic evaluation of malignant diseases. From anecdotal observations we have the impression that for antibodies to be present in the patient's serum, the tumor must express the respective antigen. Antibody titers drop and often disappear when the tumor is removed or the patient is in remission. Why only a minority of the patients with an antigen-positive tumor develops antibodies to the respective antigen also remains an open question.

Functional significance of SEREX antigens. Antigens with a known function identified by SEREX include HOM-RCC-3.1.3 (Türeci *et al.*, 1998c), which was shown to be a novel member of the carbonic anhydrase (CA) family, designated as CA XII. Overexpression of this transcript was observed in 10% of renal cell cancers (RCC), suggesting a potential significance in this tumor type. In fact, the same transcript was cloned shortly thereafter by another group based on its downregulation by the wild-type von-Hippel-Lindau tumor suppressor gene, the loss of function of which is known to be associated with an increased incidence of RCC (Ivanov *et al.*, 1998). Since the invasiveness of RCC cell lines expressing CA XII has been shown to be inhibited by acetazolamide (Parkkila *et al.*, 2000), CA XII might be exploited therapeutically.

The first cancer testis antigen to which a physiological function could be ascribed is HOM-TES-14 which is encoded by the *SCP-1* gene (Türeci *et al.*, 1998b). *SCP-1* is known to be selectively expressed during the meiotic prophase of spermatocytes and is involved in the pairing of homologous chromosomes, an essential step for the generation of haploid cells in meiosis I. The aberrant expression of this meiosis-specific gene product in the somatic cells of human tumor cells may be involved in the induction of chromosomal instabilities in cancer cells.

Reverse T-cell immunology. The main idea of cancer vaccination is to induce an effective specific cytolytic and/or T-helper immune activity against tumor cells. Serologically defined antigens are most suitable candidates for determination of such epitopes, since the isotype switching and the development of high-titered IgG *in vivo* requires cognate CD4+ T-cell help. Therefore, SEREX can be instrumentalized to analyze the CD4+ T-cell repertoire against tumor antigens. Several CD4 binding epitopes of the NY-ESO-1 antigen have been identified by us (manuscript in preparation) and by others (Jäger *et al.*, 2000). With regard to CD8+ T-lymphocytes that recognize SEREX antigens, CTL responses have been demonstrated for HOM-MEL-40 and NY-ESO-1 (Jäger *et al.*, 2000; own unpublished results).

Discussion

The SEREX approach is technically characterized by several features: (1) There is no need for established tumor cell lines and pre-characterized CTL clones; (2) the use of fresh tumor specimens restricts the analysis to genes that are expressed by the tumor cells *in vivo* and avoids *in vitro* artifacts associated with short- and long-term tumor cell culture; (3) the restriction of the screening to clones against which the patient's immune system has raised high-titered IgG or/and IgA antibody responses implies the presence of a concomitant T-helper lymphocyte response *in vivo*; (4) Since both the expressed antigenic protein and the coding cDNA are present in the same plaque of the phage immunoscreening assay, identified antigens can be sequenced immediately; (5) sequence information of excised cDNA inserts can be directly used to determine the expression spectrum of identified transcripts by Northern blot and RT-PCR to determine the specificity of the respective antigen.

SEREX allows for the identification of an entire profile of antigens using the antibody repertoire of a single cancer patient. The analysis of a variety of neoplasms demonstrated that all hitherto investigated neoplasms are immunogenic in the tumor-bearing host and that immunogenicity is conferred by multiple antigens. For the systematic documentation and archivation of sequence data and immunological characteristics of identified antigens an electronical SEREX database was initiated by the Ludwig Cancer Research Institutes, which is accessible to the public (www.licr.org/SEREX.html). By June 2001, about 2,000 entries have been made into the SEREX database, the majority of them representing independent antigens. The SEREX database is not only meant as a computational interface for discovery information management, but also as a tool for mapping the entire panel of gene products which elicit spontaneous immune responses in the tumor-bearing autologous host, for which the name "cancer immunome" has been coined by Old (LICR, New York Branch).

The cancer immunome which is defined by using spontaneously occurring immune effectors from cancer patients as probes gains increasing interest, since it has been shown that many antigens may be valuable as new molecular markers of malignant disease. The value of each of these markers or a combination of them for diagnostic or prognostic evaluation of cancer patients has to be determined by studies which correlate the presence or absence of these markers with clinical data.

The multitude of tumor-specific antigens identified by the SEREX approach has revealed that there is ample immune recognition of human tumors by the autologous host's immune system. Together with the identification of T lymphocyte epitopes a picture of the immunological profile of cancer is emerging. The knowledge of the cancer immunome provides a new basis for understanding tumor biology and for the development of new diagnostic and therapeutic strategies for cancer. The specificities of the antigens expressed by human tumors and detected by SEREX vary widely, ranging from the shared tumor antigens the expression spectrum of which is confined to tumors (and normal testis) to products of genes overexpressed in tumors, normal autoantigens, which are immunogenic in patients with other than malignant diseases, and antigens encoded by genes that are underexpressed in the malignant tumor compared to the benign tissue counterpart. This surprising finding together with the observation that some ubiquitously expressed antigens elicit immune responses only in cancer patients, suggest that the context in which a protein is presented to the immune system (e.g. the context of "danger") is more decisive for its immunogenicity (and breaking of tolerance) than its more or less tumor-restricted expression. Our results obtained with the SEREX analysis of many human tumors also show that, besides the rare tumor-specific antigens, there is a great majority of widely expressed autoantigens which are presented by the tumor and recognized by the immune system. The presentation of common autoantigens by a given tumor (presumably in the context of "danger") induces a broad range of autoimmunity, and it is only if the tumor happens to present tumor-specific molecules that tumor-specific autoimmunity can occur: specific tumor immunity is just a small part of broad autoimmunity that is commonly induced by malignant growth. The fact that tumors present a majority of molecules that are also expressed in normal tissue and only a minority of tumor-specific molecules in the context of their MHC molecules also implies that vaccines using whole tumor-cell preparations are rather unlikely to be successful, since the induction of tumor-specific immunity by such vaccines would be a quantité negligible compared to the majority of immune responses (or tolerance) induced to normal autoantigens by such a strategy.

The main goal of cancer vaccination is the induction of an effective specific cytolytic response against tumor cells which spares the cells of normal tissues. With respect to specificity several classes of antigens may be suitable targets; they include the CT antigens, differentiation antigens, tumor-associated over expressed gene products, mutated gene products and tumor-specific splice variants. Clinically the most interesting class of antigens is that of the shared tumor antigens or cancer–testis antigens. To cope with the rapidly growing number of CT antigens, a new nomenclature has been suggested for them (Old and Chen, 1998). According to the order of their initial identification the individual genes are designated by enumeration. Since individual CT antigens are expressed only in a variable proportion of tumors, only the availability of several CTA could significantly enlarge the proportion of patients eligible for vaccination studies. In this regard it is interesting that members of a given gene family tend to be expressed in a co-regulated fashion whereas different gene families are preferentially expressed in other sets of tumors. It is therefore reasonable to choose antigens from different CT families to cover as many tumors as possible. Despite the fact that SEREX enlarged the pool of available tumor antigens, the proportion of tumors for which no tumor antigen is known, is still high, particularly in frequent neoplasms such as colon and prostate cancer. Moreover, immunohistological investigations for MAGE antigens have demonstrated a heterogeneity of antigen expression even in the same tumor specimen (Hofbauer *et al.*, 1999). Thus, the combined or sequential use of a whole set of several antigens in a patient would have the potential of reducing or even preventing the *in vivo* selection of antigen loss tumor-cell variants and would also address the problem of a heterogeneous expression of a given antigen in an individual tumor specimen.

For the development of molecular vaccine strategies, the knowledge of antigen-derived peptide epitopes which are capable of priming or activating specific CTL or T-helper cells is an indispensable prerequisite. Due to the diversity of peptides presented by the highly polymorphic HLA alleles, the definition of these epitopes by "reverse T-cell immunology" represents an enormous challenge for each individual antigen. However, we and others have demonstrated that by using straightforward strategies, the identification of epitopes from SEREX defined antigens which bind and activate either CD8+ or CD4+ is feasible and has been successful for each SEREX antigen for which the definition of such epitopes has been pursued.

The knowledge and availability of a large number of human tumor antigens and their MHC binding epitopes has opened the perspective for the development of polyvalent vaccines for a wide spectrum of human cancers using pure preparations of antigenic proteins or peptide fragments. Additionally, the study and long-term follow-up of large numbers of patients will help to determine the diagnostic and prognostic relevance of tumor-related/specific autoantibodies in patients' sera and of antigen expression in tumors, as well as the correlation with CTL responses and specific T-helper cells. Finally, the knowledge of immunogenic products of cancer-associated genes will provide us with a more profound insight into genetic and molecular alterations that might be of enormous relevance for the pathogenesis and growth of cancer.

References

Boel, P., Wildmann, C., Sensei, M. L., Brasseur, R., Renauld, M., Coulie, P., Boon, T. and van der Bruggen, P. (1995) *BAGE*: a new gene encoding an antigen recognized on human melanomas by cytolytic T lymphocytes. *Immunity*, **2**, 167–175.

van der Bruggen, P., Traversari, C., Chomez, P., Lurquin, C., De Plaen, E., van den Eynde, B., Knuth, A. and Boon, T. (1991) A gene encoding an antigen recognized by cytolytic T lymphocytes on a human melanoma. *Science*, **254**, 1643–1647.

Brass, N., Heckel, D., Sahin, U., Pfreundschuh, M., Sybrecht, G. W. and Meese, E. (1997) Translation initiation factor eIF-4gamma is encoded by an amplified gene and induces an immune response in squamous cell lung carcinoma. *Hum. Mol. Genet.*, **6**, 33–39.

Chen, Y. T., Scanlan, J., Sahin, U., Türeci, Ö., Güre, A. O., Tsang, S., Williamson, S., Stockert, E., Pfreundschuh, M. and Old, L. J. (1997) A testicular antigen aberrantly expressed in human cancers detected by autologous antibody screening. *Proc. Natl. Acad. Sci. USA*, **94**, 1914–1918.

Clark, J., Roques, J. P., Crew, J., Gill, S., Shipley, J., Chand, A., Gusterson, B. and Cooper, C. S. (1994) Identification of novel genes SYT and SSX involved in the t(X,18)(p11.2,q11.2) translocation found in human synovial sarcoma. *Nat. Genet.*, **7**, 502–508.

van den Eynde, B., Peeters, O., de Backer, O., Gaugler, B., Lucas, S. and Boon, T. (1995) A new family of genes coding for an antigen recognized by autologous cytolytic T lymphocytes on a human melanoma. *J. Exp. Med.*, **182**, 689–698.

Güre, A. O., Türeci, Ö., Sahin, U., Tsang, S., Scanlan, M., Jäger, E., Knuth, A., Preundschuh, M., Old, L. J. and Chen, Y. T. (1997) SSX, a multigene family with several members transcribed in normal testis and human cancer. *Int. J. Cancer*, **72**, 965–971.

Hofbauer, G. F., Schaefer, C., Noppen, C., Boni, R., Kamarashev, J., Nestle, F. O., Spagnoli, G. C. and Dummer, R. (1997) MAGE-3 immunoreactivity in formalin-fixed, paraffin-embedded primary and metastatic melanoma: frequency and distribution. *Am. J. Pathol.*, **151**, 1549–1553.

Ivanov, S. V., Kuzmin, I., Wei, M. H., Pack, S., Geil, L., Johnson, B. E., Stanbridge, E. J. and Lerman, M. I. (1998) Down-regulation of transmembrane carbonic anhydrases in renal cell carcinoma cell lines by wild-type von Hippel–Lindau transgenes. *Proc. Natl. Acad. Sci. USA*, **95**, 12596–12601.

Jäger, E., Chen, Y. T., Drijfhout, J. W., Karbach, J., Ringhoffer, M., Jäger, D., Arand, M., Wada, H., Noguchi, Y., Stockert, E., Old, L. J. and Knuth, A. (1998) Simultaneous humoral and cellular immune response against cancer–testis antigen NY-ESO-1: definition of human histocompatibility leukocyte antigen (HLA)-A2-binding peptide epitopes. *J. Exp. Med.*, **187**, 265–270.

Jäger, E., Stockert, E., Zidianakis, Z., Chen, Y. T., Karbach, J., Jager, D., Arand, M., Ritter, G., Old, L. J. and Knuth, A. (1999) Humoral immune responses of cancer patients against "Cancer–Testis" antigen NY-ESO-1: correlation with clinical events. *Int. J. Cancer*, **84**, 506–510.

Jäger, E., Jäger, D., Kargach, J., Chen, Y.-T., Ritter, G., Nagat, Y., Gnjatic, S., Stockert, E., Arand, M., Old., L. J. and Knuth, A. (2000) Identification of NY-ESO-1 epitopes presented by human histocompatibility antigen (HAL)-DRB4*0101–0103 and recognized by $CD4^+$ T Lymphomcates of patients with NY-ESO-1-expressing melanoma. *J. Exp. Med.*, **191**, 625–630.

Lethe, B., Lucas, S., Michaux, L., De Smet, C., Godelaine, D., Serrano, A., De Plaen, E. and Boon, T. (1998) *LAGE-1*, a new gene with tumor specificity. *Int. J. Cancer*, **76**, 903–908.

Maloney, D. G., Grillo, L. A., White, C. A., Bodkin, D., Schilder, R. J., Neidhart, J. A. *et al.* (1997) IDEC-C2B8 (Rituximab) anti-CD20 monoclonal antibody therapy in patients with relapsed low-grade non-Hodgkin's lymphoma. *Blood*, **90**, 2188–2195.

Old, L. J. (1981) Cancer immunology: the search for specificity – G.H.A. Clowes Memorial lecture. *Cancer Res.*, **41**, 361–375.

Old, L. J. and Chen, Y. T. (1998) New paths in human cancer serology. *J. Exp. Med.*, **187**, 1163–1167.

Parkkila, S., Rajaniemi, H., Parkkila, A. K., Kivela, J., Waheed, A., Pastorekova, S., Pastorek, J. and Sly, W. S. (2000) Carbonic anhydrase inhibitor suppresses invasion of renal cancer cells *in vitro*. *Proc. Natl. Acad. Sci. USA*, **97**, 2220–2224.

Pierre, P., Scheel, J., Rickard, J. E. and Kreis, T. (1992) CLIP-170 links endocytic vesicles to microtubules. *Cell*, **70**, 887–892.

Sahin, U., Türeci, Ö., Schmitt, H., Cochlovius, B., Johannes, T., Schmits, R., Stenner, F., Luo, G., Schobert, I. and Pfreundschuh, M. (1995) Human neoplasms elicit multiple immune responses in the autologous host. *Proc. Natl. Acad. Sci. USA*, **92**, 11810–11813.

Sahin, U., Türeci, Ö. and Pfreundschuh, M. (1997) Serological identification of human tumor antigens. *Curr. Opin. Immunol.*, **9**, 709–716.

Scanlan, M. J., Chen, Y. T., Williamson, B., Güre, A. O., Stockert, E., Gordan, J. D., Türeci, Ö., Sahin, U., Pfreundschuh, M. and Old, L. J. (1997) Characterization of human colon cancer antigens recognized by autologous antibodies. *Int. J. Cancer*, **76**, 652–658.

Stenner-Liewen, F., Luo, G., Sahin, U., Türeci, Ö., Koslowski, M., Kautz, I., Liewen, H. and Pfreundschuh, M. (2000) Definition of tumor-associated antigens in hepatocellular carcinoma. *Cancer Epidemiol., Biomarkers, Prevention*, **9**, 285–290.

Stockert, E., Jäger, E., Chen, Y. T., Scanlan, M. J., Gout, I., Karbach, J., Arand, M., Knuth, A. and Old, L. J. (1998) A survey of the humoral immune response of cancer patients to a panel of human tumor antigens. *J. Exp. Med.*, **187**, 1349–1354.

Türeci, Ö., Sahin, U., Schobert, I., Koslowski, M., Schmitt, H., Schild, H. J., Stenner, F., Seitz, G., Rammensee, H. G. and Pfreundschuh, M. (1996) The *SSX2* gene, which is involved in the t(X,18) translocation of synovial sarcomas, codes for the human tumor antigen HOM-MEL-40. *Cancer Res.*, **56**, 4766–4772.

Türeci, Ö., Sahin, U. and Pfreundschuh, M. (1997) Serological analysis of human tumor antigens: molecular definition and implications. *Mol. Med. Today*, **3**, 342–349.

Türeci, Ö., Schmitt, H., Fadle, N., Pfreundschuh, M. and Sahin, U. (1997) Molecular definition of a novel human galectin which is immunogenic in patients with Hodgkin's disease. *J. Biol. Chem.*, **272**, 6416–6422.

Türeci, Ö., Chen, Y. T., Sahin, U., Güre, A. O., Zwick, C., Villena, C., Tsang, S., Seitz, G., Old, L. J. and Pfreundschuh, M. (1998a) Expression of *SSX* genes in human tumors. *Int. J. Cancer*, **77**, 19–23.

Türeci, Ö., Sahin, U., Zwick, C., Koslowski, M., Seitz, G. and Pfreundschuh, M. (1998b) Identification of a meiosis-specific protein as new member of the cancer/testis antigen superfamily. *Proc. Natl. Acad. Sci. USA*, **95**, 5211–5216.

Türeci, Ö., Sahin, U., Vollmar, E., Siemer, S., Göttert, E., Seitz, G., Parkkila, A. K., Shah, G., Grubb, J. H., Pfreundschuh, M. and Sly, W. S. (1998c) Carbonic anhydrase XII: cDNA cloning, expression and chromosomal localization of a novel carbonic anhydrase gene that is overexpressed in some renal cell cancers. *Proc. Natl. Acad. Sci. USA*, **95**, 7603–7613.

Chapter 11

Antibodies to human tumor oncoproteins in cancer patients

Lupe Salazar and Mary L. Disis

Summary

Antibody immunity to a number of oncogenic proteins has been identified over the last several years. While investigators are using the presence of antibody immunity as an indicator for a potential T-cell response, and, thus, identification of T-cell antigens that may be exploited for therapy, antibody immunity in and of itself may have great utility in the diagnosis and management of human malignancy. Antibodies directed against the p53 oncogenic protein have been the most completely studied. A p53 specific antibody response may eventually be developed as a diagnostic or prognostic tool in monitoring the therapy of many types of cancers. A human antibody response directed against a growth factor receptor, such as HER-2/neu, may actually have some therapeutic importance. Human antibodies directed against a number of oncogenic proteins have been defined. As we progress the assessment of antibody immunity from a laboratory to a clinical tool, assays must be better refined to allow wide scale measurement of the cancer-specific humoral immune response.

Introduction

Malignant transformation is increasingly being ascribed to the aberrant function of a set of defined cell-growth related genes. These genes, termed oncogenes, encode for proteins that are qualitatively and/or quantitatively aberrant in tumor cells relative to normal cells. Therefore, detection of these proteins, many of which are involved in the initiation of cancer, may allow for early detection of the malignant state. Many oncogenic proteins are immunogenic, that is, patients whose tumors express a particular oncogenic protein can have detectable antibody immunity directed against the protein. Over the last decade, antibody immunity directed against proteins involved in the malignant transformation has been described for almost every human tumor. Studies have progressed from observational assessments describing the ability to detect immunity to investigations designed to determine whether the measurement of tumor-specific antibody immunity might be helpful in the diagnosis and management of human cancer.

Cancer is potentially curable if diagnosed when localized. Unfortunately, many cancers are diagnosed at an advanced stage. Despite aggressive surgery and chemotherapy, the overall survival of advanced-stage cancer patients is poor. Early diagnosis is essential to make progress in the treatment and, ultimately, the survival of patients with cancer. Serologic markers, most of which are circulating cancer proteins such as carcinoembryonic antigen (CEA) and CA-125, can potentially indicate the presence of cancer. These proteins, however, are shed from the surface of growing tumors and, in general, are associated with bulky or advanced disease. A serologic marker that is prevalent in early-stage disease would be a more optimal candidate to develop as a diagnostic tool. Antibody immunity to tumor-associated proteins may be a more appropriate serologic measure of cancer exposure. The assessment of an antibody immune response to a protein is quite different from the direct measurement of the protein itself because; (1) antibody responses can be generated against proteins that are expressed on the surface of cells and do not circulate, (2) antibody responses can be detected even when small amounts of the immunogenic protein are present and (3) antibody responses can indicate exposure to tumor-related protein. Antibody immunity has been used for decades to identify individuals exposed to infectious disease proteins. The immune system can respond to minimal amounts of protein by mounting an amplified antibody response that is readily detected in serum. Immunogenic tumor antigens, such as p53, are common to a number of different cancers and, therefore, provide a logical starting point for serological studies of cancer diagnosis. Tumor antigens such as HER-2/neu have biologic relevance in the malignant transformation and offer a model system for assessing the role of antibody immunity in identifying patients at a high risk for cancer. Indeed, proteins that are aberrantly expressed in the unregulated growth of a cancer cell, such as ras and c-myc, can stimulate an antibody response. In addition, established immunological techniques are now available to improve the ability of immunologic diagnostics to discern patients with cancer from non-affected individuals.

As the detection of immunity to oncogenic proteins becomes more routine, studies, such as those described in this chapter, have evolved to begin to address specific clinical questions such as the role of antibody immunity as a marker for patients exposed to cancer, as a tool to monitor therapy, or as an indicator of disease prognosis. These clinical experimental questions now shift the experimental design of investigations from the laboratory to the clinic and underscore the problems that must be overcome in defining appropriate patient populations for study and issues surrounding obtaining well-characterized specimens for analysis. Furthermore, once antibody assays are being used to analyze multiple samples the analytic methods must be validated and have a level of reliability that will assure confidence in the interpretation of the clinical data.

The human antibody response to p53 in cancer patients

The antibody immune response to p53 has received the most attention, in part, because mutations in the *p53* gene are common, occurring in approximately 50% of all human cancers. *p53* is a tumor suppressor gene, and mutations that occur in the gene cause inactivation of encoded p53 protein. Although the malignant changes in a cell due to p53 are related to a loss of the protein's function, p53 is most likely immunogenic because mutant p53 protein accumulates in the cancer cell cytosol and nucleus due to altered intracellular trafficking. Indeed, both the wild type and mutated forms of the protein can be found in abundance in the cytosol of a cell bearing *p53* gene mutations. The *p53* gene has multiple mutational "hot spots." The majority of patients with p53-specific antibodies have associated p53 missense mutations and p53 accumulation in their tumors. Furthermore, the antibody response is directed to immunodominant epitopes common for both wild type and mutant p53 protein (Soussi, 2000). Antibody immunity to the p53 protein has been reported for nearly every human tumor with which p53 mutations are associated: colon, breast, ovarian, lung and gastric cancer to name a few (Soussi, 2000). The incidence of antibodies directed against p53 in the serum of patients with cancer ranges from 10% to 50% of patients evaluated in an individual study. Clinical investigations of the antibody response to p53 have evolved from merely defining the incidence in a particular patient population to the development of the assessment of p53 antibodies as a clinical tool. The use of the antibody response to p53 has been proposed to aid in diagnosis and to determine prognosis and response to therapy. Studies have focused on the evaluation of antibodies to p53 as a diagnostic tool due to early reports which indicated antibody responses to p53 can occur early in the course of a cancer and predict undetected malignancy or premalignancy. The first report described the evolution of the p53-antibody response in patients at high risk of developing lung cancer, such as heavy smokers (Lubin *et al.*, 1995). Although study subjects were free of cancer at the time antibody assessment started, rising titers of p53 antibodies preceded the development of early stage lung cancers bearing p53 mutations in two patients. The second indication that antibody responses to p53 may be developed as a diagnostic or screening tool lay in a report of the detection of a higher incidence of antibodies to p53 in women without breast cancer who have a significant family history of breast cancer (11%) compared to controls (1%) (Green, 1994). These two provocative early studies stimulated more detailed investigation of p53 antibodies as a cancer diagnostic.

p53 antibodies as a cancer diagnostic tool

Several recent studies highlight both the potential as well as the problems associated with using antibodies to p53 as a cancer diagnostic. Head and neck cancer is a good model for testing the utility of a biomarker, as premalignant precursor lesions are easily assessable and well characterized. Mutations in p53 have been reported in 53–93% of head and neck squamous cell carcinoma. In addition, it has been shown that p53 protein is overexpressed in both primary and recurrent oral cancers as well as premalignant lesions (Ralhan *et al.*, 1998). Ralhan and colleagues evaluated sera from 183 patients with premalignant and malignant oral lesions and normal controls for circulating p53 antibodies (Ralhan *et al.*, 1998). The results of the serum antibody assays were correlated with accumulation of p53 protein in patients' matched oral specimens. Circulating p53 antibodies were detected in 34% of cancer patients and 30% of patients with premalignant lesions. There was a significant association of the presence of p53 antibodies with increased tumor size and the anaplastic nature of the tumor, both factors indicative of a poor prognosis. The expression of p53 was analyzed in 43 matched tissue specimens, 18 premalignant and 25 oral cancers. All 18 patients with p53 antibody seropositivity, 7 premalignant and 11 cancer cases, showed p53 accumulation in their oral lesions. However, the total number of patients positive for p53 antibodies was less than that of patients with detectable p53 protein in their lesions. These data highlight one of the problems with the development of p53 antibodies as a single serologic test to diagnose malignancy: although the specificity of the approach can be high, the sensitivity is generally low. However, detection of circulating p53 antibodies in patients with premalignant oral lesions suggests that humoral immune response against p53 protein is an early event in the pathogenesis of oral cancer.

Ovarian cancer represents a tumor in which no premalignant lesions have been identified. Early diagnosis of ovarian cancer is critical in making an impact on the overall survival of the patient. Over two-thirds of all cases of ovarian cancer are diagnosed in advanced stage, and the risk of relapse after standard treatment is high. If ovarian cancer were diagnosed at an early stage when the disease could be completely surgically resected more patients could be cured. Studies have evaluated patients with ovarian cancer, borderline ovarian tumors and benign ovarian tumors for the presence of antibodies to p53 (Angelopoulou *et al.*, 1996). The prevalence of p53 antibodies in patients with invasive cancer was 19%, whereas no circulating p53 antibodies were detected in patients with borderline or benign lesions. p53 antibodies were detectable only in patients with p53 protein overexpression in their tumors, and presence of p53 antibodies correlated with tumor stage and grade and shortened overall survival and relapse-free survival. Thus, this initial study suggests that antibodies directed against p53 may have some clinical utility in identifying patients with ovarian cancer. Larger prospective studies are needed to validate p53 antibodies as an ovarian cancer biomarker.

Patients with both colorectal and lung cancer have detectable p53 antibodies which correlate with a high rate of p53 mutation in these cancers. Alterations in the *p53* gene are found in about 60% of both colorectal and lung cancer, and the prevalence of p53 antibodies in both cancers are in the range of 25–30% (Lubin *et al.*, 1995; Hammel *et al.*, 1997; Zalcman *et al.*, 1998; van der Burg *et al.*, 2001). The use of p53 antibodies as a diagnostic test in lung cancer has shown limited success due to the low sensitivity (Soussi, 2000). Rosenfeld and colleagues (Rosenfeld *et al.*, 1997) found p53 antibodies to be present in only 27 (16%) of 170 patients with small cell lung cancer (SCLC). However, none of the 50 control sera was positive for p53 antibodies, and all patients were studied at the time of SCLC diagnosis, demonstrating that p53 antibodies can be detected relatively early in the course of the disease. Investigations in colon cancer have found 14 (26%) of 54 patients with colorectal cancer to have detectable serum p53 antibodies (Hammel *et al.*, 1997). There were no detectable p53 antibodies in the 24 control patients who had non-malignant digestive disease. In this study both CEA and CA 19.9, tumor markers used to follow the course of colon cancer, had a higher sensitivity (37% and 28%, respectively) for the diagnosis of colorectal cancer than p53 antibody measurements. Of note, there was a simultaneous increase in CEA, CA 19.9 and p53 antibodies in only 20% of patients (Hammel *et al.*, 1997). Furthermore, 30% of patients with normal CEA and CA 19.9 had significant p53 antibody concentrations, suggesting that CEA, CA 19.9 and p53 antibody testing may be complementary methods in identifying patients with colorectal cancer. Indeed, the use of antibodies to p53 to detect cancer may be of greater benefit as a member of a panel of biomarkers than as a "stand alone" test.

The concept of panels of markers to diagnose malignancy, including the use of antibody response to p53, is being tested in hepatocellular carcinoma (HCC) where mutations in the *p53* tumor suppressor gene are present in up to 37% of cases (Raedle *et al.*, 1998). Alpha-fetoprotein (AFP) is the only established tumor marker with reasonable specificity for the detection of HCC. However, the sensitivity of this screening test is low at 60–69% (Edis *et al.*, 1998; Raedle *et al.*, 1998). Two recent studies have evaluated the use of p53 antibody testing in combination with AFP in detection of HCC. Raedle *et al.* (1998) found p53 antibodies to be present in 22% of HCC-positive patients with corresponding elevated AFP levels in 69% of HCC patients. By combining the two measurements, serological HCC screening was improved to 76% sensitivity, however there was a decrease in specificity from 96% to 88%. Edis *et al.* (1998) found p53 antibodies in 13% of HCC-positive patients with corresponding elevated AFP levels in 33% of patients. In this study, adding the p53 antibody test to AFP screening of patients with HCC induced by viral hepatitis increased the sensitivity from 60% to 80%. In both these studies, p53 antibodies were found in patients with chronic liver disease but without detectable HCC, 4% and 21%, respectively. The high levels of p53 antibodies in these patients without evidence of cancer might be secondary to liver cirrhosis and an increased HCC risk, or possibly these patients may have had a clinically undetectable cancer. This observation raises the possibility that the presence of p53 antibodies may precede the clinical manifestation of cancer by several months.

p53 antibodies for prognostic use and monitoring response to therapy

The use of serum p53 antibodies as a prognostic tool in various cancers has had mixed results. In SCLC, the presence of p53 antibodies in patients with newly diagnosed disease was not associated with any clinical characteristics or prognostic markers (Rosenfeld *et al.*, 1997). In univariate analysis, Angelopoulou, *et al.* (1996) found p53 antibody-positive patients with ovarian cancer to be at increased risk for relapse but not death. In multivariate analysis, however, the difference in disease-free and overall survival between patients who were p53 antibody-positive or negative was not statistically significant. Vogl *et al.* (2000) found the presence of p53 antibodies in patients with ovarian cancer to positively correlate with tumor stage and grade, as well as, decreased relapse-free and overall survival. These results were statistically significant suggesting that p53-antibody testing does have prognostic value in the clinical management of patients with ovarian cancer.

The presence of p53 antibodies in patients with colon cancer has been shown to be an indicator of poor prognosis. Two of the three studies found an association between p53 antibodies and shortened survival (Soussi, 2000). One investigation found the presence of p53 antibodies correlated with several prognostic factors including histological differentiation, grade, shape of tumor and tumor invasion into blood vessels (Houbiers *et al.*, 1995). Patients with p53 antibodies were shown to have both decreased disease-free and overall survival. In breast cancer, several studies have demonstrated the presence of p53 antibodies to be indicative of worse prognosis (Lenner *et al.*, 1999; Soussi, 2000). A recent population-based epidemiological study showed the presence of p53 antibodies to be significant for both the risk of having breast cancer and decreased overall survival suggesting that p53 antibodies be regarded as a marker for more aggressive disease (Lenner *et al.*, 1999). The association between p53 antibodies and shortened survival in oral cancer has been demonstrated in several studies (Soussi, 2000). A recent study by Ralhan and colleagues (Ralhan *et al.*, 1998) in the evaluation of oral cancers, demonstrated an association of p53 antibody seropositivity and increase in tumor size and poor differentiation of tumors, both factors indicative of poor prognosis. There was also significant decreased overall survival in the patients with p53 antibodies compared to the seronegative patients with oral cancer suggesting that detection of p53 antibodies may be a useful marker for identifying oral tumors having poor prognosis.

These studies suggest an association between p53 antibodies and poor prognosis, specifically in tumors with poor differentiation. There is also a suggestion that p53 antibodies may be of value in determining both disease-free and overall survival, however, this has not been clearly established. One of the problems with interpretation of many of these retrospective analyses is that the patient populations evaluated are associated with diseases that, in general, have an overall poor prognosis. For example, in ovarian cancer, most women are diagnosed in advanced stages of disease. Although their tumors are initially responsive to standard treatment, the majority of patients will relapse within five years and eventually succumb to their cancer. Thus, to have the statistical power to truly stratify patients for prognosis based on the p53-antibody response, the absolute number of the total population must be quite large to include a significant number of patients who do not die of their disease. The same problem occurs in the evaluation of p53 as a prognostic marker in other malignancies that are associated with an overall poor prognosis such as SCLC and oral carcinomas.

The use of p53 antibodies for monitoring cancer patients during their therapy has been studied in several types of cancer. Zalcman *et al.* (1998) monitored 32 patients with lung cancer, 16 individuals positive for p53 antibodies and 16 negative, over a period of 30 months. A decrease greater than 50%, compared to the initial titer, in p53 antibodies was seen in 12 of 16 antibody-positive patients during chemotherapy that led to partial or complete remission of disease. The specificity of these p53 antibodies was confirmed by two different ELISA procedures and immunoprecipitation. Of the patients with a decrease in antibody titer, eight had a complete response to therapy and four obtained a partial response. Of the five patients without any variation in their antibody titer, two showed a partial response and no response was seen in the other three. No patient with a complete response had a stable level of p53 antibodies, whereas patients without response always maintained an unvarying level of p53 (Zalcman *et al.*, 1998). The correlation between the specific evolution of p53 antibody titer and response to treatment suggests that p53 antibodies could be a useful tool in monitoring response to therapy and relapse before it is clinically detectable. Other studies have found the ratio of p53 antibodies in patients with colon cancer to decrease within the first months after surgery, including those with Dukes' C cancer and hepatic metastasis who underwent palliative resection (Hammel *et al.*, 1997). In two patients the variations in p53 antibody ratio strongly correlated with tumor relapse or progression suggesting that monitoring of p53 antibodies may help in the early diagnosis and treatment of relapse in asymptomatic patients (Hammel *et al.*, 1997). In breast cancer, the reappearance of p53 antibodies can be detected two years after initial therapy. These increases in p53 antibodies have been detected several months before the detection of relapse (Soussi, 2000). Thus, p53 antibodies appear to be a useful tool in monitoring the response to therapy as well as monitoring for early relapse before it becomes clinically evident in lung, colon and breast cancer.

The human antibody response to HER-2/neu in cancer patients

The HER-2/neu protein is a member of the epidermal growth factor receptor family and consists of a cysteine rich extracellular domain which functions in ligand binding, a transmembrane domain, and an intracellular domain with kinase activity. HER-2/neu is a self-protein that is expressed by some normal tissues but is often overexpressed by a variety of cancer cells (Cheever *et al.*, 1995). The *HER-2/neu* gene is present as a single copy in normal cells; however, amplification of the gene and resultant protein overexpression is seen in various cancers including breast, ovarian, colon, uterine, gastric, prostate and adenocarcinoma of the lung (Ward *et al.*, 1999; McNeel *et al.*, 2000). Overexpression of this oncogenic protein is associated with more aggressive disease and higher risk of relapse in breast cancer patients (Slamon *et al.*, 1987; Press *et al.*, 1997). In colorectal cancer, overexpression of the HER-2/neu oncogenic protein is seen in 30–50% of cases. The presence of endogenous HER-2/neu-specific antibodies has been identified in breast, ovarian, prostate and colorectal cancer patients (Ward *et al.*, 1999, Disis *et al.*, 2000; McNeel *et al.*, 2000).

Investigations of the antibody response to HER-2/neu in patients with cancer are less developed than studies of the antibody response to p53. However, similar to investigations of antibody response to p53, endogenous humoral immunity to HER-2/neu directly correlates to overexpression of the protein by the patient's tumor. Similarly, antibody responses to the HER-2/neu protein can be detected in patients with early stage disease indicating that the presence of antibodies are not simply a reflection of tumor burden. HER-2/neu antibodies at titers of >1 : 100 were detected in 12 of 107 (11%) breast cancer patients versus 0 of 200 (0%) controls ($p < 0.01$) (Disis *et al.*, 1997). Detection of antibodies to HER-2/neu also correlated to overexpression of HER-2/neu protein in the patient's primary tumor. Nine of 44 (20%) patients with HER-2/neu-positive tumors had HER-2/neu-specific antibodies, whereas 3 of 63 (5%) patients with HER-2/neu-negative tumors had detectable antibodies ($p = 0.03$). The presence of HER-2/neu-specific antibodies in breast cancer patients and the correlation with HER-2/neu-positive tumors implies that immunity to HER-2/neu develops as a result of exposure of patients to HER-2/neu protein expressed by their own cancer. A later study evaluated 45 patients with advanced stage (III/IV) HER-2/neu-overexpressing breast and ovarian cancer for detection of pre-existent humoral immunity to HER-2/neu (Disis *et al.*, 2000). All patients were documented to be immune competent prior to study by DTH testing. Three of 45 patients (7%) had detectable HER-2/neu-specific IgG antibodies suggesting that patients with advanced-stage HER-2/neu-overexpressing breast and ovarian cancer can mount an antibody immune response to their tumor, however this endogenous immune response was found only in a minority of patients.

Similarly, antibodies to HER-2/neu have been found in the sera of patients with colon cancer; and, again, their presence correlates with overexpression of protein in the primary tumor ($p < 0.01$) (Ward *et al.*, 1999). Most recently, HER-2/neu was demonstrated to be a shared tumor antigen in patients with prostate cancer. Antibody immunity to HER-2/neu was significantly higher in patients with prostate cancer (15.5%, 31/200) compared with controls (2%, 2/100, $p = 0.0004$), and titers greater than 1 : 100 were most prevalent in the subgroup of patients with androgen-independent disease (16%, 9/56) (McNeel *et al.*, 2000). Studies such as those described here provide the basis for evaluating antibodies to HER-2/neu as a potential tool for cancer diagnostics, but also underscore the questionable utility of antibodies to stand alone as a single marker. Although the specificity of the approach may be significant, that is, few responses are found in non-cancer bearing individuals, the sensitivity of antibodies to identify all patients with HER-2/neu-overexpressing tumors is low.

HER-2/neu antibodies as a cancer diagnostic tool

Of note, the HER-2/neu protein may be a unique immunologic target in which to evaluate antibody immunity as a cancer diagnostic tool. Extensive studies of antibodies to the HER-2/neu protein have not been reported in patients with ductal carcinoma *in situ* (DCIS). While the HER-2/neu oncogenic protein is overexpressed in 30% of invasive breast cancers, it is overexpressed in 60–80% of DCIS. If antibody responses are proved to occur at the *in situ* stage of breast cancer, the detection of HER-2/neu antibodies could serve as an early indicator of exposure to the malignant phenotype. Thus, the detection of pre-existent human antibody responses against HER-2/neu protein in patients with DCIS might be clinically beneficial in early diagnosis and therapy.

As stated, the prevalence of HER-2/neu-specific antibodies in DCIS and their use in early diagnosis of breast cancer has not been established. However, many studies have evaluated the amplification and/or overexpression of the HER-2/neu oncogenic protein, itself, in DCIS (Allred *et al.*, 2001). HER-2/neu is overexpressed in 60–80% of DCIS with protein overexpression seen in approximately 10% of non-comedo subtypes and 60% of comedo subtype (Allred *et al.*, 2001). Histologic studies have evaluated the role of HER-2/neu in the development and progression of breast cancer by measuring its overexpression in a series of hyperplastic, dysplastic and malignant neoplastic lesions (Allred *et al.*, 1992, 2001). The neoplasms evaluated included DCIS and infiltrating ductal carcinoma (IDC). The latter were stratified into IDC combined with or not combined with DCIS. Overexpression of HER-2/neu was not detected in the hyperplastic or dysplastic lesions, however it was present in 56% of the pure DCIS (77% of the comedo subtype). Only 15% of the IDC overexpressed HER-2/neu, yet 22% of the IDC with combined DCIS had HER-2/neu protein overexpression. This observation suggests that HER-2/neu plays a more important role in the initiation of ductal carcinoma than it does in progression, an ideal characteristic for a diagnostic marker. HER-2/neu protein overexpression is associated with DCIS and unfavorable prognostic factors that include a tendency for high nuclear-grade and comedo-type necrosis. Therefore, the evaluation of HER-2/neu overexpression in DCIS lesions may prove beneficial in being able to stratify patients into low-risk groups which can be followed conservatively or high-risk groups that may require more extensive post-biopsy surgical procedures to prevent recurrence. As antibody responses to the HER-2/neu protein have been shown to correlate with protein overexpression by the tumor it remains to be shown whether the detection of serum antibody response can aid in the identification and monitoring of patients with DCIS.

Amplification of the *HER-2/neu* oncogene and overexpression of HER-2/neu protein have also been studied in benign breast disease (BBD). A recent investigation of women with benign breast biopsies demonstrated both *HER-2/neu* gene amplification and a proliferative histopathologic diagnosis that suggested an increased risk for subsequent breast cancer. In this case control study, 6,805 women who were diagnosed with BBD were followed and observed for the development of breast cancer for a median period of 10 years between 1967 and 1981 (Stark *et al.*, 2000). Benign breast parenchyma with fibrosis and fibroadenoma were two of the most common conditions. Archival tissue blocks, both benign and tumorous, were studied. *HER-2/neu* gene amplification occurred more often in the malignant tumors if the corresponding prior biopsies exhibited amplification. Overexpression of HER-2/neu protein was not detected in any of the benign tissues with or without *HER-2/neu* gene amplification. Previous studies similarly found a lack of HER-2/neu protein overexpression in benign breast biopsies, suggesting that overexpression of HER-2/neu occurs at the transition from hyperplasia to DCIS (Stark *et al.*, 2000). A significant correlation was found between *HER-2/neu* gene amplification and protein overexpression in the malignant tumors (Stark *et al.*, 2000). This study suggests that genetic alterations, including HER-2/neu amplification occur as a relatively early event in the development of breast cancer. Given that histopathologic risk classification identifies only a small portion of breast cancer cases that develop among women with BBD, that is, less than 4% of women with biopsy-diagnosed BBD have atypical hyperplasia, amplification of the *HER-2/neu* oncogene in BBD may serve as a sensitive marker of susceptibility for subsequent invasive breast cancer. If antibodies to HER-2/neu could be detected in selected high-risk populations with BBD, the finding would have important implications in the use of antibodies in the screening and early intervention of breast cancer.

Antibodies against cancer-related proteins important in malignant transformation or progression

Studies described in the previous section demonstrate the potential of harnessing the evaluation of antibody immunity against oncogenic proteins as a clinical tool. The choice of protein for evaluation is critical. Ideal proteins to be studied would be those that play a role in the malignant transformation or maintenance of the malignant phenotype. Investigations by several groups have demonstrated that many self-proteins can be recognized by the humoral immune system. As an example, antibodies have been detected to proteins as ubiquitous in cancer as ras, c-myc and bcr-abl.

Human antibodies directed against ras protein

Three ras genes, *H-ras*, *K-ras* and *N-ras*, have been defined, and these genes encode a 21 kDa protein designated as p21 ras (Takahashi *et al.*, 1995; Schneider *et al.*, 2000). Point mutations in codons 12, 13 or 61 of one of the three ras genes result in activation of ras. Activating amino acid substitutions in the p21 protein impair the GTPase activity of the ras protein and generate constitutively activated signaling complexes with transforming activity (Takahashi *et al.*, 1995; Anderson *et al.*, 1998). Mutated ras proteins are implicated in malignant transformation of various tumor types. The highest incidence of ras mutations are found in adenocarcinoma of the pancreas (90%), colon (50%), lung (30%), thyroid tumors (50%) and in myeloid leukemia (30%) (Anderson *et al.*, 1998). The *K-ras* gene is the predominant mutated ras gene in adenocarcinoma of the lung, pancreas and colon, whereas *N-ras* mutations are implicated in myeloid leukemia (Anderson *et al.*, 1998). Lung cancer patients with *K-ras* mutations have been found to have a significantly worse disease-free and overall survival when compared to patients without ras mutations; however, some tumors, such as thyroid cancer, lack any specificity of a particular ras mutation with histopathology or outcome of disease (Anderson *et al.*, 1998). It has been suggested that the type of ras mutation at position 12, valine versus aspartic acid or others such as cysteine or serine, may determine whether the cancer will undertake a benign or more aggressive form (Anderson *et al.*, 1998).

The biologic expression of ras protein in particular disease states makes evaluation of ras-specific markers an attractive biologic monitoring tool. Overexpression of p21 ras proteins has been detected in plasma from workers with heavy occupational carcinogen exposure. Elevated levels of p21 in serum, five-fold over controls, were initially detected in 3 of 16 (19%) workers (Brandt-Rauf and Pincus, 1998). During an 18-month period of follow-up, one of the seropositive workers developed a villous adenoma or premalignant colonic tumor. Upon resection of the tumor, the patient's serum p21 returned to normal, suggesting that the adenoma was the source of the serum elevation and detection before evident clinical disease was possible (Brandt-Rauf and Pincus, 1998). More recent studies have demonstrated plasma p21 overexpression to increase with increasing size of adenoma and increasing stage of carcinoma. In addition, there was a statistically significant correlation between the overexpression of p21 in the plasma and in the corresponding tumor tissue (Brandt-Rauf and Pincus, 1998). In a recent two-year follow-up study by Weissfeld and colleagues, ras-related proteins were measured in the serum of 37 patients who subsequently developed fatal cancers, 59 patients who developed non-fatal cancers, 58 patients who developed benign tumors, and 94 healthy controls (Brandt-Rauf and Pincus, 1998). Detectable serum levels of ras-related proteins were statistically more common in patients who subsequently developed fatal malignancies (10.8%) which included lung, gastrointestinal, brain, prostate, ovary, cervical cancers, lymphomas and leukemias, compared with controls (1.1%). The use of increased ras p21 proteins in plasma as a biomarker for the carcinogenic process or for the general disease state was evaluated in patients with chronic obstructive pulmonary disease (COPD) (Anderson *et al.*, 1998). Nine of twenty COPD patients had increased p21 protein levels. Eighteen of forty lung cancer patients had increased ras p21 protein levels. COPD patients had lower ras p21 protein values than cancer patients. Thus, the detection of ras protein in blood may be a useful molecular marker of carcinogenesis. An antibody response to p21 ras may be a more sensitive measure of exposure to the protein than direct measurement of protein in serum.

Antibodies to p21 ras proteins have been detected in patients with colon cancer. Takahashi *et al.* (1995) examined sera from 160 colon cancer patients and 60 normal controls to determine whether antibodies to mutated p21 ras protein were present. Antibodies of IgA subtype against mutated p21 ras protein were detected in 51 of 160 (32%) colon cancer patients and in 1 of 40 (2.5%) of normal controls. The greater incidence of p21 ras antibodies in the cancer patients suggested that immunization to ras proteins occurred as a result of the malignancy. Antibodies in the majority of colon cancer patients recognized normal p21 K-ras-G12 protein. In only a small number of colon cancers was the serum antibody reactivity against only the mutated p21 ras-D12 protein. Antibody responses to the non-mutated segment may provide a method to identify individuals who have been exposed to mutated p21 ras protein and have yet to show an early indication of incipient or occult malignancy undetectable by current conventional methods (Takahashi *et al.*, 1995). In this study, antibody reactivity did not correlate with patient age, sex, histology, stage or serum CEA, but did correlate with lymphocyte count prior to surgery. Although the presence of p21 ras overexpression is associated with worse prognosis in terms of disease-free and overall survival, the association of p21 ras antibodies and prognosis is not known and needs to be further evaluated.

Human antibodies directed against c-myc protein

The c-myc protein belongs to the myc oncoprotein family which also includes N-myc and L-myc. The gene product encoded by the *c-myc* oncogene is a transcription factor which plays a key role in transcriptional transactivation, proliferation and transformation of cells (Hoffman and Liebermann, 1998; Nesbit *et al.*, 1999). By influencing cellular proliferation, differentiation and apoptosis, c-myc is a critical factor in both the positive and negative growth of cells. c-myc is expressed in almost all proliferating normal cells and its expression is strictly dependent on mitogenic stimuli and is downregulated in many cells when they are induced to terminally differentiate (Hoffman and Liebermann, 1998). Forced expression of c-myc has been shown to inhibit differentiation and associated growth arrest in several cell types which include myeloid, erythroid, myogenic, preadipocyte and nerve cells (Hoffman and Liebermann, 1998). Alterations in c-myc expression or protein structure are associated with many human malignancies. Several genetic alterations which include chromosomal alterations, proviral insertion, retroviral transduction and gene amplification have been shown to activate *myc* genes, however, in many cases the basis for altered myc expression is not understood (Hoffman and Liebermann, 1998).

Overexpression of c-myc results in intracellular accumulation and increased levels of the oncoprotein can be detected in human tumors. By mechanisms that are not well defined, the oncoprotein can also accumulate in the extracellular environment. c-myc oncoprotein has been detected in a variety of cancers which include colon, breast, lung, osteosarcoma, myeloid leukemia and lymphoma including Burkitt's lymphoma (BL). c-myc dysregulation is an essential aspect of Burkitt's lymphoma and virtually all cases of BL involve c-myc rearrangements. Both rearranged *c-myc* genes and somatic mutations are felt to contribute significantly to the phenotype and/or progression of the disease. In breast cancer, earlier studies have shown c-myc amplification to be an independent prognostic indicator of early relapse, more powerful than either estrogen receptor status or tumor size (Nesbit *et al.*, 1999). More recent studies have also shown c-myc amplification in 28% of tumors; however, no significant association with tumor stage or the presence of nodal disease has been shown (Nesbit *et al.*, 1999). Most breast cancer studies have tended to indicate that a correlation exists between *c-myc* gene amplification and disease progression or recurrence that ultimately influences long-term disease-free survival. Serial measurements of myc related proteins in serum of patients with resected colorectal carcinoma have showed a gradual return to normal following resection (Nesbit *et al.*, 1999).

Circulating antibodies to the myc protein have been detected in cancer patients. Serum antibodies to myc were first described in 4 of 6 (67%) of patients with colon cancer, 12 of 125 (10%) of breast cancer patients, 1 of 2 (50%) osteosarcoma patients, 1 of 9 (11%) ovarian cancer patients, and 3 of 3 (100%) of patients with cancer of unknown origin (Brandt-Rauf and Pincus, 1998). In a later study, c-myc antibodies were detected in the serum of 25 of 44 (57%) cases of colorectal cancer compared with 8 of 46 (17%) of normal controls (Brandt-Rauf and Pincus, 1998). These studies did not establish an association between the presence of c-myc antibodies and prognosis. A study by Yamamoto *et al.* (1999) examined serum c-myc antigens and antibodies against c-myc in 68 lung cancer patients and 30 healthy volunteers. Anti-c-myc antibodies were detected in 9 of 68 (13.2%) patients with lung cancer and 1 of 30 (3.3%) of normal controls. Circulating c-myc antigen was not detected in any individuals with lung cancer or normal controls.

Human antibodies directed against bcr-abl

The molecular hallmark of chronic myelogenous leukemia (CML) is the Philadelphia (Ph) chromosome, which results from a reciprocal translocation of human c-abl proto-oncogene from chromosome 9 to the bcr region on chromosome 22. This t(9;22) (q34;q11) translocation results in chimeric *bcr-abl* genes, which encode hybrid protein, $p210^{Bcr\text{-}Abl}$, which has abnormal tyrosine kinase activity (Oka *et al.*, 1998; Talpaz *et al.*, 2000). The bcr-abl protein can stimulate the growth of hematopoietic progenitor cells and is essential in the pathogenesis of CML. The $p210^{Bcr\text{-}Abl}$ protein is found in more than 95% of patients with CML and some acute lymphocytic leukemia patients (Cheever *et al.*, 1995).

Although several studies have established specific cellular immunity to bcr-abl oncogene-derived proteins, a role for B cells or antibodies in CML immunity has yet to be defined (Talpaz *et al.*, 2000). A recent study demonstrated the presence of antibodies against $p210^{Bcr\text{-}Abl}$ in both Ph-positive and Ph-negative leukemia patients and in healthy volunteers (Talpaz *et al.*, 2000). Plasma from 18 of 31 (58%) individuals was able to immunoprecipitate $p210^{Bcr\text{-}Abl}$ including 14 of 20 patients (70%) with Ph-positive CML. Plasma of 2 of 7 (29%) normal subjects also contained p210 specific antibodies. The antibodies detected in the patient plasma were also able to precipitate normal bcr and abl proteins, suggesting that recognition of $p210^{Bcr\text{-}Abl}$ is most likely attributable to reactivity with epitopes found in normal abl and/or normal bcr proteins as opposed to a bcr-abl-specific epitope.

Measurement of antibodies to oncogenic proteins evolves from a research to a clinical tool

An optimal method for detecting cancer-specific antibodies must be defined. The presence of antibodies directed against oncogenic proteins may be a specific, but not yet sensitive, test for the presence of malignancy. Published data in the studies described in the previous section suggests that a minority of patients with cancer can be distinguished from a control population using oncogenic protein antibody immunity as a serologic screen. Clearly, additional experimental development of the antibody detection systems must be performed to improve the assay sensitivity. Several strategies have been explored in viral model systems to improve the sensitivity of detecting the antibody response directed against a virus associated with the development of a malignancy. "Second generation" antibody assays are currently being studied for the potential to identify patients with viral-related malignancies such as nasopharyngeal carcinoma associated with Epstein–Barr Virus (EBV) and cervical cancer associated with Human Papilloma Virus (HPV). Review of the experimental development of these viral antibody assays demonstrates strategies for increasing assay sensitivity which can be applied to antibody assays detecting tumor specific responses: (1) evaluating the antibody response to a panel of proteins rather than an individual tumor antigen, (2) evaluating the level of the immune response either by titer or absolute amount of immunoglobulin, (3) discerning the classes of the antibody involved in the immune response IgG, IgA or IgM and (4) determining the most appropriate form of the antigen to be used in the assay, for example, "captured" protein, recombinant protein, or even peptides.

Evaluating immune sera against a panel of immunogenic proteins is feasible. Antibodies directed against several tumor-associated proteins should be evaluated in each individual as the tumor-specific immune response in cancer is, most likely, directed against multiple antigens. The field of tumor immunology has undergone many changes in the last several years due to advances in molecular biology (van der Bruggen, 1991; Coulie *et al.*, 1994). Previously a major avenue of investigation had been to discern if human tumors were immunogenic. This is no longer an open experimental question (Disis and Cheever, 1996). We now know that cancer patients do mount an immune response against their cancer and several tumor antigens have been defined and are continuing to be discovered. Melanoma has been studied in great detail. Many antigens have been determined for melanoma, and this tumor remains a model for characterizing the human cancer-specific immune response due to its inherent immunogenicity and the ease with which surface tumor samples can be obtained for analysis. Evaluation of the melanoma-specific immune response, at a T-cell level, indicates that an individual patient has immunity towards several proteins expressed by its tumor, not just one dominant protein (Van den Eynde *et al.*, 1989; de-Vries *et al.*, 1997). Extrapolating from the melanoma model, cancer patients have the potential to be immune to many proteins expressed in their tumor. This theory has not been explored extensively with the antibody immune response to solid tumors. The analysis of multiple proteins in discerning the tumor-specific immune response, however, has been studied in malignancies associated or caused by oncogenic viruses. Viruses are implicated in the pathogenesis of many human tumors, and immune responses against viral proteins have been studied extensively as an aid in early diagnosis. Unfortunately, in many cases, the measurement of the antibody response to a single viral protein is not predictive of malignancy as many normal individuals are seropositive for proteins in viruses such as HPV and EBV (Littler *et al.*, 1991; Lehtinen *et al.*, 1993; Lennette *et al.*, 1993). However, the proteins eliciting an immune response in patients with EBV-related nasopharyngeal carcinomas can be distinct from the proteins generating immunity in a control population (Littler *et al.*, 1991; Gadducci *et al.*, 1996). To potentially distinguish the "normal" EBV-specific immune response from one associated with the later development of malignancy, a large panel of proteins had to be analyzed (Lennette *et al.*, 1993). The evaluation of immunity to multiple oncogenic proteins may increase the sensitivity of the approach in cancer.

The level, or titer, of the immune response can be helpful in improving both the sensitivity and specificity of potential cancer diagnosis. Titer of antibodies directed against the EBV capsid antigen are 10× higher in EBV-related cancer than patients with other head and neck cancers or a normal seropositive population (Lennette *et al.*, 1993). Similarly, increasing titers of antibodies to human herpes virus 6, a virus implicated in Hodgkin's disease, was associated with relapse when antibody responses were measured over time (Levine *et al.*, 1992). Several laboratories measure antibody responses to cancer-related proteins in terms of Western blot analysis which is a non-quantitative measure of an antibody response. Quantitative methods are essential to allow a measurement of antibody response. Quantitative detection of antibody class can also potentially improve the sensitivity of antibody diagnosis. One study evaluated the HPV antibody response to a panel of HPV proteins in sera from 64 patients with anal cancer and 79 blood donor controls. There was no difference in the IgG HPV antibody response, but IgA reactivity to HPV was quite different (Heino *et al.*, 1993). Serum IgA HPV-specific antibodies were found in 89% of patients with anal cancer compared to 24% of controls ($p = 0.0001$). Therefore, an accurate, controlled, and sensitive measure of the antibody class of the tumor-specific immune response may play an important role in determining the sensitivity and specificity of antibody diagnosis.

Finally, the source of antigen used in the assay can affect the predictive value of the test in detecting malignancy. The two major methods for detection of tumor-specific antibodies are immunoblotting (Western blot) and ELISA systems. Western blot analysis involves separating tumor proteins by gel electrophoresis, transferring to nitrocellulose, and then probing with patient serum samples. While Western blotting is useful for identifying new immunogenic tumor proteins, it is not a quantitative assay. By contrast, ELISA methodology allows a quantitative evaluation of the antibody response and is capable of allowing the analysis of a great number of patients with little effort. Use of ELISA requires a source of reasonably pure protein for detection of antibodies. The type of ELISA used to detect antibody responses, such as "sandwich" or "indirect" ELISA, may affect the sensitivity of the test. A "sandwich" ELISA allows an impure source of protein to be used. Briefly, the method entails the binding of a protein-specific monoclonal or polyclonal antibody to the surface of a plate followed by incubation with the protein of interest in solution. Protein will bind to the antibody, impurities are washed away, and the antibody-bound protein can then be probed with patient sera. An "indirect" ELISA requires a purified source of protein, either recombinant protein or chemically synthesized fragments of protein termed "peptides." The pure antigen can be directly bound to the surface of a plate and probed with patient sera. The sensitivity of each approach may differ depending upon experimental conditions. It will be important to explore several strategies with a series of candidate antigens to determine the most optimal method to further develop for use in screening for the presence of cancer in a high-risk or asymptomatic population.

As the measurement of antibodies to oncogenic proteins is being tested as a clinical tool, the assays used to measure responses have to be validated. Clinicians have an expectation with each laboratory test result they receive that the result is backed by a rigorous analysis of a reference population to define baseline responses as well as validation of the reproducibility of the result. This type of data is rarely published or discussed in the context of reporting results from studies of antibody immunity in specific cancer populations; however, the steps needed to determine validation data are well defined. Quality control monitoring and assay validation is composed of several analytical measures. These measures are routine, for the most part, for serologic studies. First is an assessment of accuracy. Accuracy refers to the correctness and exactness of the test result. It is defined as the closeness of a test result to the true value and can only be calculated by comparison to a standard. Many of the immunogenic oncogenic proteins described here have readily available commercial monoclonal antibodies for use in immunohistochemical staining. Although in a murine background, these antibodies are useful to use as positive controls in both ELISA-based platforms as well as Western blot analyses. A second important measure is precision, or the reproducibility of the test. Precision is defined as the closeness of the test results to one another when using the same specimen. Precision is expressed as a standard deviation and coefficient of variation of multiple sample runs. The same sample can be run multiple times on the same plate or on the same day, intraassay precision, or the same sample can run multiple times over several days, interassay precision (within run and within day). Identification of known positive specimens during screening will allow the estimate of precision to be made for any particular assay.

Sensitivity is the limit of detection of a method or the capacity of the method to detect small amounts of a substance with some assurance. Sensitivity can be determined by adapting an ELISA to a standard curve using purified human IgG or IgA depending on the antibodies of interest to be detected. By estimating the antigen-specific antibody responses in ug/ml of Ig a quantitative analysis can be performed. Sensitivity is often linked with specificity, and specificity is the ability of a method to measure only the substance being tested. Whether specificity of the antibody assay will be determined by the ability of the marker to discriminate between cancer-affected individuals or individuals whose tumors express the protein of interest is dependent on the ultimate use of the test. The challenge before us is to develop reproducible assays for the quantitation of immunity when there is no "gold standard" with which to compare them. In addition, as these laboratory assays become increasingly adapted as clinical tools, publications of clinical studies should include assay validation parameters to aid in the interpretation of the clinical results.

Analysis of antibody responses to human oncoproteins in control populations

Prior to large-scale studies of antibody immunity to cancer-related proteins as a potential diagnostic or prognostic tool, evaluations in well-defined reference populations must be performed. Many of the analyses cited above compared the development of an oncogenic antibody response in cancer patients to a control population of volunteer individuals without cancer. However, there are many other disease states that are common in volunteers that may be associated with significant levels of antibody immunity to oncogenic proteins.

As an example, antibodies against p53 have been detected in the sera of patients with autoimmune diseases, including systemic lupus erythematosus (SLE), rheumatoid arthritis (RA), Sjogren syndrome (SS), and autoimmune thyroid disease. In a study by Kovac *et al.* (1997) the frequency of p53 specific IgG antibodies were studied in sera from 50 patients with SLE, 20 autoimmune disease (non-SLE) control patients, and 20 healthy controls (Kovacs *et al.*, 1997). p53 antibodies were detected in 15 (32%) SLE patients, 3 (15%) disease control patients, and none of the healthy controls. Of the three antibody positive disease control patients, two had SS and one had systemic sclerosis. The frequency of p53 mutations were studied in T-cell lines derived from 18 SLE patients and four normal controls. No mutations could be detected in either the control or SLE T cells, suggesting that in SLE, the high frequency of p53 antibodies is not due to mutations in the T-cell derived p53 gene. In contrast, another study determined p53 antibodies are rarely found in patients with RA and SS (Mariette *et al.*, 1999). Serum p53 antibodies were detected in 2 of 106 patients with RA and the synovial fluid from one of these patients was also positive for p53 antibodies. p53 antibodies were not found in 72 patients with SS but were detected in 2 of 14 patients with lymphoma complicating SS. A recent study by Fenton *et al.* (2000) detected p53 antibodies in 2 of 48 (4.2%) patients with autoimmune thyroid disease, including one patient with Hashimoto's thyroiditis and one with Graves' disease. A third patient with pseudohypoparathyroidism but without thyroid disease was also seropositive. None of the 19 patients with differentiated thyroid cancer had p53 antibodies. This data showed that p53 antibodies could be detected in sera from 4% of patients with autoimmune thyroid disease which could be indicative of DNA damage and increased apoptosis associated with autoimmune disease.

In nearly all models of oncogenic protein studies, the development of an antibody response to the protein was correlated with aberrant expression of that protein in the tumor. Many of these oncogenic proteins can be aberrantly expressed in other non-malignant disease states as well. As an example, overexpression of several proto-oncogenes, including *c-myc, c-raf, bcl-2* and *n-ras* have been detected in PBMC, and B and T cells of patients with SLE when compared to normal controls (Rapoport *et al.*, 1997). In contrast, expression of another proto-oncogene, c-fos, has been shown to be decreased or unchanged in SLE, suggesting that the up- and down-regulation of proto-oncogenes in autoimmune disease depends on the specific proto-oncogene being examined (Rapoport *et al.*, 1997). Expression of c-myc in SLE patients is directly associated with disease activity, with the highest levels detected in the most active patients (Rapoport *et al.*, 1997). Increase in the levels of proto-oncogenes may precede clinical exacerbation and remission secondary to immunosuppressive therapy is associated with downregulation of the proto-oncogene levels, suggesting that these levels may be used to monitor disease activity in SLE patients (Rapoport *et al.*, 1997). However, other proto-oncogenes such as bcl-2 do not correlate with SLE disease activity and remain persistently elevated supporting their primary role in production and maintenance of lymphoid hyperactivity associated with SLE (Rapoport *et al.*, 1997). Increased levels of proto-oncogenes, including c-myc, c-myb and c-raf have also been detected in patients with RA (Rapoport *et al.*, 1997). Overexpression of the proto-oncogenes *jun-b, c-fos* and *c-ras* was also found in cultured synovial cells from RA patients, and this overexpression is thought to play a key role in the upregulation and *in situ* activation of synovial mesenchymal cells, which leads to formation of an invasive tumor-like pannus resulting in joint destruction (Rapoport *et al.*, 1997). The proto-oncogene *p21 ras* has been associated with autoimmune insulin dependent diabetes mellitus (IDDM). Normal expression of p21 ras and its regulatory elements is seen in IDDM, however a decreased receptor-mediated activation of the p21 ras pathway has been shown to correlate with disease progression, and correction of this defect prevented disease (Rapoport *et al.*, 1999).

Aberrant expression of several proto-oncogenes, including c-myc, c-raf, c-myb and n-ras, has been reported in a small number of patients with SS, progressive systemic sclerosis, glomerulonephritis, Henoch-Schonlein purpura, and dermatomyositis (Rapoport *et al.*, 1997).

In angioblastic lymphadenopathy, high levels of n-ras and low levels of c-fos correlate with disease activity and are normalized by immunosuppressive therapy. These limited findings suggest that aberrant expression of proto-oncogenes in lymphoid cells is a common finding in various autoimmune diseases, thus, extensive and thoughtful evaluation of a reference population to define baseline values will be essential in the clinical development of antibodies to oncogenic proteins as a clinical tool for diagnosing and monitoring human malignancy.

Conclusion

The last decade of research in tumor immunology has demonstrated the immunogenicity of human cancer. Much investigation has centered on developing oncogenic proteins as therapeutic immune targets; however, antibody immunity can be useful as a marker of disease presence and progression. The challenge that is before us is to design population-based experiments focused on answering clinical questions with the same scientific rigor as laboratory-based analyses. However, the complexities of defining the appropriate patient populations, obtaining quality clinical material, and having the statistical power and number of patients to definitively answer the specific clinical question can be daunting. Furthermore, assay systems need to be refined to meet clinical standards, both in the ease of use and in the validation of the reproducibility of results. Antibody immunity is our most powerful tool in infectious disease systems for identifying individuals exposed to pathogens. In the next few years studies will show whether antibody immunity to oncogenic proteins can have similar impact in the diagnosis and treatment of human malignancy.

Acknowledgments

This work is supported for MLD by NCI grants K24 CA85218 and U54 CA090818. We thank Ms Marsha Weese for reviewing and editing the manuscript and Ms Chalie Livingston for assisting with manuscript preparation.

References

Allred, D., Clark, G., Molina, R., Tandon, A., Schnitt, S., Gilchrist, K. *et al.* (1992) Overexpression of HER-2/neu and its relationship with other prognostic factors change during the progression of *in situ* to invasive breast cancer. *Hum. Pathol.*, **23**, 974–979.

Allred, D. C., Mohsin, S. K. and Fuqua, S. A. (2001) Histological and biological evolution of human premalignant breast disease. *Endocr. Relat. Cancer*, **8**, 47–61.

Anderson, D., Hughes, J. A., Cebulska-Wasilewska, A., Nizankowska, E. and Graca, B. (1998) Ras p21 protein levels in human plasma from patients with chronic obstructive pulmonary disease (COPD) compared with lung cancer patients and healthy controls. *Mutat. Res.*, **403**, 229–235.

Angelopoulou, K., Rosen, B., Stratis, M., Yu, H., Solomou, M. and Ep, D. (1996) Circulating antibodies against p53 protein in patients with ovarian carcinoma. Correlation with clinicopathologic features and survival. *Cancer*, **78**, 2146–2152.

Brandt-Rauf, P. W. and Pincus, M. R. (1998) Molecular markers of carcinogenesis. *Pharmacol. Ther.*, **77**, 135–148.

van der Bruggen, P., Traversari, C., Chomez, P., Lurquin, C., De Plaen, E., Van den Eynde, B. *et al.* (1991) A gene encoding an antigen recognized by cytolytic T lymphocytes on a human melanoma. *Science*, **254**, 1643–1647.

van der Burg, S. H., de Cock, K., Menon, A. G., Franken, K. L., Palmen, M., Redeker, A. *et al.* (2001) Long lasting p53-specific T cell memory responses in the absence of anti-p53 antibodies in patients with resected primary colorectal cancer. *Eur. J. Immunol.*, **31**, 146–155.

Cheever, M. A., Disis, M. L., Bernhard, H., Gralow, J. R., Hand, S. L., Huseby, E. S. *et al.* (1995) Immunity to oncogenic proteins. In *Immunol. Rev.*, (Ed. Moller, G.) pp. 33–59. Copenhagen: Munksgaard Intl Pub.

Coulie, P. G., Brichard, V., Van-Pel, A., Wolfel, T., Schneider, J., Traversari, C. *et al.* (1994) A new gene coding for a differentiation antigen recognized by autologous cytolytic T lymphocytes on HLA-A2 melanomas. *J. Exp. Med.*, **180**, 35–42.

Disis, M. L. and Cheever, M. A. (1996) Oncogenic proteins as tumor antigens. *Curr. Opin. Immunol.*, **8**, 637–642.

Disis, M. L., Pupa, S. M., Gralow, J. R., Dittadi, R., Menard, S. and Cheever, M. A. (1997) High-titer HER-2/neu protein-specific antibody can be detected in patients with early-stage breast cancer. *J. Clin. Onc.*, **15**, 3363–3367.

Disis, M. L., Knutson, K. L., Schiffman, K., Rinn, K. and McNeel, D. G. (2000) Pre-existent immunity to the HER-2/neu oncogenic protein in patients with HER-2/neu overexpressing breast and ovarian cancer. *Breast Cancer Res. Treat.*, **1766**, 1–8.

Edis, C., Kahler, C., Klotz, W., Herold, M., Feichtinger, H., Konigsreiner, A. *et al.* (1998) A comparison between alpha-fetoprotein and p53 antibodies in the diagnosis of hepatocellular carcinoma. *Transplant Proc.*, **30**, 780–781.

Fenton, C. L., Patel, A., Tuttle, R. M. and Francis, G. L. (2000) Autoantibodies to p53 in sera of patients with autoimmune thyroid disease. *Ann. Clin. Lab. Sci.*, **30**, 179–183.

Gadducci, A., Ferdeghini, M., Buttitta, F., Fanucchi, A., Annicchiarico, C., Prontera, C. *et al.* (1996) Preoperative serum antibodies against the p53 protein in patients with ovarian and endometrial cancer. *Anticancer Res.*, **16**, 3519–3523.

Green, J. A., Mudenda, B., Jenkins, J., Leinster, S. J., Tarunina, M., Green, B. *et al.* (1994) Serum p53 auto-antibodies: incidence in familial breast cancer. *Eur. J. Cancer*, **30A**, 580–584.

Hammel, P., Boissier, B., Chaumette, M. T., Piedbois, P., Rotman, N., Kouyoumdjian, J. C. *et al.* (1997) Detection and monitoring of serum p53 antibodies in patients with colorectal cancer. *Gut*, **40**, 356–361.

Heino, P., Goldman, S., Lagerstedt, U. and Dillner, J. (1993) Molecular and serological studies of human papillomavirus among patients with anal epidermoid carcinoma. *Int. J. Cancer*, **53**, 377–381.

Hoffman, B. and Liebermann, D. A. (1998) The proto-oncogene *c-myc* and apoptosis. *Oncogene*, **17**, 3351–3357.

Houbiers, J. G., van-der-Burg, S. H., van-de-Watering, L. M., Tollenaar, R. A., Brand, A., van-de-Velde, C. J. *et al.* (1995) Antibodies against p53 are associated with poor prognosis of colorectal cancer. *Br. J. Cancer*, **72**, 637–641.

Kovacs, B., Patel, A., Hershey, J. N., Dennis, G. J., Kirschfink, M. and Tsokos, G. C. (1997) Antibodies against p53 in sera from patients with systemic lupus erythematosus and other rheumatic diseases. *Arthritis Rheum.*, **40**, 980–982.

Lehtinen, T., Lumio, J., Dillner, J., Hakama, M., Knekt, P., Lehtinen, M. *et al.* (1993) Increased risk of malignant lymphoma indicated by elevated Epstein–Barr virus antibodies – a prospective study. *Cancer Causes Contl.*, **4**, 187–193.

Lenner, P., Wiklund, F., Emdin, S. O., Arnerlov, C., Eklund, C., Hallmans, G. *et al.* (1999) Serum antibodies against p53 in relation to cancer risk and prognosis in breast cancer: a population-based epidemiological study. *Br. J. Cancer*, **79**, 927–932.

Lennette, E. T., Rymo, L., Yadav, M., Masucci, G., Merk, K., Timar, L. *et al.* (1993) Disease-related differences in antibody patterns against EBV-encoded nuclear antigens EBNA 1, EBNA 2 and EBNA 6. *Eur. J. Cancer*, **11**, 1584–1589.

Levine, P. H., Ebbesen, P., Ablashi, D. V., Saxinger, W. C., Nordentoft, A. and Connelly, R. R. (1992) Antibodies to human herpes virus-6 and clinical course in patients with Hodgkin's disease. *Int. J. Cancer*, **51**, 53–57.

Littler, E., Baylis, S. A., Zeng, Y., Conway, M. J., Mackett, M. and Arrand, J. R. (1991) Diagnosis of nasopharyngeal carcinoma by means of recombinant Epstein–Barr virus proteins. *Lancet*, **337**, 685–689.

Lubin, R., Zalcman, G., Bouchet, L., Trédaniel, J., Legros, Y., Cazals, D. *et al.* (1995) Serum p53 antibodies as early markers of lung cancer. *Nat. Med.*, **1**, 701–702.

Mariette, X., Sibilia, J., Delaforge, C., Bengoufa, D., Brouet, J. C. and Soussi, T. (1999) Anti-p53 antibodies are rarely detected in serum of patients with rheumatoid arthritis and Sjogren's syndrome. *J. Rheumatol.*, **26**, 1672–1675.

McNeel, D. G., Nguyen, L. D., Storer, B. E., Vessella, R., Lange, P. H. and Disis, M. L. (2000) Antibody immunity to prostate cancer-associated antigens can be detected in the serum of patients with prostate cancer. *J. Urol.*, **164**, 1825–1829.

Nesbit, C. E., Tersak, J. M. and Prochownik, E. V. (1999) *MYC* oncogenes and human neoplastic disease. *Oncogene*, **18**, 3004–3016.

Oka, T., Sastry, K. J., Nehete, P., Schapiro, S. J., Guo, J. Q., Talpaz, M. *et al.* (1998) Evidence for specific immune response against P210 BCR-ABL in long-term remission CML patients treated with interferon. *Leukemia*, **12**, 155–163.

Press, M., Bernstein, L., Thomas, P., Meisner, L., Zhou, J., Ma, Y. *et al.* (1997) *HER-2/neu* gene amplification characterized by fluorescence in situ hybridization: poor prognosis in node-negative breast carcinomas. *J. Clin. Oncology*, **15**, 2894–2904.

Raedle, J., Oremek, G., Truschnowitsch, M., Lorenz, M., Roth, W. K., Caspary, W. F. *et al.* (1998) Clinical evaluation of autoantibodies to p53 protein in patients with chronic liver disease and hepatocellular carcinoma. *Eur. J. Cancer*, **34**, 1198–1203.

Ralhan, R., Nath, N., Agarwal, S., Mathur, M., Wasylyk, B. and Shukla, N. K. (1998) Circulating p53 antibodies as early markers of oral cancer: correlation with p53 alterations. *Clin. Cancer Res.*, **4**, 2147–2152.

Rapoport, M., Mor, A., Bistritzer, T., Ramot, Y., Levi, O., Slavin, S. *et al.* (1997) Protooncogenes and autoimmunity. *Isr. J. Med. Sci.*, **33**, 262–266.

Rapoport, M. J., Mor, A., Amit, M., Rosenberg, R., Ramot, Y., Mizrachi, A. *et al.* (1999) Decreased expression of the p21ras stimulatory factor hSOS in PBMC from inactive SLE patients. *Lupus*, **8**, 24–28.

Rosenfeld, M. R., Malats, N., Schramm, L., Graus, F., Cardenal, F., Vinolas, N. *et al.* (1997) Serum anti-p53 antibodies and prognosis of patients with small-cell lung cancer. *J. Natl. Cancer Inst.*, **89**, 381–385.

Schneider, J., Presek, P., Braun, A., Loffler, S. and Woitowitz, H. J. (2000) Serum ras (p21) as a marker for occupationally derived lung cancer? *Clin. Chem. Lab. Med.*, **38**, 301–305.

Slamon, D. J., Clark, G. M., Wong, S. G., Levin, W. J., Ullrich, A. and McGuire, W. L. (1987) Human breast cancer: correlation of relapse and survival with amplification of the *HER-2/neu* oncogene. *Science*, **235**, 177–182.

Soussi, T. (2000) p53 antibodies in the sera of patients with various types of cancer: a review. *Cancer Res.*, **60**, 1777–1788.

Stark, A., Hulka, B. S., Joens, S., Novotny, D., Thor, A. D., Wold, L. E., Schell, M. J., Melton, L. J., 3rd, Liu, E. T., Conway, K. (2000) HER-2/neu amplification in benign breast disease and the risk of subsequent breast cancer. *J. Clin. Oncol.*, **18**(2), 267–274.

Takahashi, M., Chen, W., Byrd, D., Disis, M. L., Huseby, E., McCahill, L. *et al.* (1995) Antibody to ras proteins in patients with colon cancer. *Clin. Cancer Res., Advances in Brief*, **1**, 1071–1077.

Talpaz, M., Qiu, X., Cheng, K., Cortes, J. E., Kantarjian, H. and Kurzrock, R. (2000) Autoantibodies to Abl and Bcr proteins. *Leukemia*, **14**, 1661–1666.

Van den Eynde, B., Hainaut, P., Herin, M., Knuth, A., Lemoine, C., Weynants, P. *et al.* (1989) Presence on a human melanoma of multiple antigens recognized by autologous CTL. *Int. J. Cancer*, **44**, 634–640.

de-Vries, T. J., Fourkour, A., Wobbes, T., Verkroost, G., Ruiter, D. J. and van-Muijen, G. N. (1997) Heterogeneous expression of immunotherapy candidate proteins gp100, MART-1, and tyrosinase in human melanoma cell lines and in human melanocytic lesions. *Cancer Res.*, **57**, 3223–3229.

Vogl, F. D., Frey, M., Kreienberg, R. and Runnebaum, I. B. (2000) Autoimmunity against p53 predicts invasive cancer with poor survival in patients with an ovarian mass. *Br. J. Cancer*, **83**, 1338–1343.

Ward, R. L., Hawkins, N. J., Coomber, D. and Disis, M. L. (1999) Antibody immunity to the HER-2/neu oncogenic protein in patients with colorectal cancer. *Hum. Immunol.*, **60**, 510–515.

Yamamoto, A., Shimizu, E., Takeuchi, E., Houchi, H., Doi, H., Bando, H. *et al.* (1999) Infrequent presence of anti-c-Myc antibodies and absence of c-Myc oncoprotein in sera from lung cancer patients. *Oncology*, **56**, 129–133.

Zalcman, G., Schlichtholz, B., Tredaniel, J., Urban, T., Lubin, R., Dubois, I. *et al.* (1998) Monitoring of p53 autoantibodies in lung cancer during therapy: relationship to response to treatment. *Clin. Cancer Res.*, **4**, 1359–1366.

Chapter 12

Antibody and T-cell responses to the NY-ESO-1 antigen

Elke Jäger, Dirk Jäger and Alexander Knuth

Summary

NY-ESO-1 is one of the most immunogenic tumor antigens known to date. NY-ESO-1 belongs to the category of "Cancer–Testis" CT antigens according to the gene expression patterns in different types of cancer and normal germ cells. NY-ESO-1 was identified with the SEREX method from an esophageal cancer cDNA library (Chen *et al.*, 1997). Later, it was shown that high-titered antibody responses against NY-ESO-1 were found in patients with NY-ESO-1+ cancers. A close positive correlation was observed for antibody titers and the clinical development of NY-ESO-1+ disease (Jäger *et al.*, 1999). The identification of peptide epitopes recognized by CD4+ and CD8+ T lymphocytes in the context of different MHC class I and II alleles has set the basis for the specific monitoring of spontaneous and vaccine-induced cellular immune responses against NY-ESO-1 (Wang *et al.*, 1998; Gnjatic *et al.*, 2000; Zarour *et al.*, 2000; Jäger *et al.*, 2000b; Zeng *et al.*, 2001). Integrated immune responses with detectable serum antibody and T-cell reactivity against NY-ESO-1 are found in approximately 50% of patients with NY-ESO-1+ cancers (Stockert *et al.*, 1998; Jäger *et al.*, 2000d). Immunization with HLA-A2 restricted NY-ESO-1 peptides has led to strong primary CD8+ T-cell responses in patients with different NY-ESO-1+ cancers (Jäger *et al.*, 2000a). The high frequency of spontaneous NY-ESO-1 immunity, the close correlation between humoral and cellular immune responses against NY-ESO-1 and the high immunogenicity of NY-ESO-1 derived peptides renders NY-ESO-1 to be a model antigen in cancer immunology today.

Cancer–testis antigen NY-ESO-1

NY-ESO-1 was identified from a squamous cell esophageal carcinoma using a tumor cDNA library screened for specific recognition of recombinant gene products with autologous patient serum. The coding region for NY-ESO-1 contains 543 bp, and the full-length NY-ESO-1 protein consists of 180 amino acids. Similar to other members of the CT family of genes, *NY-ESO-1* is mapped to chromosome Xq28 (Chen *et al.*, 1997). *LAGE-1*, *CAMEL* and *CAG-3* are related genes displaying high degrees of homologies to *NY-ESO-1*. Immune responses against gene products derived from alternative open reading frames of these genes were identified simultaneously and with equivalent efficiency in patients with melanoma and breast cancer (Lethe *et al.*, 1998; Wang *et al.*, 1998; Aarnoudse *et al.*, 1999).

NY-ESO-1 is expressed in different types of cancer, and not in normal tissues except male germ cells (spermatogonia). Non-small cell and small cell lung cancer, bladder cancer, melanoma, ovarian cancer, breast and head and neck cancer were found to be NY-ESO-1+ in 30–60% of all cases tested as assessed by RT-PCR and by immunohistochemistry.

NY-ESO-1 detection at the cellular protein level showed a highly heterogeneous expression in the majority of tissues evaluated (Jungbluth *et al.*, 2001). Expression levels of NY-ESO-1 showed no correlation with other known tumor antigens and/or MHC class I/II. Different metastases of individual patients displayed varying patterns of NY-ESO-1 expression independent of the metastatic site. Follow-up assessment of single metastases showed a loss of NY-ESO-1 expression along with disease progression or during specific immunotherapy with NY-ESO-1 derived peptides (Jäger *et al.*, 2000a).

Antibody reactivity against NY-ESO-1

IgG1 type antibody responses against NY-ESO-1 are detectable in approximately 50% of patients with NY-ESO-1+ cancers (Stockert *et al.*, 1998). Antibody detection is not dependent on specific histopathological features of tumors, that is, cellular infiltrates, MHC class I/II expression, homogeneity and intensity of NY-ESO-1 expression. An association between NY-ESO-1 serum antibody and HLA-DP4 positivity was described by Zeng *et al.* (2001) suggesting efficient stimulation of NY-ESO-1 specific CD4+ T cells in the context of DP4 molecules.

NY-ESO-1 serum antibody is dependent on the presence of NY-ESO-1+ cancer *in vivo*. Surgical resection of NY-ESO-1+ tumors, or regression of irresectable disease subsequent to chemo- and/or immunotherapy induced a decrease of NY-ESO-1 serum antibody within 4–6 months. Progression of disease or upregulation of NY-ESO-1 in pre-existing lesions was associated with increasing NY-ESO-1 serum antibody titers (Jäger *et al.*, 1999).

The analysis of spontaneous NY-ESO-1 immunity revealed a close correlation between detectable NY-ESO-1 serum antibody and CD8+ T-cell response. This suggests that NY-ESO-1 induces an integrated immune response in a proportion of patients with NY-ESO-1+ cancers (Jäger *et al.*, 2000d). Future studies will show whether spontaneous immunity against NY-ESO-1 is associated with defined patient characteristics (i.e. HLA type, type of disease, treatment etc.) or tumor-related features (i.e. location, metastatsis, intratumoral cellular infiltrates, MHC class I/II expression etc.) characteristics. In a population of patients with transitional cell carcinoma, NY-ESO-1 serum antibody was associated with tumor grading, detectable only in patients with G3 cancers (Kurashige *et al.*, 2001). Thus far, spontaneous NY-ESO-1 serum antibody is indicative for the presence of NY-ESO-1+ disease and associated with detectable CD8+ T-cell reactivity.

In the first NY-ESO-1 peptide vaccination trial reported so far, pre-existing NY-ESO-1 serum antibody did not change during immunization in five patients. However, in one of seven NY-ESO-1 antibody negative patient, a sero-conversion was observed during peptide vaccination. NY-ESO-1 serum antibody was first detectable after three months, subsequent to the *de novo* induction of a strong NY-ESO-1 peptide specific CD8+ T-cell response (Jäger *et al.*, 2000a). Since NY-ESO-1 antibodies do not interact with MHC class I restricted NY-ESO-1 peptides used for vaccination, the humoral NY-ESO-1 response was most likely the consequence of the specific interaction of CD8+ T cells induced by vaccination with tumor cells leading to tumor necrosis and secondary stimulation of CD4+ T cells. This clinical example shows that NY-ESO-1 specific CD8+ T-cell responses may be induced in the absence of or prior to NY-ESO-1 serum antibody.

NY-ESO-1 specific CD4+ T-cell reactivity

Detectable NY-ESO-1 serum antibody of the IgG1 type suggests the presence of cognate CD4+ T-cell reactivity. The analysis of two patients with NY-ESO-1+ metastatic melanoma and strong serum antibody reactivity against NY-ESO-1 has led to the identification of HLA-DRB4*0101–0103 restricted peptide epitopes that are specifically recognized by CD4+ T lymphocytes. CD4+ T-cell lines and clones generated by repeated stimulation with the respective NY-ESO-1 peptides efficiently recognized NY-ESO-1 protein-loaded autologous dendritic cells and tumor cells, indicating that these peptides represent naturally processed epitopes (Jäger *et al.*, 2000b). Later, further MHC class II restricted NY-ESO-1 peptides were identified by different groups, which specifically induce CD4+ T-cell responses *in vitro* in cancer patients and normal individuals (Zarour *et al.*, 2000; Zeng *et al.*, 2001). Spontaneous NY-ESO-1 serum antibody and CD4+ T-cell reactivity was systematically correlated with MHC class II genotypes in cancer patients by Zeng *et al.* (2001). The results of this study show that the majority of NY-ESO-1 antibody positive patients (16/17) were HLA-DP4+, but only 2/8 NY-ESO-1 antibody negative patients. A peptide was identified that elicits strong CD4+ T-cell responses in the context of HLA-DP4 in 5/6 NY-ESO-1 antibody positive patients (Zeng *et al.*, 2001). Larger studies in patients with NY-ESO-1+ cancer will have to be performed, however, to confirm any association between a specific HLA phenotype and NY-ESO-1 specific immunity.

NY-ESO-1 specific CD8+ T-cell reactivity

CD8+ T lymphocytes were identified as the mediators of cytotoxicity against tumor cells *in vitro* and of tumor regressions in individual patients (Jäger *et al.*, 2000c). Therefore, the characterization of peptides inducing CD8+ T-cell responses in the context of different MHC class I alleles have been a long-standing goal in cancer immunology. The first step towards the identification of NY-ESO-1 peptides was based on the observation that a melanoma cell line, NW-MEL-38, was efficiently lyzed by an autologous T-cell line, NW-38-IVS 1, generated by repetitive *in vitro* sensitization with NW-MEL-38 cells. T-cell mediated lysis was restricted by HLA-A2, but the target antigen was unknown. Since the patient NW38 had a high-titered NY-ESO-1 serum antibody, it was hypothetized that the CD8+ T-cell reactivity was directed against gene products of *NY-ESO-1*. Co-transfection of NY-ESO-1 and HLA-A2 cDNA into antigen presenting cells confirmed that the target antigen for the T-cell line NW-38-IVS 1 was indeed NY-ESO-1. Computer assisted analysis of the NY-ESO-1 protein for potential HLA-A2 binding peptide motifs has helped identifying 28 candidate peptides of 9 or 10 amino acids in length. These were tested for specific recognition by the T-cell line NW-38-IVS 1. Three peptides with overlapping sequences were identified initially: NY-ESO-1 p155–163 (QLSLLMWITQC), NY-ESO-1 p157–167 (SLLMWITQCFL) and NY-ESO-1 p157–165 (SLLMWITQC) (Jäger *et al.*, 1998). Later, another epitope was found to be recognized in the context of HLA-A2, NY-ESO-1 p158–166 (LLMWITQCF), as well.

The identification of HLA-A2 restricted peptide epitopes has led to large-scale analyses of spontaneous NY-ESO-1 specific CD8+ T-cell reactivity in patients with NY-ESO-1+ cancer and detectable or absent NY-ESO-1 serum antibody. Complemetary methods of high sensitivity were established and standardized for the monitoring of CD8+ T cells for NY-ESO-1 specific-binding capacity (recombinant HLA-A2 tetramer–peptide complexes), specific cytokine release (ELISPOT and cytospot assays) and specific cytotoxicity (^{51}Chromium release assay). Collaborative efforts were initiated to correlate different methods of T-cell assessment to identify the most sensitive way of monitoring CD8+ T-cell responses for clinical studies. It was demonstrated that tetramer and ELISPOT assays showed closely corresponding results, which were confirmed by detectable cytotoxicity in the majority of cases tested (Jäger *et al.*, 2000d). Rarely, it was observed that NY-ESO-1 specific CD8+ T cells showed high levels of cytokine release, but a low binding capacity for recombinant NY-ESO-1 peptide–tetramer complexes, or no cytotoxicity against NY-ESO-1 peptide-pulsed antigen presenting cells or NY-ESO-1+ tumor cell lines. As a consequence, tetramer and ELISPOT assays are currently regarded as standard methods for the monitoring of spontaneous or vaccine-induced NY-ESO-1 specific CD8+ T-cell responses.

The relative *in vitro* immunogenicity of the HLA-A2 binding NY-ESO-1 p157–165 peptide was analyzed against several peptide analogs modified by single amino acid substitutions. Exchange of cysteine at the COOH-terminus with valine, leucine, isoleucine or alanine resulted in a 100-fold increased binding affinity to the HLA-A* 0201 molecule, but also to an altered panel of T-cell receptors activated by these peptide analogs. Further studies have been initiated to identify the most immunogenic peptide analogs, which efficiently stimulate T-cell receptors interacting with naturally presented NY-ESO-1 epitopes. Modified peptides with enhanced immunogenicity may represent promising candidates for active immunotherapy in patients with NY-ESO-1+ cancers (Chen *et al.*, 2000; Romero *et al.*, 2001).

The analysis of spontaneous CD8+ T-cell responses in patients with NY-ESO-1+ cancer and detectable NY-ESO-1 serum antibody has led to the identification of NY-ESO-1 peptide epitopes presented by different MHC class I alleles. A 10-mer peptide was found to be recognized by CD8+ T lymphocytes of a breast cancer patient in the context of HLA-A31 (Wang *et al.*, 1998). Adenoviral vector transfection of NY-ESO-1 to antigen presenting cells was used to identify NY-ESO-1 peptide epitopes recognized by CD8+ T lymphocytes in the context of HLA-Cw3 and -Cw6 (Gnjatic *et al.*, 2000). Ongoing studies analyze the expression levels of these MHC class I alleles in different tumors and correlate these with the efficacy of recogntion by NY-ESO-1 specific CD8+ T lymphocytes *in vitro* and *in vivo*. Differential expression or regulation of MHC class I alleles may represent an important mechanism to render tumor cells susceptible for the interaction with epitope-specific subsets of NY-ESO-1 reactive CD8+ T lymphocytes.

Primary induction of NY-ESO-1 specific CD8+ T-cell responses *in vivo*

HLA-A2 restricted NY-ESO-1 peptides were used in a clinical trial first to induce peptide specific CD8+ T-cell responses *in vivo*. Twelve patients were enrolled in the study, five were NY-ESO-1 antibody positive (three melanoma, one ovarian cancer, one breast cancer), seven (six melanoma, one ovarian cancer) were NY-ESO-1 antibody negative before immunization. NY-ESO-1 peptides were injected intradermally at a dose of 100 μg per injection weekly for four weeks. After a treatment-free interval of four weeks, injections were repeated and systemic GM-CSF was added as a systemic adjuvant. Delayed-type hypersensitivity (DTH) reactions were observed after peptide immunization with increasing intensity in the NY-ESO-1 antibody negative patients, whereas NY-ESO-1 antibody positive patients showed strong DTH reactions already after the first immunization. In parallel, peptide specific CD8+ T-cell responses were induced in 4/7 NY-ESO-1 antibody negative patients during immunization. NY-ESO-1 antibody positive patients had spontaneous NY-ESO-1 specific CD8+ T-cell reactivity that did not change significantly during immunization. Although major clinical remissions were not observed in the patient population studied, partial and complete regressions of single metastases were observed in NY-ESO-1 antibody negative and positive patients. In one melanoma patient, a primary CD8+ T-cell response against NY-ESO-1 peptides p157–167 and p157–165 was generated. Subsequently, a necrotic transformation of a large inguinal lymph node metastasis was observed that showed an increased infiltrate of intratumoral CD8+ T lymphocytes. After three months of immunization, the patient developed increasing titers of NY-ESO-1 serum antibody, suggesting the concurrent activation of B cells after the primary CD8+ T-cell response to the vaccine (Jäger *et al.*, 2000a).

Considering the high rate of primary CD8+ T-cell responses to peptide vaccination in patients with advanced NY-ESO-1+ cancer, vaccine approaches will be extended to other NY-ESO-1 derived MHC class I and II restricted epitopes to study their immunogenicity in the context of different immunization schedules and adjuvants.

Perspectives

NY-ESO-1 is one of the most immunogenic tumor antigens known to date. Corresponding to the CT pattern of expression, NY-ESO-1 is found in different types of cancer including hematologic malignancies, that is, multiple myeloma, Non-Hogdkin's lymphoma. Integrated spontaneous immune responses with detectable serum antibody and CD4+ and CD8+ T-cell reactivity are measurable in approximately 50% of patients with NY-ESO-1+ cancers. The high frequency of spontaneous immune responses may be the result of a very efficient presentation of NY-ESO-1 epitopes on MHC class I and class II molecules, inducing strong specific CD8+ and CD4+ T-cell responses. In contrast to other tumor antigens identified so far, spontaneous immune responses against NY-ESO-1 almost exclusively include cellular and humoral effectors simultaneously. Based on this observation, serum antibody reactivity against NY-ESO-1 is considered a sensitive marker for the presence of NY-ESO-1+ disease and detectable NY-ESO-1 specific T-cell reactivity in cancer patients. Further, changes of NY-ESO-1 antibody titers over time reflect the clinical development of NY-ESO-1+ disease in cancer patients under treatment.

The identification of MHC class I and class II restricted NY-ESO-1 peptide epitopes recognized by CD8+ and CD4+ T lymphocytes has opened new perspectives for specific

Table 12.1 HLA-restriction of NY-ESO-1 peptides

HLA-restriction	*HLA-frequency (%)*	*Peptide*	*Position*	*Reference*
A2	44	SLLMWITQC	157–165	Jäger *et al.* (1998), Chen *et al.* (2000)
A2	44	MLMAQEALAFL	alt ORF	Aarnoudse *et al.* (1999)
A31	5	ASGPGGGAPR	53–62	Wang *et al.* (1998)
A31	5	LAAQERRVPR	alt ORF	Wang *et al.* (1998)
Cw3	24	LAMPFATPM	92–100	Gnjatic *et al.* (2000)
Cw6	16	ARGPESRLL	80–88	Gnjatic *et al.* (2000)
DR4	24	PGVLLKEFTVSGNILTIRL	119–138	Jäger *et al.* (2000)
DR4	24	AADHRQLQLSISSCLQQ	139–156	Jäger *et al.* (2000)
DP4	70	WITQCFLPVFLAQPPSG	161–180	Zeng *et al.* (2001)

Note
NY-ESO-1 peptide sequences recognized by CD4+ and CD8+ T lymphocytes in the context of different HLA class I and class II alleles.

immunotherapeutic approaches targeting NY-ESO-1 in cancer patients. HLA-A2 restricted NY-ESO-1 peptides have been used in a clinical study. Primary NY-ESO-1 specific CD8+ T-cell responses were induced in the majority of cancer patients without detectable immunity against NY-ESO-1 at the start of treatment. Subsequent studies are being designed to evaluate the effects of concurrent immunization with MHC class I and class II restricted NY-ESO-1 epitopes to induce both CD8+ and CD4+ T-cell responses at the same time. The results of these studies will provide further insight whether the activation of NY-ESO-1 specific CD4+ T cells affects the kinetics and maintenance of NY-ESO-1 specific CD8+ T-cell responses to NY-ESO-1 peptide vaccination *in vivo* (Table 12.1).

Viral constructs, that is, vaccinia-NY-ESO-1 and fowl pox-NY-ESO-1, will be used for vaccination of patients with NY-ESO-1+ cancers to study the potential of full protein expression to induce both CD8+ and CD4+ T-cell responses against any NY-ESO-1 epitope presented in the context of the individual HLA alleles. The results will provide further evidence for the efficacy of processing MHC class I and class II restricted epitopes after viral transfer of the entire *NY-ESO-1* gene into antigen presenting cells. If integrated immune responses were observed in the majority of patients after immunization with viral constructs, NY-ESO-1 specific cancer vaccination would become available for a larger patient population independent of any individual HLA genotype.

References

Aarnoudse, C. A., van den Doel, P. B., Heemskerk, B. and Schrier, P. L. (1999) Interleukin-2-induced, melanoma-specific T cells recognize CAMEL, an unexpected translation product of LAGE-1. *Int. J. Cancer*, **82**, 442–8.

Chen, J. L., Dunbar, P. R., Gileadi, U., Jäger, E., Gnjatic, S., Nagata, Y., Stockert, E., Panicali, D. L., Chen, Y. T., Knuth, A., Old, L. J. and Cerundolo, V. (2000) Identification of NY-ESO-1 peptide analogues capable of improved stimulation of tumor-reactive CTL. *J. Immunol.*, **165**, 948–55.

Chen, Y.-T., Scanlan, M. J., Sahin, U., Türeci, Ö., Gure, A. O., Tsang, S., Williamson, B., Stockert, E., Pfreundschuh, M. and Old, L. J. (1997) A testicular antigen aberrantly expressed in human cancers detected by autologous antibody screening. *Proc. Natl. Acad. Sci. USA*, **94**, 1914–18.

Gnjatic, S., Nagata, Y., Jäger, E., Stockert, E., Shankara, S., Roberts, B. L., Mazzara, G. P., Lee, S. Y., Dunbar, P. R., Dupont, B., Cerundolo, V., Ritter, G., Chen, Y. T., Knuth, A. and Old, L. J. (2000) Strategy for monitoring T cell responses to NY-ESO-1 in patients with any HLA class I allele. *Proc. Natl. Acad. Sci. USA*, **97**, 10917–22.

Jäger, E., Gnjatic, S., Nagata, Y., Stockert, E., Jäger, D., Karbach, J., Neumann, A., Rieckenberg, J., Chen, Y. T., Ritter, G., Hoffman, E., Arand, M., Old, L. J. and Knuth, A. (2000a) Induction of primary NY-ESO-1 immunity: CD8+ T lymphocyte and antibody responses in peptide-vaccinated patients with NY-ESO-1+ cancers. *Proc. Natl. Acad. Sci. USA*, **97**, 12198–203.

Jäger, E., Jäger, D., Karbach, J., Chen, Y. T., Ritter, G., Nagata, Y., Gnjatic, S., Stockert, E., Arand, M., Old, L. J. and Knuth, A. (2000b) Identification of NY-ESO-1 epitopes presented by human histocompatibility antigen (HLA)-DRB4*0101-0103 and recognized by CD4(+) T lymphocytes of patients with NY-ESO-1-expressing melanoma. *J. Exp. Med.*, **191**, 625–30.

Jäger, E., Maeurer, M., Hohn, H., Karbach, J., Jäger, D., Zidianakis, Z., Bakhshandeh-Bath, A., Orth, J., Neukirch, C., Necker, A., Reichert, T. E. and Knuth, A. (2000c) Clonal expansion of Melan A-specific cytotoxic T lymphocytes in a melanoma patient responding to continued immunization with melanoma-associated peptides. *Int. J. Cancer*, **86**, 538–47.

Jäger, E., Nagata, Y., Gnjatic, S., Wada, H., Stockert, E., Karbach, J., Dunbar, P. R., Lee, S. Y., Jungbluth, A., Jäger, D., Arand, M., Ritter, G., Cerundolo, V., Dupont, B., Chen, Y. T., Old, L. J. and Knuth, A. (2000d) Monitoring CD8 T cell responses to NY-ESO-1: correlation of humoral and cellular immune responses. *Proc. Natl. Acad. Sci. USA*, **97**, 4760–5.

Jäger, E., Stockert, E., Zidianakis, Z., Chen, Y.-T., Karbach, J., Jäger, D., Arand, M., Ritter, G., Old, L. J. and Knuth, A. (1999) Humoral immune responses of cancer patients against 'Cancer-Testis' antigen NY-ESO-1: correlation with clinical events. *Int. J. Cancer.*, **84**, 506–10.

Jäger, E., Chen, Y.-T., Drijfhout, J. W., Karbach, J., Ringhoffer, M., Jäger, D., Arand, M., Wada, H., Noguchi, Y., Stockert, E., Old, L. J. and Knuth, A. (1998) Simultaneous humoral and cellular immune response against Cancer-Testis Antigen NY-ESO-1: definition of human histocompatibility leucocyte antigen (HLA)-A2-binding peptide epitopes. *J. Exp. Med.*, **187**, 265–9.

Jungbluth, A. A., Chen, Y. T., Stockert, E., Busam, K. J., Kolb, D., Iversen, K., Coplan, K., Williamson, B., Altorki, N. and Old, L. J. (2001) Immunohistochemical analysis of NY-ESO-1 antigen expression in normal and malignant human tissues. *Int. J. Cancer*, **92**, 856–60.

Kurashige, T., Noguchi, Y., Saika, T., Ono, T., Nagata, Y., Jungbluth, A., Ritter, G., Chen, Y. T., Stockert, E., Tsushima, T., Kumon, H., Old, L. J. and Nakayama, E. (2001) Ny-ESO-1 expression and immunogenicity associated with transitional cell carcinoma: correlation with tumor grade. *Cancer Res.*, **61**, 4671–4.

Lethe, B., Lucas, S., Michaux, L., De Smet, C., Godelaine, D., Serrano, A., De Plaen, E. and Boon, T. (1998) LAGE-1, a new gene with tumor specificity. *Int. J. Cancer*, **76**, 903–8.

Romero, P., Dutoit, V., Rubio-Godoy, V., Lienard, D., Speiser, D., Guillaume, P., Servis, K., Rimoldi, D., Cerottini, J. C. and Valmori, D. (2001) CD8+ T-cell response to NY-ESO-1: relative antigenicity and in vitro immunogenicity of natural and analogue sequences. *Clin. Cancer Res.*, **7**, 766s–72s.

Stockert, E., Jäger, E., Chen, Y.-T., Scanlan, M. J., Gout, I., Karbach, J., Knuth, A. and Old, L. J. (1998) A survey of the humoral immune response of cancer patients to a panel of human tumor antigens. *J. Exp. Med.*, **187**, 1349–54.

Wang, R. F., Johnston, S. L., Zeng, G., Topalian, S. L., Schwartzentruber, D. J. and Rosenberg, S. A. (1998) A breast and melanoma-shared tumor antigen: T cell responses to antigenic peptides translated from different open reading frames. *J. Immunol.*, **161**, 3598–606.

Zarour, H. M., Storkus, W. J., Brusic, V., Williams, E. and Kirkwood, J. M. (2000) NY-ESO-1 encodes DRB1*0401-restricted epitopes recognized by melanoma-reactive CD4+ T cells. *Cancer. Res.*, **60**, 4946–52.

Zeng, G., Wang, X., Robbins, P. F., Rosenberg, S. A. and Wang, R. F. (2001) CD4(+) T cell recognition of MHC class II-restricted epitopes from NY-ESO-1 presented by a prevalent HLA DP4 allele: association with NY-ESO-1 antibody production. *Proc. Natl. Acad. Sci. USA*, **98**, 3964–9.

Index